METHODS IN MOLECULAR BIOLOGY™

Series Editor
John M. Walker
School of Life Sciences
University of Hertfordshire
Hatfield, Hertfordshire, AL10 9AB, UK

For other titles published in this series, go to
www.springer.com/series/7651

Plant Chromosome Engineering

Edited by

James A. Birchler

Division of Biological Sciences
University of Missouri
Columbia, MO, USA

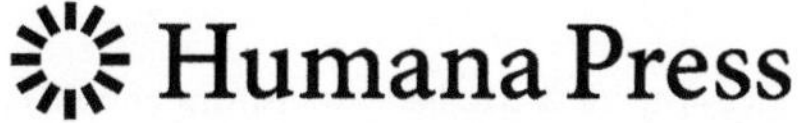

Editor
James A. Birchler, Ph.D.
Division of Biological Sciences
University of Missouri
Columbia, MO, USA
BirchlerJ@Missouri.edu

ISSN 1064-3745 e-ISSN 1940-6029
ISBN 978-1-61737-956-7 e-ISBN 978-1-61737-957-4
DOI 10.1007/978-1-61737-957-4
Springer New York Dordrecht Heidelberg London

Printed on acid-free paper

Humana Press is part of Springer Science+Business Media (www.springer.com)

Preface

With a stable or decreasing amount of arable land on earth but with a continuing increase in human population, creative improvements in agriculture will be needed in the coming decades to maintain or improve the standard of living. Novel approaches to the production of food, feed, fuel, fiber, and pharmaceuticals will be needed, and the modification of crops and other plant species is one means to achieve such a goal. This volume on plant chromosome engineering includes reviews and protocols for transformation procedures, chromosome painting, production of engineered minichromosomes, gene targeting and mutagenesis, site-specific integration, gene silencing, protein expression, chromosome sorting and analysis, protocols for generating chromosomal rearrangements, enhancer trapping, and means of studying chromosomes in vivo. Collectively, these chapters touch upon the spectrum of tools currently available for modifying plant genomes and chromosomes and provide the foundation for future developments.

Columbia MO *James A. Birchler*

Contents

Contributors

PATRICE S. ALBERT • *Division of Biological Sciences, University of Missouri, Columbia MO, USA*

ANIÇA AMINI • *Biotechnology Center for Agriculture and the Environment, Rutgers, The State University of New Jersey, New Brunswick NJ, USA*

SHYAMKUMAR BARAMPURAM • Division of Plant Sciences, University of Missouri, Columbia MO, USA

JAMES A. BIRCHLER • *Division of Biological Sciences, University of Missouri, Columbia MO, USA*

SIVANANDAN CHUDALAYANDI • *Division of Biological Sciences, University of Missouri, Columbia MO, USA*

JARMILA ČÍHALÍKOVÁ • *Laboratory of Molecular Cytogenetics and Cytometry, Institute of Experimental Botany, Olomouc, Czech Republic*

TATIANA DANILOVA • *Division of Biological Sciences, University of Missouri, Columbia MO, USA*

KEITH R. DAVIS • *Owensboro Cancer Research Program, Owensboro KY, USA; Department of Pharmacology and Toxicology and the James Graham Brown Cancer Center, University of Louisville School of Medicine, Louisville KY, USA*

JAROSLAV DOLEŽEL • *Laboratory of Molecular Cytogenetics and Cytometry, Institute of Experimental Botany, Olomouc, Czech Republic*

ANDREW L. EAMENS • *School of Microbial and Molecular Sciences, University of Sydney, Sydney, Australia*

TAKASHI R. ENDO • *Laboratory of Plant Genetics, Graduate School of Agriculture, Kyoto University, Kyoto, Japan*

SETH FINDLEY • *Division of Biological Sciences, University of Missouri, Columbia MO, USA*

ROBERT T. GAETA • *Division of Biological Sciences, University of Missouri, Columbia MO, USA*

ZHI GAO • *Division of Biological Sciences, University of Missouri, Columbia MO, USA*

FANGPU HAN • *Division of Biological Sciences, University of Missouri, Columbia MO, USA*

ANDREAS HOUBEN • *Leibniz Institute of Plant Genetics and Crop Plant Research (IPK), Gatersleben, Germany*

JIMING JIANG • *Department of Horticulture, University of Wisconsin, Madison WI, USA*

AKIO KATO • *Division of Biological Sciences, University of Missouri, Columbia MO, USA*

LAKSHMINARASIMHAN KRISHNASWAMY • *Department of Biological Sciences, University of Missouri, Columbia MO, USA*

DAL-HOE KOO • *Department of Horticulture, University of Wisconsin, Madison WI, USA*

MARIE KUBALÁKOVÁ • *Laboratory of Molecular Cytogenetics and Cytometry, Institute*

of Experimental Botany, Olomouc, Czech Republic
RALF G. KYNAST • *Jodrell Laboratory, Royal Botanic Gardens, Kew, Richmond Surrey, UK*
ERIC LAM • *Biotechnology Center for Agriculture and the Environment, Rutgers, The State University of New Jersey, New Brunswick NJ, USA*
JONATHAN C. LAMB • *Division of Biological Sciences, University of Missouri, Columbia MO, USA*
AVRAHAM A. LEVY • *Department of Plant Sciences, The Weizmann Institute of Science, Rehovot, Israel*
MICHAL LIEBERMAN-LAZAROVICH • *Department of Plant Sciences, The Weizmann Institute of Science, Rehovot, Israel*
CHONGYUAN LUO • *Biotechnology Center for Agriculture and the Environment, Rutgers, The State University of New Jersey, New Brunswick NJ, USA*
PAL MALIGA • *Waksman Institute of Microbiology, Rutgers, The State University of New Jersey, Piscataway NJ, USA*
NOBUYUKI MATOBA • *Owensboro Cancer Research Program, Owensboro KY, USA; Department of Pharmacology and Toxicology and the James Graham Brown Cancer Center, University of Louisville School of Medicine, Louisville KY, USA*
SHUHEI NASUDA • *Laboratory of Plant Genetics, Graduate School of Agriculture, Kyoto University, Kyoto, Japan*
DAVID W. OW • *Plant Gene Expression Center, USDA-ARS & Plant & Microbial Biology, University of California-Berkeley, Albany CA, USA*
KENNETH E. PALMER • *Owensboro Cancer Research Program, Owensboro KY, USA; Department of Pharmacology & Toxicology and the James Graham Brown Cancer Center, University of Louisville School of Medicine, Louisville KY, USA*
THOMAS PETERSON • *Department of Genetics, Development, and Cell Biology and Department of Agronomy, Iowa State University, Ames IA, USA*
OSCAR RIERA-LIZARAZU • *Department of Crop and Soil Science, Oregon State University, Corvallis OR, USA*
HANA ŠIMKOVÁ • *Laboratory of Molecular Cytogenetics and Cytometry, Institute of Experimental Botany, Olomouc, Czech Republic*
PAVLA SUCHÁNKOVÁ • *Laboratory of Molecular Cytogenetics and Cytometry, Institute of Experimental Botany, Olomouc, Czech Republic*
ZORA SVAB • *Waksman Institute of Microbiology, Rutgers, The State University of New Jersey, Piscataway NJ, USA*
JAMES G. THOMSON • *Crop Improvement Unit, Western Regional Research Center, USDA-ARS, Albany CA, USA*
DANIEL F. VOYTAS • *Department of Genetics, Cell Biology, and Development and Center for Genome Engineering, University of Minnesota, Minneapolis MN, USA*
YUEJU WANG • *Plant Gene Expression Center, USDA-ARS & Plant & Microbial Biology, University of California-Berkeley, Albany CA, USA*
PETER M. WATERHOUSE • *School of Microbial and Molecular Sciences, University of Sydney, Sydney, Australia*
CHUNHUI XU • *Department of Biology, The Chinese University of Hong Kong, Shatin, N.T. Hong Kong, China*
YUAN-YEU YAU • *Plant Gene Expression Center, USDA-ARS & Plant & Microbial*

Biology, University of California-Berkeley, Albany CA, USA
WEICHANG YU • *Department of Biology, The Chinese University of Hong Kong, Shatin, N.T. Hong Kong, China*
CHUANHE YU • *Department of Genetics, Development, and Cell Biology and Department of Agronomy, Iowa State University, Ames IA, USA*
FENG ZHANG • *Department of Genetics, Cell Biology, and Development and Center for Genome Engineering, University of Minnesota, Minneapolis MN, USA*
JIANBO ZHANG • *Department of Genetics, Development, and Cell Biology and Department of Agronomy, Iowa State University, Ames IA, USA*
ZHANYUAN J. ZHANG • *Division of Plant Sciences, University of Missouri, Columbia MO, USA*

Chapter 1

Recent Advances in Plant Transformation

Shyamkumar Barampuram and Zhanyuan J. Zhang

Abstract

Plant genetic engineering has become one of the most important molecular tools in the modern molecular breeding of crops. Over the last decade, significant progress has been made in the development of new and efficient transformation methods in plants. Despite a variety of available DNA delivery methods, *Agrobacterium*- and biolistic-mediated transformation remain the two predominantly employed approaches. In particular, progress in *Agrobacterium*-mediated transformation of cereals and other recalcitrant dicot species has been quite remarkable. In the meantime, other transgenic-enabling technologies have emerged, including generation of marker-free transgenics, gene targeting, and chromosomal engineering. Although transformation of some plant species or elite germplasm remains a challenge, further advancement in transformation technology is expected because the mechanisms of governing the regeneration and transformation processes are now better understood and are being creatively applied to designing improved transformation methods or to developing new enabling technologies.

Key words: *Agrobacterium*, Biolistic, Genetic engineering, Marker free, Transgenic plant

1. Introduction

The world's agriculture and farming are heavily dependent on crops that provide food and fibers for human use, either directly or through livestock. For the past two centuries, modern technology has improved agricultural practices, thereby augmenting conventional plant breeding methods to achieve improved yield and quality of crops. However, multiple factors such as population growth, environmental stress, ecological considerations, and demand for renewable energy have led to the demand for further improvements in the quality and quantity of crops. Plant genetic engineering offers new avenues in this regard and has become one of the most important molecular tools in the modern molecular breeding of crops (1).

James A. Birchler (ed.), *Plant Chromosome Engineering: Methods and Protocols*, Methods in Molecular Biology, vol. 701, DOI 10.1007/978-1-61737-957-4_1,

Advancements in plant genetic engineering have made it possible to transfer genes into crop plants from unrelated plants and even from nonplant organisms; as a result, many crop species are being genetically modified for better agronomical traits, including disease resistance, insect tolerance, better nutritional values, and other desirable qualities (2). Presently, foreign genes from various origins and production of products in transgenic plants represent a new aspect of the molecular agriculture revolution. In addition, transgenic plants have great impact on nonagricultural applications and represent an alternative for the production of medicinally useful and recombinant proteins and vaccines (3). However, in some crop plants, the lack of efficient transformation methods to introduce foreign DNA remains an obstacle to the application of plant genetic engineering.

Over the last decade, some significant achievements have been made in the development of new and efficient transformation methods in plants. Methods for delivering exogenous DNA to plant cells and gene transformation in general can be divided into two major categories: indirect and direct DNA deliveries. In the former approach, genes of interest are introduced into the target cell via bacteria, e.g., *Agrobacterium tumefaciens* or *Agrobacterium rhizogenes* (4). In contrast, the latter approach does not employ bacterial cells as mediators to transfer DNA to plant cells. Although various delivery methods have been reported, including the use of other bacterial strains (i.e., TransBacter™ Technology) (5), *Agrobacterium*-mediated transformation remains the method of choice for plant transformation. The enduring success of the *Agrobacterium*-mediated approach is primarily attributable to the continuous improvements in plant tissue culture and T-DNA transfer processes. As a result, reproducible and efficient *Agrobacterium*-mediated protocols have been developed for many dicot and some monocot crops (6–8). *Agrobacterium*-mediated transformation possesses intrinsic advantages over direct DNA delivery systems. These advantages include the ability to transfer large intact segments of DNA, simple transgene insertions with defined ends and low copy number, stable integration and inheritance, and consistent gene expression over the generations.

The lower rates of success achieved with *Agrobacterium* in monocots and recalcitrant plant species have led to the development of specific direct DNA transfer methods, one of which is microparticle bombardment. This technology, first developed by Sanford and coworkers (9), is often termed biolistics or gene gun. Particle bombardment has a high success rate in monocot species in which agroinfection is limited. However, biolistic technology possesses several intrinsic disadvantages, including a low transformation efficiency as compared with *Agrobacterium* (when agroinfection works), a high frequency integration of the vector backbone and a loss of transgene cassette integrity, and transgene silencing due to multicopy insertions (10). Alternative delivery systems have also been used for gene transfer in plants, including electroporation, microinjection, silicon carbide, and chloroplast transformation (11). Of these, silicon carbide-mediated transformation represents one of the least complicated methods of plant transformation. Yet, each of these methods has their own limitations in the successful production of transgenic plants.

To date, *Agrobacterium*-mediated and particle bombardment transformation are the most commonly used methods for plant transformation. Nevertheless, the resulting transgenic plants are of course subject to biosafety issues related to the presence of vector backbone sequences and/or selectable marker genes, irrespective of the delivery method used. Transgenic plants produced by *Agrobacterium*-mediated transformation are likely subject to integration of vector backbone sequences. This leads to multiple transgene copies in the transgenic plants, complicating the regulatory process of genetically engineered plants. In recent years, there has been an important advancement in generation of vector-backbone- and selectable-marker-free transgenic plants while still enabling the use of marker genes to select and identify transgenic plants (12).

Several strategies have been proposed and are in use for the production of marker-free transgenic plants. The methods that have been developed include simultaneous transformation of two marker genes (cotransformation), the movement of a transgene segment within the genome (transposition), and recombination between two specific sequences that are not necessarily homologous (site-specific recombination) (13). Currently, there is a great demand for simultaneous expression of multiple genes for expression of complex traits in plants. With recent advancement in molecular biology and vector construction technology, it is possible to achieve stable expression of multiple transgenes in a single genome. The recent development of minichromosome technology might represent a strategy for gene stacking in plants. The minichromosome technique can be used to incorporate desirable traits such as insect, bacterial, or fungal resistance, herbicide tolerance, and increased crop quality (14). Hence, new techniques are in demand to boost yield or to improve crop traits. Finally, the genetic engineering of plants has already begun to play a crucial role in the production of biofuels. This chapter discusses various methods and recent advances in plant transformation technology.

2. Plant Transformation Methods

Recent advances in genetic transformation have made it possible to transfer genes of both academic and agronomic importance into various crop species. A prerequisite for successful transformation system is an efficient regeneration protocol when tissue culture-based transformation process is employed. The very basis of plant regeneration relies on the realization that plant somatic cells are totipotent and can be stimulated to regenerate into whole plants. However, this insight is limited because, in reality, only a limited number of plant species and certain types of explant tissues have been found to be capable of regenerating whole plants under appropriate culture conditions. Therefore, much effort has been aimed at establishing and improving plant regeneration systems. Yet, efficient regeneration alone does not necessarily lead to efficient transformation.

There is a need to develop advanced transformation methods that would not only incorporate the required characteristics (stable and desirable transgene integration and expression) into plants but also enable generation of transgenic events in a high-throughput manner. These requirements are particularly relevant now in the crop post-genome era in which ever-increasing amounts of genome sequence information, BAC clones, ESTs, and full-length cDNAs are available. This situation presents both new challenges and opportunities for plant transformation research. At present, *Agrobacterium* and microprojectile are the commonly used methods for this purpose; other methods, such as electroporation and microinjection, are still used only rarely. The following sections discuss in detail the recent advances in each of the plant transformation methods.

2.1. Agrobacterium-Mediated Transformation

In this method, *A. tumefaciens* or *A. rhizogenes* is employed to introduce foreign genes into plant cells. *A. tumefaciens* is a soil-borne gram-negative bacterium that causes crown-gall, a plant tumor. The tumor-inducing capability of this bacterium is due to the presence of a large Ti (tumor-inducing) plasmid in its virulent strains. Similarly, Ri (root-inducing) megaplasmids are found in virulent strains of *A. rhizogenes*, the causative agent of "hairy root" disease. Both Ti- and Ri-plamids contain a form of "T-DNA" (transferred DNA). The T-DNA contains two types of genes: oncogenic genes, encoding enzymes involved in the synthesis of auxins and cytokinins (causing tumor formation), and genes involved in opine production. The T-DNA element is flanked by two 25-bp direct repeats called the left border (LB) and right border (RB), respectively, which act as a *cis* element signal for the T-DNA transfer (15). Both oncogenic and opine catabolism genes are located inside the T-DNA of the Ti plasmid whereas the virulence (*vir*) genes are situated outside the T-DNA on the Ti plasmid and bacterial chromosome. These *vir* genes are organized into several operons (*virA, virB, virC, virD, virE, virF, virG, and virH*) on the Ti-plasmid and other operons (*chvA, chvB, and chvF*) that are chromosomal and are essential for T-DNA transfer.

The mechanism of gene transfer from *A. tumefaciens* to plant cells involves several steps, which include bacterial colonization, induction of the bacterial virulence system, generation of the T-DNA transfer complex, T-DNA transfer, and integration of the T-DNA into the plant genome. The process of T-DNA transfer is initiated upon receipt of specific signals (e.g., phenolic compounds) received from host cells. Previous observations suggested that wounding or vigorous cell division also promotes T-DNA transfer, presumably due to induction by phenolic compounds produced during cell repair or during the formation of new cells. In response, a signal received by *virA* activates a cascade of other *vir* protein machinery genes. However, very little is known about the nature and function of the factors that *Agrobacterium* utilizes, for instance, specific receptors on the host cell surface and/or cell wall. Subsequently, *virD1* and *virD2* proteins nick both the left and right borders on the bottom strand of the T-DNA. The resulting single-stranded T-DNA molecule (T-strand), together with several vir proteins, is then exported into the host cell cytoplasm through a channel formed by the *Agrobacterium* VirD4 and VirB protein complex (16). Before its entry into the host cell cytoplasm, the VirD2–T-strand conjugate is most likely coated by VirE2, forming the T-complex. VirE2 is a single-stranded DNA-binding *Agrobacterium* protein that is transported into the plant cell, where it presumably functions to protect the T-DNA from degradation.

The *Agrobacterium* T-complex is likely transported through the host cell cytoplasm by a cellular-motor-assisted mechanism. In a recent report, a dynein-like *Arabidopsis protein* (DLC3), coupled with another protein (VIP1), has been proposed to function in the intracellular transport of the *Agrobacterium* T-complex (17). Recently, an additional *Arabidopsis* protein, VIP2 (VirE2 interacting protein2), has been demonstrated to play a major role in T-DNA integration into the plant genome (18). The T-complex then enters the cell nucleus by an active mechanism mediated by the nuclear import machinery of the host cell. This facilitates integration of the T-strand into the host genome at random positions by a process of nonhomologous, or more precisely, illegitimate recombination.

Many recent reviews have addressed mechanisms related to T-DNA transfer (19–23). Characterization of the mechanisms governing the T-DNA transfer process is very important for plant transformation studies and should facilitate the identification of conditions to maximize T-DNA transfer. The best example of this is the use of a phenolic compound (e.g., acetosyringone) as well as a low-pH media and temperature to induce T-DNA transfer during the *Agrobacterium* infection stage.

Advancement in molecular biology techniques have enabled the development of binary Ti vectors that are compatible with utilization of both *Agrobacterium* strains and *Escherichia coli*. Development of the binary vector and bacterial strain systems for plant transformation is achieved by placing virulence genes on a separate plasmid (the large Ti-plasmid) and the gene to be transferred on separate vector (the small binary vector) (24). Since most gene manipulations are carried out in *E. coli*, the binary plasmids are designed to replicate in both *E. coli* and *Agrobacterium*. Recent advancements in vector-cloning techniques have led to the development of *bi*nary *b*acterial *a*rtificial *c*hromosome (BIBAC) vectors (25, 26) that enhance the frequency of T-DNA transfer of large-sized DNA fragments. The key features that make BIBAC vectors useful include its extremely low copy number and high stability when they are replicated in either *Agrobacterium* or *E. coli* cells.

The development of superbinary vectors made it possible to transform monocot plants for the first time. A superbinary vector represents an improved version of a binary vector; the vector carries a 14.8 kb *Kpn*I fragment containing the *virB, virG, vir C* genes derived from pTiBo542. These genes are responsible for the supervirulence phenotype of *A. tumefaciens* strain A281 (27). The superbinary vector has been highly efficient in transforming various plants, particularly recalcitrant species, such as important cereal crops.

The integration and enhancement of gene expression in the plant genome greatly depends on the promoter that is fused at the 5′ end of the gene of interest. The most widely used foreign regulatory elements include the 35S promoter of the cauliflower mosaic virus and the transcriptional terminator of the *Agrobacterium* nopaline synthase gene (*nos*), which together promote high-level gene expression in transgenic plants (28). The 35S promoter is a constitutive promoter that is used in vector constructs to drive target gene expression in many plant species. Recently, a new, stronger promoter has been developed. This "super promoter" is a hybrid construct combining a triple repeat of the octopine synthase (*ocs*) activator sequence plus the mannose synthase (*mas*) activator elements fused to the *mas* promoter. An initial study performed with this construct in maize (*Zea mays*) and tobacco (*Nicotiana tabacum*) (29, 30) confirmed the stable expression of superpromoter – GUS fusion gene in both the plant species (31). In tobacco, activity of the superpromoter is higher in mature leaves than young leaves, whereas in maize, the activity differed little among the tested aerial portions of the plant.

In order to achieve efficient *Agrobacterium*-mediated T-DNA transfer, several factors must be taken into consideration, including the plant genotypes, sources of explants, *Agrobacterium* strains, medium salt strength and pH, duration and temperature of *Agrobacterium*–explant interactions (inoculation and cocultivation), and use of T-DNA-inducing compounds (32, 33). *Agrobacterium*-mediated transformation of higher plants is now well established for dicotyledonous species. In recent years, the frequency of gene transfer to monocotyledonous species has also been greatly improved (Table 1). A variety of explants can be used as target material for *Agrobacterium*-mediated transformation,

Table 1
Major gene transfer methods used for transformation of monocot and dicotyledonous plant species

Plant species	Types of explants	Transfer method	Gene transferred	Transformation efficiency (%)	Reference
Monocots					
Oryza sativa	C	*Ag*	*cry1Ac*	2	(152)
	C	MpB	*shGH*	79.5	(153)
	SA	EP	*npt* II, *ppt*	13.8	(154)
Hordeum vulgare	ImE	*Ag*	*hpt* II, *luc*	25	(155)
	EC	MpB	*AtNDPK2*	0.15	(156)
Saccharum sp.	AxB	*Ag*	*npt* II, *bar*, *gusA*	50	(157)
	C	EP	*gusA*	80	(158)
Sorghum bicolor	C	*Ag*	*Man*, *gfp*	8.3	(159)
Triticum sp.	ImE	*Ag*	*bar*, *gusA*	9.7	(160)
Zea mays	ImE	*Ag*	*gusA*, *bar*	12.2	(41)
	ImE	MPB	*gusA*, *hpt* II	31	(73)
Dicots					
Arabidopsis thaliana	S	*Ag*	T-DNA	26	(161)
	L	MPB	*gusA*	–	(162)
Arachis hypogea	CN	*Ag*	*gusA*	38	(163)
	SE	MPB	*VP2*, *gusA*	12.3	(164)
	EL	EP	*gusA*	3	(165)
Brassica oleracea	ML	Cl-MPB	*cry1Ab*	11.1	(166)
Cajanus cajan	CN	*Ag*	*npt* II, *H*	51	(167)
Eucalyptus sp.	ApS	*Ag*	*gusA*	9	(168)
Glycine max	CN	*Ag*	*bar*, *gusA*	5.5	(169)
	SE	MPB	*Os-mALS*	60	(170)
	Fl	PTP	*phyA*	13	(171)
Gossypium hirsutum	EC	*Ag*	*cry1Ia5*	83	(7)
	EC	SCW	*AVP1*, *npt* II	64	(172)
Malus domestica	IVS	*Ag*	*Lc*	50	(173)
Pinus sp.	EC	*Ag*	*npt* II, *bar*, *gusA*	65–98	(174)

Ag Agrobacterium-mediated transformation, *AtNDPK2 Arabidopsis* nucleoside diphosphate kinase gene, *ApS* apical shoot, *AVP1 Arabidopsis* vacuolar pyrophosphatase, *AxB* axilllary bud, *bar* bialaphos-resistance gene, *C* callus, *Cl-MPB* chloroplast-mediated MPB, *CN* cotyledonary node, *cry1Ia5 Bacillus thuringiensis* toxins, *EC* embryogenic callus, *EL* embryonic leaflets, *EP* electroporation, *Fl* flower, *gusA* β-glucuronidase, *H* hemagglutinin protein, *hGH* human growth hormone, *hpt* II hygromycin phosphotransferase II, *ImE* immature embryo, *IVS* in vitro shoot, *L* leaf, *Lc* maize leaf color regulatory gene, *ML* mature leaf, *MPB* microprojectile bombardment-mediated transformation, *npt* II neomycin phosphotransferase II, *Os-mALS* acetolactate synthase derived from rice, *PTP* pollen tube pathway transformation, *phyA* phytase A, *S* seed, *SA* shoot apex, *SCW* silicon-carbide-whiskers-mediated transformation, *SE* somatic embryo, *T-DNA* transfer DNA including embryonic cultures, immature or mature zygotic embryos, mature seed-derived calli, meristems, shoot apices, primary leaf nodes, excised leaf blades, roots, cotyledons (including or excluding nodal areas), stem segments, and callus suspension cultures. These explant tissues have been able to regenerate through either a somatic embryogenesis or organogenesis regime. As an alternative to the organogenesis regeneration regime, somatic embryogenesis offers the advantage of single cell regeneration. However, the types and physiological conditions of explants used are critical to successful regeneration and subsequent recovery of whole transgenic plants. For example, in sorghum, a higher transformation efficiency was achieved in immature embryos taken from field grown plants than in the immature embryos from greenhouse-grown plants (34).

Competence for transformation can be enhanced in recalcitrant explants by phytohormone treatments. The maximal percent of calli showed higher transient GUS activity when picloram was used in *Typha latifolia* (35). Similarly, in *Hibiscus cannabinus* (Kenaf), preculturing of explants for 2 days in benzyladenine (BA) containing medium enhanced transient GUS expression (36). Phytohormone treatment activates cell division and dedifferentiation in many tissues. The stimulation of cell division by phytohormones suggests that efficient *Agrobacterium* transformation may occur at a particular stage of the plant cell cycle (37).

The procedures that promote *Agrobacterium* cells to come into close contact with the plant cells around wounded tissue sites have been found to enhance T-DNA transfer. For example, "dip-wounding," which is prewounding of the explants prior to cocultivation with blade dipped in *Agrobacterium* suspension, increases transformation frequency as high as 10-fold (Table 2, Xinlu Chen, Xiujuan Su, and Zhanyuan J. Zhang, unpublished data). When "dip-wounding" is combined with the use of phenolic compounds in inoculation and cocultivation media, the attraction of *Agrobacterium* is presumably enhanced at wounded sites, which facilitates increased access of bacteria to plant cells.

Table 2
Impact of dip-wounding on soybean regeneration and transformation[a]

Inoculation procedure	Number of explants			Percentage of explants		Number of events	Percentage of events
	To start	Good	Green	Good	Green		
Standard	350	172	129	49.0	37.0a	1	0.3a
Dip-wounding	350	198	236	57.0	67.0b	12	3.4b

[a]Data were collected from three independent experiments using the soybean genotype "Maverick." Percentages within the column followed by different letters indicate significant difference as detected by Duncan's multiple range test at $\alpha = 0.05$ level. Good: the explant showing multiple bud or shoot formation without axillary shoot; Green: the explant in which over half of the multiple buds or shoots are green (alive)

In a number of plant species, explants are hypersensitive to *Agrobacterium* infection, forming necrotic barriers; this can be overcome by the use of antioxidants to reduce the oxidative burst. Tissue browning/necrosis associated with *Agrobacterium*-mediated transformation has been reported in various types of explants of both dicotyledonous and monocotyledonous species (38). Antioxidants such as polyvinylpyrrolidone (PVPP), dithiothreitol (DTT), cysteine, glutathione, lipoic acid, ascorbic acid, and citric acid are now commonly used to reduce tissue browning/necrosis of explants during plant transformation. In sugarcane, pretreatment of explants on media containing ascorbic acid and cysteine prior to transformation results in higher transformation efficiency (39). In grapes and rice, the addition of antioxidants, such as polyvinylpyrrolidone and dithiothreitol also increases the transformation efficiency. In soybean, which is difficult to transform, a higher transformation rate was achieved by including L-cysteine, DTT, and sodium thiosulphate in cocultivation media (40). These antioxidants also enhanced the transformation in maize Hi-II (41, 42). Such increased T-DNA transfer enabled the use of standard binary vectors to routinely achieve efficient transformation without the use of a superbinary vector. In contrast, lipoic acid was found to enhance GUS transient expression and transformation efficiency in tomato (43, 44).

In addition to the strategies discussed above, other treatment conditions have been devised more recently to promote *Agrobacterium*-mediated transformation. Desiccation of explants prior to *Agrobacterium* infection enhances transformation efficiency in sugarcane, whereas addition of surfactants such as Silwet-L77 or pluronic acid F68 enhances transformation in wheat (45). Such desiccation helps to reduce cell damage due to the reduction of cell turgidity, whereas the use of surfactants may induce wounds and thinner cell walls in the explant tissues, thereby promoting *Agrobacterium* attachment to explants and ultimately T-DNA transfer. Another method is treatment of plant tissues and *Agrobacterium* to brief sonication, which allows *Agrobacterium* and T-DNA entry into the tissues. In loblolly pine, sonication was found to enhance not only transient transformation but also the recovery of hygromycin-resistant lines (46).

Plant transformation frequency is also associated with cell division or dedifferentiation of the host explants. Recent studies have revealed the phase of the plant cell cycle at the time of transformation to be a major determinant of transformation and regeneration efficiency. To achieve a stable transgenic event, the differentiated cell for regeneration should adopt a "stem-cell-like" state for pluripotentiality to renter the S phase of the cell cycle (47, 48). The transformation competence of the cells is high in S and G2 phase/M phase, and lower in G0 and G1 phases (49). A cell cycle study identified the RepA, HP1, E2Fa, CycD3, and CycD1 genes to be involved in the pluripotency of cells in G1-S phase and further progression through S and G2 phases (50, 51). Gordon-Kamm et al. demonstrated substantial improvement in maize transformation by overexpressing the RepA gene (52). *Arabidopsis* histone H2A, which is expressed at higher levels in tissues that are more susceptible to *Agrobacterium* infection, is essential for T-DNA integration in somatic cells (53). Co-overexpressing E2Fa together with its dimerization partner, DPa, resulted in increased cell proliferation in cotyledons, leading to approximately twice the number of cells as wild type (54).

Transcriptome analysis of E2Fa–DPa-overexpressing plants showed upregulation of 14 genes that are involved in DNA replication and S phase onset (54). VIP1 (VirE2 interacting protein 1) in *Arabidopsis*; those genes were also found to be upregulated during the pluripotent stage. Overexpression of VIP1 increased the rate of transient and stable plant transformation and predicted its role in interacting with histone proteins (17). A recent study in *Phaseolous coccineus* has identified the proteins PIN and CUC to be involved in shoot apical meristem formation (55, 56). The use of transgenic marker genotypes, such as *WUSCHEL* (WUS)-reporter or *STM*-reporter (*SHOOTMERISTEMLESS*) construct should be useful in identifying meristemoids in early stages in development.

Agrobacterium rhizogenes strains contain a T-DNA region located on the Ri plasmid that carries genes involved in root initiation, which are essential for production of hairy roots. Studies on the function of Ri T-DNA-encoded genes performed using the agropine-type Ri plasmids (57) led to the identification of 18 ORF, including rolA, rolB, rolC, and rolD genes. It was evident that these genes also participate in the production of hairy roots. Several of the experiments in this study were carried out to inactivate or overexpress various *rol* genes, generating stable transgenic lines with various alterations in plant phenotypes and root morphology.

In general, however, *A. rhizogenes*-mediated root transformation has received considerably less attention than *A. tumefaciens* transformation. The main reason for this is the difficulty in regenerating plants from hairy roots transformed by *A. rhizogenes.* Therefore, this delivery system has been predominantly used to generate transgenic roots for transient assays. One of the most advanced systems in such type of assay is the production of "composite plants" (58). The important characteristic feature of hairy roots is their ability to grow in plant hormone-free media. These growth characteristics have made hairy roots a useful tool for secondary metabolite production, use in metabolic engineering, and studies of root biology in general (59, 60). Recently, it has been demonstrated that hairy-root cultures can be adapted for T-DNA-activation tagging studies (61). Hairy roots are also used to genetically investigate root–bacterial interactions in soybean. For example, various studies have revealed that hairy roots derived from various soybean cultivars maintain their cognate nematode resistance or susceptible phenotypes (62). Recently, hairy roots have been used to produce recombinant proteins (63). This system is also suitable for high-throughput analysis of root-related transgene expression in *Medicago* or soybean root tissues and is now expected to be applied to root transformation for high-throughput functional analysis of certain gene expressions in other plant species (64).

In summary, *Agrobacterium*-mediated transformation has been very successfully employed recently in transformation of both dicots and particularly monocots, the latter of which had long been thought to be unable to host *Agrobacterium*. These successes are attributable to the development of the superbinary vector, the use of antioxidants, and optimization of the composition of inoculation and cocultivation media. This trend of success in transformation of various plant species will continue not just because of new ideas and approaches in improving *Agrobacterium* transformation but also because of the obvious advantages of such a natural gene delivery system.

2.2. Microprojectile (Particle) Bombardment Transformation

Microprojectile bombardment is one of the direct gene transfer methods for development of transgenics. This method was developed in 1980s to genetically engineer plants that were recalcitrant to transformation with *Agrobacterium*. Subsequently, the technique has been widely used to produce transgenic plants in a wide range of plant species (65). The first particle delivery method was developed by Sanford and coworkers. The Sanford device was extensively modified to produce the PDS-1000/He machine, which was licensed to DuPont. The technique involves coating microcarriers (gold or tungsten particles approx 0.6–1.0 μm in diameter) with the DNA of interest and then accelerating them at high velocities, to penetrate into the cell of essentially any organism.

Briefly, the microcarriers are spread evenly on circular plastic film (macrocarrier). The entire unit is then placed below the rupture disk in the main vacuum chamber of the biolistic device. A variety of rupture disks are available that burst at pressures ranging from 450 to 2,200 psi. Below the macrocarrier is a stopping screen, in which a wire-mesh is designed to retain the macrocarrier, while allowing the microcarriers to pass through. The target tissue is placed below the launch assembly unit. Under a partial vacuum, the microprojectile is fired, and helium is then allowed to fill the gas-acceleration tube. The helium pressure builds up behind a rupture disk, which bursts at a specific pressure, thus releasing a shock wave of helium that forces the macrocarriers down onto the stopping screen. The microcarriers leave the circular plastic film and continue flying down the chamber to hit and penetrate the target tissue, thus delivering the DNA.

Several factors must be considered for successful gene transfer using particle bombardment technology. These factors include the design of a suitable vector with a small size and high copy number, as well as the quantity and quality of the delivered DNA. The entire process must be performed under sterile conditions to prevent contaminations of target tissue during subsequent tissue culture. The types and sizes of microcarriers are important choices because they affect the depth of penetration of the accelerated microcarrier as well as the degree of damage to the target cells. Gold particles ranging from 0.6 to 3.0 μm in diameter are commercially available. The degree of penetration required will depend on the thickness of the cell wall, the type of tissue being transformed, and the depth of the target cell layers. Variation in the helium pressures, the level of vacuum generated, the size of the particles, and the position of target tissues will dictate the momentum and penetrating power at which the microprojectiles strike the tissue. All of these parameters are under the experimenter's control and must be optimized for a given target tissue (10).

Treatment of the target tissues prior to and after particle bombardment has a significant effect on the frequency of recoverable transgenic cell lines and plants. An attractive feature of particle bombardment is its ability to transfer foreign DNA into any cell or tissue type whose cell wall and plasma membrane can be penetrated. Embryogenic and meristematic tissues are the most commonly employed target tissues for the production of genetically transformed plants. Particle bombardment of embryogenic tissues has been successfully exploited to produce transgenic plants in a wide range of agronomically important plants, including legumes, tuber crops, starchy staples, trees, commodity crops, and all of the major cereals (66, 67). In the case of bombarding apical meristems, the treatment, physiological status, and age of the mother plants prior to excision of the explants must be taken into consideration. Use of an osmotic pretreatment or partial drying of the target cells prior to bombardment is a commonly used strategy to increase the frequency of successful transformation (68).

One of the advantages of particle bombardment is the possible expression of multiple transgenes in the target tissue, which can be achieved by fusion of genes within the same plasmid that is then bombarded into the target tissues. In recent years, multiple independent gene expression cassettes have been successfully transferred using particle bombardment in wheat, rice, and soybean (67, 69, 70). The use of microprojectile bombardment has made it easy to transfer large DNA fragments into the plant genome, although the integrity of the DNA is a concern. Integration of yeast artificial chromosomes (YACs) into the plant genome by particle bombardment has been successful with inserts of up to 150 kb (71, 72). Although requiring further development, integration of large DNA fragments promises to be an important tool in future plant research and crop biotechnology.

In summary, over the last two decades, many published studies have utilized microprojectile technology in both monocot and dicot plant species (Table 1). In the process, a diverse range of agronomic traits have been transferred and imparted to crop plants via particle bombardment, which remains the most promising technique to genetically engineer plastids. However, this technology is limited due to several drawbacks, such as the integration of multiple copies of the desired transgene, in addition to superfluous DNA sequences that are associated with the plasmid vector. Multicopy integrations and superfluous DNA can lead to silencing of the gene of interest in the transformed plant. This problem was overcome by transferring the desired coding region only with its control elements into the target cells of plant genome (73).

2.3. Electroporation-Mediated Transformation

Electroporation-mediated transformation requires the application of strong electric field pulses to cells and tissues and is known to cause some type of structural rearrangement of the cell membrane. *In vitro* introduction of DNA into cells is now the most common application of electroporation. The technique was originally developed for protoplast transformation but has subsequently been shown to work with intact plant cells as well. A voltage of 25 mV and an amperage of 0.5 mA for 15 min are the most often used parameters. However, factors such as surface concentration of DNA and tolerance of cells to membrane permeation may affect electroporation efficiency. Using the electroporation method, successful transformation has been achieved with protoplasts of both monocot and dicot plants. The first report of fertile transgenic rice utilized electroporation of DNA into embryogenic protoplasts (74). However, using protoplasts as explants for regeneration of transformants limits the use of electroporation for stable transformation because the protoplast-to-plant regeneration system has not been developed in most plant species. The electroporation of plant cells and tissues is very similar in its principles to the electroporation of protoplasts. This approach enabled the recovery of the first transgenic plants in barley (75). In sugarcane, a gene was transferred into intact meristem tissue using electroporation-mediated transformation (Table 1). While electroporation was proposed as an alternative to biolistics, it is not nearly as efficient. Compared to biolistics, it is inexpensive and simple, but the technique has only been successful in a few plant species. The thick cell walls of intact tissues represent key physical barriers to electroporation.

2.4. PEG/Liposome-Mediated Transformation

Polyethylene glycol (PEG)-mediated transformation is a method used to deliver DNA using protoplasts as explants. The method is similar to electroporation in that the DNA to be introduced is simply mixed with the protoplast, and uptake of DNA is then stimulated by the addition of PEG, rather than an electrical pulse. PEG-mediated transformation has several advantages because it is easy to handle and no specialized equipment is required. However, the technique is rarely used due to the low frequency of transformation and because many species cannot be regenerated into whole plants from protoplasts. In addition, fertility may be a concern because of the somaclonal variation of the transgenic plants derived from protoplast cultures. Nonetheless, using this method, transgenic maize and barley have been produced (76). Thus, protoplast transformation is feasible in cereals, even though fertility problems in the regenerants are often encountered. In cotton, transformation was achieved using combined polybrene–spermidine-based callus treatment (77).

Related to PEG-mediated transformation is the liposome-mediated transformation technique. In this method, DNA enters protoplasts via endocytosis of liposomes. Generally, this process involves three steps: adhesion of liposomes to the protoplast surface, fusion of liposomes at the site of adhesion, and release of the plasmid inside the cell. Liposomes are microscopic spherical vesicles that form when phospholipids are hydrated. Liposomes being positively charged tend to attract negatively charged DNA and cell membrane (78). In this process, the engulfed DNA is free to integrate into the host genome. However, there have been very few successful reports on the application of this technique in plant species because the technique is very laborious and is associated with low efficiency. In tobacco, intact YACs were transformed via lipofection-PEG technique (79).

2.5. Silicon Carbide-Mediated Transformation (SCMT)

Kaeppier et al. first reported the use of silicon-mediated transformation, which is one of the least complicated methods. In this method, small needle-type silicon carbide whiskers are mixed with plant cells and the gene of interest, and the mixture is then vortexed (80). In the process, the whiskers pierce the cells, permitting DNA entry into the cells. The fibers most often used in this procedure have an elongated shape, a length of 10–80 mm and a diameter of 0.6 mm and show high resistance to expandability. The method is simple, inexpensive, and effective on a variety of cell types (81). The efficiency of SCMT depends on fiber size, vortexing, the shape of vortexing vessel, as well as the plant material used for transformation. The SCMT technique has been used in a variety of plants, including maize, rice, wheat, tobacco, etc. Furthermore, silicon carbide fibers have been found to improve the efficiency of *Agrobacterium*-mediated transformation (82). The main disadvantages of SCMT include low transformation efficiency and damage to cells, thereby negatively affecting their regeneration capacity. Furthermore, this method imposes health hazards due to fiber inhalation if not performed properly.

More recently, two related technologies have been developed: silicon fibers have been reported to increase callus transformation by 30–50% in rice (83), and mesoporous silica nanoparticles have been used to deliver DNA and chemicals into both plant cells and intact leaves (84). Mesoporous silica nanoparticles are synthesized from a reaction between tetraethyl orthosilicate and a template made of micellar rods (85).

2.6. Microinjection

Microinjection is the direct mechanical introduction of DNA into the nucleus or cytoplasm using a glass microcapillary injection pipette. Using a microscope, cells (protoplasts) are immobilized in low-melting-point agar with a holding pipette and gentle suction; DNA is then injected into the cytoplasm or nucleus (86). The microinjection technique requires relatively expensive technical equipment for the micromanipulation of single cells under a microscope and involves precise injection of small amounts of DNA solution; the procedure is also very time-consuming. The injected cells or clumps of cells are subsequently cultured in tissue culture systems and regenerated into plants. Successful regeneration of microinjected rapeseed (87), tobacco, and barley (88) has produced genome-integrated, stable transformants. However, microinjection has achieved only limited success in plant transformation due to the thick cell walls of plants and, more challengingly, to a lack of availability of a single-cell-to-plant regeneration system in most plant species.

2.7. Chloroplast-Mediated Transformation

In genetically modified plants, the gene of interest usually integrates into the nucleus; however, it is also possible to transfer the gene into the plastid. The chloroplast genome is highly conserved among plant species and typically consists of double-stranded DNA of 120–220 kb, arranged in monomeric circles or in linear molecules. In most higher plant species, the chloroplast genome has two similar inverted repeat (IR_A and IR_B) regions of 20–30 kb, that separate a large single copy (LSC) region and small single copy (SSC) region (89). Both the microprojectile or protoplast-mediated transformation methods are capable of delivering DNA to plastids (90), but to achieve successful transformation, chloroplast-specific vectors are required. The basic plastid transformation vector is comprised of chloroplast-specific expression cassettes and target-specific flanking sequences. Integration of the transgene into the chloroplast occurs via homologous recombination of the flanking sequences used in the chloroplast vectors (91). The first successful chloroplast transformation (of the *aada* gene, which confers spectinomycin resistance) was reported in *Chlamydomonas* (92). In higher plants, plastid transformation has been accomplished in tobacco with various foreign genes. In recent years, several crop chloroplast genomes have been transformed through organogenesis, and maternal inheritance has been observed (93). In economically important crops such as cotton, efficient plastid transformation has been achieved through somatic embryogenesis by bombarding embryogenic cell cultures.

Several transgenes engineered through chloroplast transformation have conferred valuable agronomic traits in plants, including insect and pathogen resistance, and both drought and salt resistance. In soybean expressing the Cry1Ab gene, insecticidal activity against velvet bean caterpillar was conferred to the transformed plant (Table 1). Advancement in chloroplast engineering has made it possible to use chloroplasts as bioreactors for the production of recombinant proteins and biopharmaceuticals. Because plastid genes are maternally inherited, transgenes inserted into these plastids are not disseminated by pollen. Additional advantages of this transformation system include the ability to express several genes as a polycistronic unit, thereby potentially eliminating position effects and gene silencing in chloroplast genetic engineering (94, 95).

2.8. Native Gene Transfer

Transformation of native genes (including regulatory elements) into plants without a selectable marker is highly desirable to overcome consumers' concerns about GM crops. Historically, the development of transgenics with important agronomic traits depended on the use of genes derived from other organisms. Over the past decade, however, rapid advances in plant molecular biology have resulted in a major shift from bacteria and viruses to plants as important gene sources. A broad variety of plant genes associated with agronomically important traits have now been identified (96). For example, plants containing modified acetolactate synthase (ALS) genes displayed the same high levels of sulfonylurea tolerance as transgenic plants that expressed bacterial ALS tolerance genes (97). Likewise, the occurrence of glyphosate tolerance in a goosegrass (*Eleusine indica*) biotype has been associated with a mutated 5-enolpyruvylshikimate-3-phosphate synthase (*EPSPS*) gene (98). With respect to native gene transfer research, the application of transposon tagging and map-based cloning methods have resulted in the identification of more than 50 functionally active resistance (R) genes, several of which are currently being used as viable alternatives to foreign antimicrobial genes in crop improvement programs. One of the most agronomically important R-genes isolated is the *Solanum bulbocastanum* RB gene, which provides resistance to the potato late blight fungus *Phytophthora infestans* (99).

Rapid progress has also been made in the development of plant-based gene alternatives and recovery of various insecticidal proteins that are involved in insect resistance. It has been suggested that a 30-kDa maize cysteine protease can be used to enhance maize tolerance to caterpillars and armyworms (100). Alternatively, silencing or overexpressing key biosynthetic genes can enhance a plant's ability to produce insecticidal secondary metabolites. In tobacco, suppression of a P450 hydroxylase gene resulted in a 19-fold increase in cembratrieneol levels in trichomes, dramatically enhancing aphid resistance (101). About 40 diverse plant genes have been used to enhance tolerance to abiotic stresses. In *Arabidopsis*, overexpression of C-repeat-binding transcription factor (CBF) increased survival resistance to freezing, drought, and salt (102).

Various methods such as promoter trapping and RNA fingerprinting have facilitated the identification of hundreds of plant promoters (e.g., ubiquitin and actin gene promoters), many of which contain regulatory elements that support high-level gene expression in most tissues of transgenic plants. Use of the same genetic material available in the plant benefits genetic engineering approaches into existing plant breeding programs.

3. Marker Genes and Methods to Remove Selectable Markers from Transgenics

After explants are transformed with the requisite vector having the gene of interest, the transformed cells/tissues need to be selected, which requires a selectable marker gene in the vector. In current transformation systems, a selectable marker gene is codelivered with the gene of interest to identify and resolve rare transgenic cells from nontransgenic cells (103). However, during transformations, only a few plant cells accept the integration of foreign DNA; most cells remain nontransgenic. Several selectable marker genes are currently in use, and most utilize antibiotic or herbicide selection (Table 3). Of these, the *npt* II gene (encoding neomycin phosphotransferase II), which imparts kanamycin resistance, is commonly used in the development of transgenics. Visual observation of gene expression is achieved using reporter genes that can represent important components for transient and stable expression studies. The most commonly used reporter genes in plant transformation are the β-glucuronidase (GUS) gene and the green fluorescent protein *gfp*. Selectable marker genes are required to recover stably transformed plants. However, due to environmental concerns, as well as human health risks, the use of nonselectable markers needs to be promoted. In recent years, the following techniques have been used for the production of selectable-marker-free plants.

3.1. Cotransformation

Cotransformation is a strategy that utilizes two plasmid vectors: one containing the gene of interest and the other containing a selectable marker gene. With this approach, integration of genes may be either within a single locus or at unlinked loci (104). The method achieves a high cotransformation efficiency and is usually

Table 3
Various selectable markers and reporter genes commonly used in transgenic plants

Gene	Enzymes encoded	Substrate	Gene source	Reference
Selectable markers				
bar	Phosphinothricin acetyl transferase	Phosphinothricin	*Streptomyces hygroscopicus*	(175)
BADH	Betaine aldehyde dehydrogenase	Betaine aldehyde	*Spinacia oleracea*	(176)
bxn	Bromoxynil nitrilase	Oxynils	*Klebsiella pneumonia*	(177)
cat	Chloramphenicol acetyl transferase	Chloramphenicol	*Escherichia coli* Tn5	(178)
dhfr	Dihydrofolate reductase	Methotrexate	*Candida albicans*	(179)
EPSPS	5-Enolpyruvyl shikimate-3-phosphate synthase	Glyphosate	*Petunia* × *hybrida*	(180)
gox	Glyphosate oxidoreductase	Glyphosate	*Ochrobactrum anthropi*	(181)
hpt II	Hygromycin phospho-transferase II	Hygromycin B	*E. coli*	(182)
ManA	Phosphomannose isomerase	D-Mannose	*E. coli*	(183)
npt II	Neomycin phosphotrans-ferase II	Kanamycin	*E. coli* Tn5	(184)
xylA	Xylose isomerase	D-Xylose	*Streptomyces rubignosus*	(185)
Reporter genes				
uidA/GUS	β-Glucuronidase	X-gluc	*E. coli*	(186)
gfp	Geen fluorescent protein		*Aequorea victoria*	(187)
lacZ	Galactosidase	X-gal	*E. coli*	(188)
luc	Luciferase	Luciferin	*Photinus pyralis*	(189)
	Oxalate oxidase	Oxalic acid	*Triticum aestivum*	(190)

carried out using biolistic or *Agrobacterium*-mediated transformation. The approach may permit the simultaneous introduction of many genes, independent of gene sequence, with a limited number of selectable marker genes. For example, in one report, nine genes were transferred into the rice genome by biolistics. Because the cotransformed genes are integrated at a single locus, they cosegregate. However, *Agrobacterium*-mediated cotransformation has the advantage over biolistic transformation that cotransformed genes often integrate into different loci, which results in the segregation of unlinked selectable marker genes from the gene of interest, thereby permitting production of marker-free transgenic plants (105). Cotransformation via *Agrobacterium* uses either two mixed *Agrobacterium* populations (each carrying a different binary vector) or a single *Agrobacterium* population (carrying two different binary vectors) (Fig. 1).

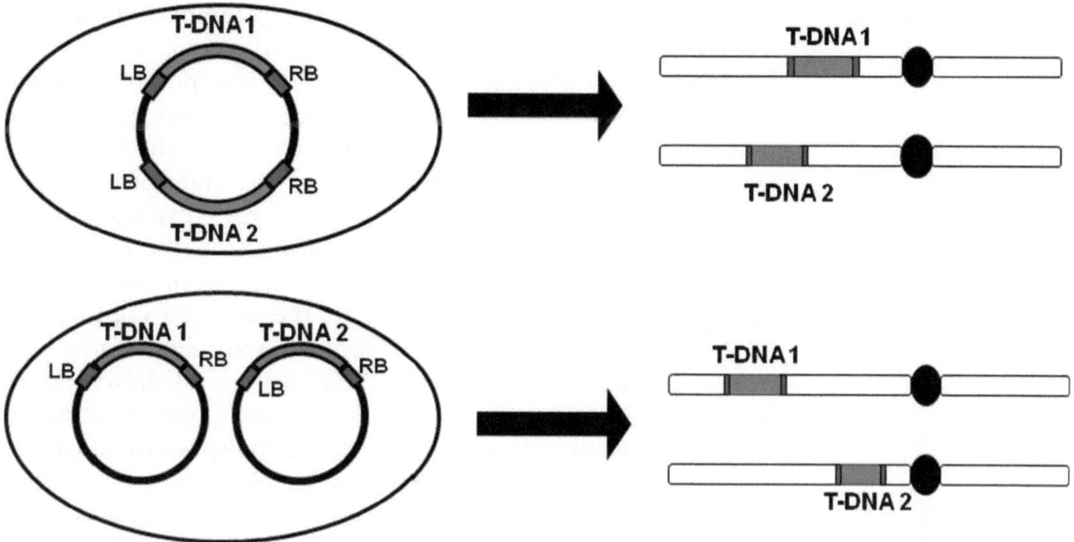

Fig. 1. Schematic representation of two T-DNA on single or two separate vectors in *Agrobacterium*. LB and RB: T-DNA left and right borders, respectively. After transformation, the cotransformed T-DNA integrates, along with its flanking sequences (the LB and RB), into the plant genome at different loci.

3.2. Transposon-Mediated Transformation

Recent developments in genetic engineering have made it possible to use transposon-based transformation to eliminate selectable marker genes in transgenics; for example, the maize *Ac/Ds* transposable element system has been used to excise selectable marker genes from plants. In this approach, the marker gene is flanked by the inverted repeat sequences of the *Ds* element. Subsequent to the transformation and T-DNA integration, expression of the *Ac* transposase from within the T-DNA results in excision of the gene of interest from the T-DNA insert containing the selectable marker gene. As a result, the gene of interest is transferred from the T-DNA site to another chromosomal location. Successful application of the system therefore requires the activity of the *Ac* transposase for the development of marker-free transgenic plants (106).

Repositioning within the genome can also enhance the expression profile of the gene of interest. In tomato, transposition of the GUS reporter gene and subsequent generation of *npt* II-free plants has been accomplished with both single and multiple T-DNA insertions. For example, *hpt* II-free rice plants were created that expressed *Bt* endotoxin encoded by the *cry1B* gene excision; reinsertion of the transgene occurred at very high frequencies (25–37%), preserving high levels of resistance to striped stem borer (107). However, because this technology relies on crossing plants to segregate the gene of interest from the marker gene and transposase, it is of limited use in plants that are vegetatively propagated or which have a long reproductive cycle.

3.3. Multiautosystem

To improve the selectable-marker-free plant technology, an *IPT*-type MAT (multiautotransformation) vector system was developed as an alternative to the *Ac*/Ds transposable element system. A MAT vector system uses the *IPT* (isopentenyltransferase) gene and *Ac* element under the control of the CaMV35S promoter. However, *IPT* expression in the transformed plant generates an abnormal phenotype called the "extreme shooty phenotype." Thus, subsequent to transformation, the *IPT* gene is removed using the *Ac* transposable element from the T-DNA, leaving only the gene of interest in the inserted copy of the T-DNA (108). This results in marker-free transgenic plants with a normal phenotype. However, there are several drawbacks of using a MAT system for marker gene removal (109). Specifically, there is the possibility of variable rates of transposition efficiency and also of reinsertion of the transposable element.

Because of potential biosafety concerns, attempts have been made to generate transgenic plants devoid of selectable marker genes and vector backbone sequences from the T0 generation (110). Thus, methods have been devised to reduce the frequency of vector backbone sequence integration during plant transformation. In maize, the ovary-drip method was used to increase transformation frequency using linear green fluorescent protein (GFP) cassettes (Ubi-GFP-nos) flanked by 25-bp T-DNA borders as the transfer gene (111, 112). In another approach, multiple tandem LB repeats were used to suppress the transfer of vector backbone in rice transformation. And more recently, soybean transformation experiments have been conducted to reduce vector backbone sequences by using nonlethal genes that interfere with plant development (113), resulting in a high frequency of vector-backbone-free transgenics. Multiple border sequences can also be used to generate vector-backbone-free transgenics (114, 115). And finally, in maize, backbone-free transgenics were developed using vectors having a selectable marker gene in the vector backbone and the gene of interest in the T-DNA (116).

3.4. Site-Specific Recombination

The most widely used site-specific recombination system is the Cre–*lox* system from bacteriophage P1, which is very effective in the generation of marker-free plants (117). In this system, the plant is transformed with a T-DNA vector carrying the gene of interest with *lox* sites (34 bp repeats in direct orientation) flanking the selectable marker. In the second round of transformation, Cre recombinase is introduced to achieve precise excision of the marker gene. Specifically, Cre catalyzes the recombination between the *lox* repeat sequences, thereby eliminating the marker gene in the progeny (118). This system has been used in various plant species to generate marker-free transgenics. Using the Cre–*lox* system, it is also possible to resolve multiple transgene copies into only a single copy per recipient genome (119). Recent advances in this technology have led to the use of different inducible promoter systems; for example, a strategy has been developed using the B estradiol-inducible promoter system in which an artificial transcription factor, XVE, was constructed for use in plants with its cognate promoter (120). Using the Cre/*lox* system under the control of an inducible promoter, successful marker-free transgenic plants were developed. Recently, successful excision of *lox*P-flanked selectable marker has been achieved using a flower-specific promoter in rice (121). However, to achieve a high transformation efficiency, more refinement is needed to improve this technology.

3.5. Two T-DNA-Mediated Transformation

Using two T-DNAs, one bearing a selectable marker and the other containing the gene of interest, in a single vector, marker-free transgenic plants have been produced. This approach yields higher frequencies of cotransformation than a strategy using a mixture of *A. tumefaciens* strains carrying separate vectors. In one study, GUS and *hpt* II genes cotransformed into tobacco showed segregation of both the genes at unlinked loci, resulting in *hpt* marker-free plants (122). In another report, a 100% cotransformation frequency was achieved in tobacco when the selected T-DNA was twice as large as the nonselected T-DNA. In maize, cotransformation with an octopine strain carrying a binary vector with two T-DNAs yielded cotransformation frequencies of 93% for the *bar* and GUS genes in the T0 generation (123). In barley, a similar approach using smaller vectors yielded a cotransformation frequency of 66%, but only 24% of these segregated as marker-free plants (124). These studies clearly demonstrate that marker-free plants can be generated with variable efficiencies using *Agrobacterium*-mediated cotransformation followed by segregation of the genes in the subsequent sexual generations.

4. Gateway Plant Transformation Vectors

Gateway cloning technology offers a fast and reliable high-throughput, restriction-enzyme-free cloning strategy for plasmid construction. The Gateway technology is based on the site-specific recombination reaction mediated by bacteriophage λ DNA fragments flanked by recombination sites (*att*). These sites can be transferred into vectors containing compatible recombination sites (*att*L *att*R or *att*B, *att*P) in a reaction mixture mediated by the Gateway clonase mix (125). The backbone of all described Gateway-compatible plant transformation vectors is the plasmid pPZP200 (115). Two recombination reactions, catalyzed by LR and BP recombinases (clonases), respectively, are used in Gateway cloning. The first step, catalyzed by LR, inserts the gene of interest into the Gateway vector at the *att*L and *att*R sites. The resulting construct is called the entry clone. All entry clones have *att*Ls flanking the gene of interest. These are necessary in the Gateway system because these *att*L sites are cut to form sticky ends by the Gateway clonase. These sticky ends match with the sticky ends of the destination vector, which contains *att*R restriction sites. This process is called a LR reaction and is mediated by LR clonase mix, which contains the recombination proteins necessary for excision and incision. The product formed in the LR reaction is called the expression clone, which represents a subclone of the starting DNA sequence, correctly positioned in a new vector backbone. The second Gateway step is the BP reaction, which is the reverse of the LR reaction. In the BP reaction, the DNA insert flanked by 25 bp *att*B sites is transferred from the expression clone into a vector donated by a plasmid containing the *att*P sites. The final product is termed the destination clone and contains the transferred DNA sequence. Alternatively, these two sequential steps can be reversed to meet specific cloning needs (Fig. 2) (Olga V. Karpova and Zhanyuan J. Zhang, unpublished). The BP reaction thus allows rapid, efficient, directional PCR Cloning (126, 127).

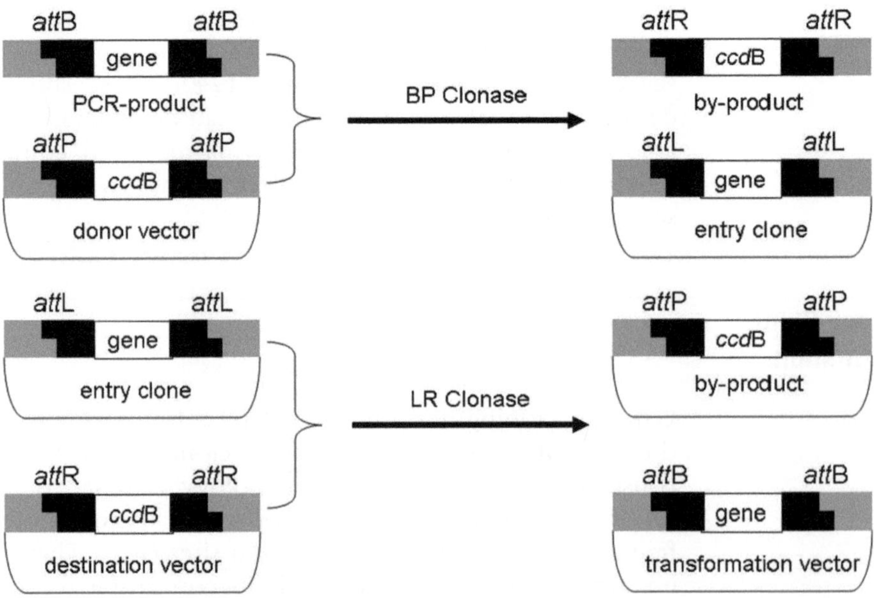

Fig. 2. Diagram of gateway cloning technology to clone a PCR-product containing the gene of interest into plant transformation vector, without the use of restriction or ligation enzymes. *ccd*B, a bacterial suicide gene whose replacement by the incoming transgene-containing PCR product, will allow the transformed bacterial cells to survive, ensuring the presence of the transgene in either the entry or destination clone.

With advances in Gateway technology, it is possible to create many vectors with fused or directly linked multiple transgenes (128). Recently, for RNA interference, a high throughput cloning system has been designed using pHELLSGATE vectors. In these vectors, the recombination sites do not affect gene silencing performance, in contrast to the conventional restriction enzyme cloning vectors. In *Arabidopsis*, overexpression of large DNA fragments was found to be effective using Gateway technology. Molecular analysis of T0 and T1 generations confirmed that Gateway-compatible constructs were active and also that the *att* recombination sites did not inhibit transgene activity or interfere with enhancer activity *in vivo* (129, 130).

However, the use of Gateway-RNAi transformation vectors can be a challenge due to the requirement for generating double-stranded RNA via inverted repeats. Thus, because the *att* recombination sites need to be duplicated, there is an increased chance of altering the orientation of the intron spacer whose splicing efficiency may be adversely impacted, causing reduced RNAi-mediated silencing efficacy. One solution to this problem is to use a double-intron spacer in the opposite orientation (131). However, the effectiveness of such a design needs to be validated in more experiments with different plant species.

Recently, the "pEarleyGate" vectors have been designed for *Agrobacterium*-mediated plant transformations; these vectors translationally fuse FLAG, HA, cMyc, AcV5, or tandem affinity purification epitope tags onto target proteins, with or without an adjacent fluorescent protein (132, 133). A high-throughput protocol has recently been developed using the Gateway binary vector R4pGWBs for transformation of *Arabidopsis thaliana*. This vector is designed for the one-step construction of chimeric genes between any promoter and any cDNA (134). Also, autofluorescent protein tags are useful due to their ability to visualize cellular processes in vivo (135). For example, the pSAT vectors are useful for both autofluorescent protein tagging and multiple gene transfer (136). In many plant laboratories, Gateway binary vector technology has enabled *in planta* expression of recombinant proteins fused to fluorescent tags (137). Efficient expression of fluorescent protein has been achieved in *Nicotiana benthamiana* (138). For easy manipulation and efficient cloning of DNA fragments for gene expression studies, a new Gateway expression vector has been developed by combining the Gateway system and a recombineering system. The recombineering system uses bacteriophage-based homologous recombination in which genomic DNA in a bacterial artificial chromosome (BAC) is modified or subcloned without restriction enzymes or DNA ligase (139), thereby permitting the direct cloning of gDNA fragments from BACs to plant transformation vectors. The construct is converted into a novel Gateway Expression vector that incorporates cognate

5′ and 3′ regulatory regions, using recombineering, to replace the intervening coding region with the Gateway Destination cassette. Using this approach, efficient transformation has been achieved in *Arabidopsis* (140).

5. Gene Targeting (Zinc-Finger)

Zinc-finger endonucleases (ZFNs), or "molecular scissors," have recently been developed to target genes in plant systems. ZFNs are engineered proteins that have highly specific ZF domains fused to a sequence-independent nuclease domain. The current generation of ZFNs combines the nonspecific cleavage endonuclease domain of the Fok I restriction enzyme with several (usually three) zinc fingers domains that provide cleavage specificity. Subsequent to transformation, ZFNs introduce targeted double-stranded breaks in genomic DNA, thereby inducing recombination and repair processes at specific sites. In this process, homologous recombination will occur using the homologous sequence of the transferred gene to repair the double-strand DNA break (141, 142) (Fig. 3).

ZFN-mediated gene targeting promises to be a powerful tool in the development of novel crop species possessing beneficial

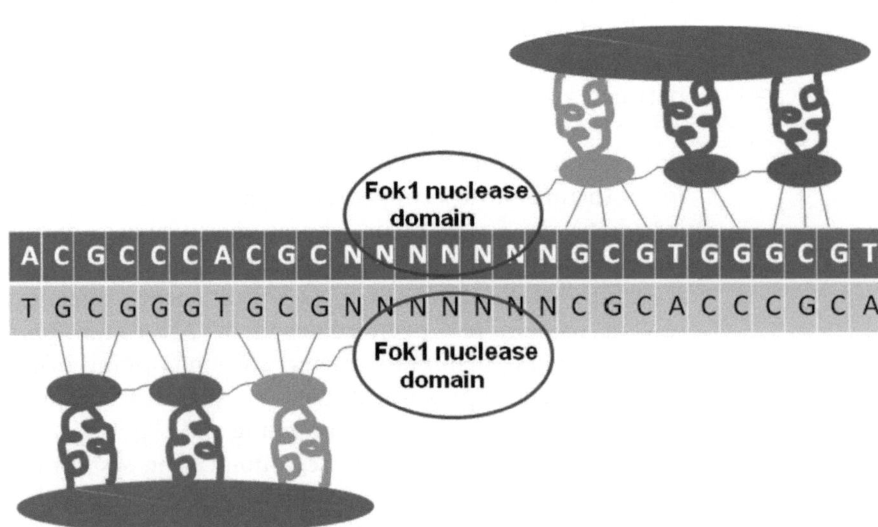

Fig. 3. Schematic representation of a zinc finger nuclease (ZFN) binding to its target site. The three different zinc finger domain indicate different sequences. The zinc-finger domains are fused to the cleavage domain of endonuclease FokI to create the ZFN. The three-finger ZFN requires two copies of the 9 bp recognition site in an inverted orientation to produce a double-stranded break (DSB).

agronomic traits. ZFNs have been utilized successfully in both tobacco and *Arabidopsis*. In *Arabidopsis*, the ZFN have been shown to cleave and stimulate mutation at specific genomic sites, resulting in a mutation frequency of 0.2 per target (143). Site-specific cleavage and transgene integration has also been demonstrated in an engineered tobacco cell culture system (144). Such studies demonstrate the utility of zinc finger technology in plants and hold much promise for the development of novel transgenic plants. One major limitation in ZFN technology is the potential for cellular toxicity due to off-target DNA cleavage; thus, ZFNs are being modified to reduce this effect (145).

6. Chromosome Truncation

The recent development of plant artificial chromosome technology provides an opportunity for the stable expression and maintenance of multiple transgenes in a single genome. Plant artificial chromosomes are produced through telomere-mediated truncation of endogenous chromosomes. Telomeres are comprised of arrays of tandemly repeat sequences present at each end of a plant chromosome. To facilitate *in vivo* telomere-mediated truncation, an array of telomere sequences is cloned into a vector and then delivered into the genome via *Agrobacterium*-mediated T-DNA transformation or particle bombardment. Upon integration of the construct into the genome, the telomeric sequences can at some frequency be recognized as a bona fide chromosome terminus in the recipient chromosome, thereby truncating the chromosome at the insertion site (146). Recently, a 2.6-kb telomere repeat array isolated from *Arabidopsis* has been successfully transformed in maize to generate telomere-truncated chromosomes whose structure was verified by marker gene expression and FISH (fluorescent in situ hybridization)-based karyotyping (147, 148). Plant minichromosome technology can also be combined with site-specific recombination systems to facilitate the stacking of multiple transgenes. This strategy for construction of engineered chromosomes should easily be extended to other plant species because it does not rely on species-specific cloned centromere sequences.

7. Future Prospects

Over the last two decades numerous transformation techniques have been developed for plants. Both *Agrobacterium*-mediated and microprojectile bombardment-based transformations are now standard laboratory techniques in plant labs. Recent advancements in genetic engineering techniques have led to the identification of genes and proteins involved in the DNA transfer mechanism accomplished by *Agrobacterium* transformation. Yet, despite this progress, many economically important crops and tree species remain highly recalcitrant to *Agrobacterium* infection. Thus, much effort is being made to develop regeneration protocols that can efficiently integrate exogenous genes to develop stable transgenics using both *Agrobacterium*- and biolistic-mediated transformation. Novel techniques are also under development to develop genetically transformed plants with desired characteristics (149).

The generation of transgenic plants requires the use of various selectable marker genes that are introduced together with the gene of interest. Yet, it is abundantly clear that marker-free transgenic plants will be required in the future, thereby requiring more progress in genomics, cloning technology, and vector design, so as to eliminate the need for residual bacterial selectable marker genes in the future. Genetically engineered plants also play an important role in the ongoing study of gene function and metabolic pathways. Promising research has led to the identification of genes that control organogenesis or somatic embryogenesis which may function as selectable marker genes. The reintroduction of native genes and regulatory elements into plants therefore represents a viable means to achieve exogenous marker-free plants. A wide variety of plant genes associated with agronomically important traits have now been identified (150).

Chloroplast transformation technology is gaining importance due to its unique advantage of gene stacking without concern about gene silencing and of creating an opportunity to produce multivalent vaccines in a single transformation step. The recently developed zinc finger nuclease technology holds much promise in both basic and applied agricultural biotechnology. ZFN-assisted gene targeting and chromatin remodeling studies should also aid in characterizing gene function in plants. Resources are now available for engineering ZFNs in numerous plant species. Advancements in genetic engineering have also led to the development of minichrosome technology, which may provide a solution to gene stacking. Large DNA sequences, including multiple genes, could be introduced into the genome using this technology. Therefore, advances in transgenic technology would provide a solution for production of improved crop species to meet the world's demands for food, feed, fiber, and fuel (151). Hence, it is hoped that in the near future, GM plants with minimal genomic modifications can be developed.

Acknowledgments

The authors would like to thank Dr. Seth Findley (University of Missouri-Columbia, USA) for his critical proofreading. This work is supported by National Science Foundation.

References

1. Job, D. (2002) Plant biotechnology in agriculture. *Biochimie* **84**, 1105–1110.
2. Vain, P. (2007) Thirty years of plant transformation technology development. *Plant Biotechnol. J.* **5**, 221–229.
3. Fischer, R., Stoger, E., Schillberg, S., Christou, P., and Twyman, R.M. (2004) Plant based production of biopharmaceuticals. *Curr. Opin. Plant Biol.* 7, 152–158.
4. Tzfira, T., and Citovsky, V. (2006) *Agrobacterium*-mediated genetic transformation of plants: biology and biotechnology. *Curr. Opin. Biotechnol.* **17**, 147–154.
5. Broothaerts, W., Mitchell, H.J., Weir, B., Kaines, S., Smith, L.M., Yang, W., Mayer, J.E., Roa-Rodríguez, C., and Jefferson, R.A. (2005) Gene transfer to plants by diverse species of bacteria. *Nature* **433**, 629–633.
6. Hiei, Y., Ohta, S., Komari, T., and Kumashiro, T. (1994) Efficient transformation of rice (*Oryza sativa*) mediated by *Agrobacterium* and sequence analysis of the boundaries of the T-DNA. *Plant J.* **6**, 271–282.
7. Leelavathi, S., Sunnichan, V.G., Kumria, R., Vijaykanth, G.P., Bhatnagar, R.K., and Reddy, V.S. (2004) A simple and rapid *Agrobacterium*-mediated transformation protocol for cotton (*Gossypium hirsutum* L.): embryogenic calli as a source to generate large numbers of transgenic plants. *Plant Cell Rep.* **22**, 465–470.
8. Zhang-Hua, H., Jin-Qing, C., Guan-Ting, W., Wei, J., Chun-Xiu, L., Rui-Zhi, H., Fu-Lin, W., Zhi-Hong, L., and Xiao-Yun, C. (2005) Highly efficient transformation and plant regeneration of tall fescue mediated by *Agrobacterium tumefaciens. J. Plant. Physiol. Mol. Biol.* **31**, 149–159.
9. Sanford, J.C. (1990) Biolistic plant transformation. *Physiol. Plant.* **79**, 206–209.
10. Taylor, N.J., and Fauquet, C.M. (2002) Microparticle bombardment as a tool in plant science and agricultural biotechnology. *DNA Cell Biol.* **21**, 963–977.
11. Rakoczy-Trojanowska, M. (2002) Alternative methods of plant transformation – a short review. *Cell Mol. Biol. Lett.* 7, 849–858.
12. Scutta, C.P., Zubko, E., and Meyer, P. (2002) Techniques for the removal of marker genes from transgenic plants. *Biochimie* **84**, 1119–1126.
13. Darbani, B., Eimanifar, A., Stewart, C.N., Jr., and Camargo, W.N. (2007) Methods to produce marker-free transgenic plants. *Biotechnol. J.* **2**, 83–90.
14. Yu, W., Lamb, J.C., Han, F., and Birchler, J.A. (2006) Telomere-mediated chromosomal truncation in maize. *Proc. Natl. Acad. Sci. USA* **103**, 17331–17336.
15. Zupan, J., Muth, T.R., Draper, O., and Zambryski, P. (2000) The transfer of DNA from *Agrobacterium tumefaciens* into plants: a feast of fundamental insights. *Plant J.* **23**, 11–28.
16. Christie, P.J. (1997) *Agrobacterium tumefaciens* T-complex transport apparatus: a paradigm for a new family of multifunctional transporters in Eubacteria. *J. Bacteriol.* **179**, 3085–3094.
17. Tzfira, T., Vaidya, M., and Citovsky, V. (2002) Increasing plant susceptibility to *Agrobacterium* infection by overexpression of the *Arabidopsis* nuclear protein VIP1 *Proc. Natl. Acad. Sci. USA* **99**, 10435–10440.
18. Anand, A., Krichevsky, A., Schornack, S., Lahaye, T., Tzfira, T., Tang, Y., Citovsky, V., and Kirankumar, S.M. (2007) *Arabidopsis* VIRE2 INTERACTING PROTEIN2 is required for *Agrobacterium* T-DNA Integration in plants. *Plant Cell* **19**, 1695–1708.
19. Tzfira, T., and Citovsky, V. (2002) Partners-in-infection: host proteins involved in the transformation of plant cells by *Agrobacterium*. *Trends Cell Biol.* **12**, 121–129.
20. Bako, L., Umeda, M., Tiburcio, A.F., Schell, J., and Koncz, C. (2003) The VirD2 pilot protein of *Agrobacterium*-transferred DNA interacts with the TATA box-binding protein and a nuclear protein kinase in plants. *Proc. Natl. Acad. Sci. USA* **100**, 10108–10113.
21. Gelvin, S.B. (2000) *Agrobacterium* and plant genes involved in T-DNA transfer and integration. *Annu. Rev. Plant. Physiol.* **51**, 223–256.

22. Veena, J.H., Doerge, R.W., and Gelvin, S.B. (2003) Transfer of T-DNA and vir proteins to plant cells by *Agrobacterium tumefaciens* induces expression of host genes involved in mediating transformation and suppresses host defense gene expression. *Plant J.* **35**, 219–236.
23. Lacroix, B., Kozlovosky, S.V., and Citovsky, V. (2008) Recent patents on *Agrobacterium*-mediated gene and protein transfer for research and biotechnology. *Recent Pat. DNA Gene Seq.* **2**, 69–81.
24. Hoekema, A., Hirsch, P.R., Hooykaas, P.J.J., and Schilperoort, R.A. (1983) A binary plant vector strategy based on separation of *vir* and T-region of the *Agrobacterium tumefaciens* Ti plasmid. *Nature* **303**, 179–180.
25. Hamilton, C.M. (1997) A binary-BAC system for plant transformation with high-molecular-weight DNA. *Gene* **200**, 107–116.
26. Rui-Feng, H., Yuan-Yuan, W., Bo, D., Ming, T., Ai-Qing, Y., Li-Li, Z., and Guang-Cun, H. (2006) Development of transformation system of rice based on binary bacterial artificial chromosome (BIBAC) vector. *Acta Gen. Sin.* **33**, 269–276.
27. Komari, T. (1990) Transformation of cultured cells of *Chenopodium quinoa* by binary vectors that carry a fragment of DNA from the virulence region of pTiBo542. *Plant Cell Rep.* **9**, 303–306.
28. Lee, L., and Gelvin, S.B. (2008) T-DNA binary vectors and systems. *Plant Physiol.* **146**, 325–332.
29. Ishida, Y., Saito, H., Ohta, S., Hiei, Y., Komari, T., and Kumashiro, T. (1996) High efficiency transformation of maize (*Zea mays*) mediated by *Agrobacterium tumefaciens. Nat. Biotechnol.* **14**, 745–750.
30. Khanna, H.K., and Daggard, G.E. (2003) *Agrobacterium tumefaciens*-mediated transformation of wheat using a super binary vector and a polyamine supplemented regeneration medium. *Plant Cell Rep.* **21**, 429–436.
31. Lee, L.-Y., Kononov, M.E., Bassuner, B., Frame, B.R., Wang, K., and Gelvin, S.B. (2007) Novel pant transformation vectors containing the superpromoter. *Plant Physiol.* **145**, 1294–1300.
32. Cheng, M., Lowe, B.A., Spencer, T.M., Ye, X., and Armstrong, C.L. (2004) Factors influencing *Agrobacterium*-mediated transformation of monocotyledonous species. *In Vitro Cell Dev. Biol. Plant* **40**, 31–45.
33. Opabode, J.T. (2006) *Agrobacterium* mediated transformation of plants: emerging factors that influence efficiency. *Biotech. Mol. Biol. Rev.* **1**, 12–20.
34. Zhao, Z., Cai, T., Tagliani, L., Miller, M., Wang, N., Pang, H., Rudert, M., Schroeder, S., Hondred, D., Seltzer, J., and Pierce, D. (2000) *Agrobacterium*-mediated sorghum transformation. *Plant Mol. Biol.* **44**, 789–798.
35. Nandakumar, R., Chen, L., and Rogers, S.M.D. (2004) Factors affecting the *Agrobacterium*-mediated transient transformation of the wetland monocot, *Typha latifolia. Plant Cell Tiss. Organ Cult.* **79**, 31–38.
36. Herath, S.P., Suzuki, T., and Hattori, K. (2005) Factors influencing *Agrobacterium*-mediated genetic transformation of kenaf. *Plant Cell Tiss. Organ Cult.* **82**, 201–206.
37. Chateau, S., Sangwa, R.S., and Sangwan-Norreel, B.S. (2000) Competence of *Arabidopsis thaliana* genotypes and mutants for *Agrobacterium tumefaciens*-mediated gene transfer: role of phytohormones. *J. Exp. Bot.* **51**, 1961–1968.
38. Dan, Y. (2008) Biological functions of antioxidants in plant transformation. *In Vitro Cell Dev. Biol. Plant* **44**, 149–161.
39. Enriquez-Obregon, G.A., Vazquez-Padron, R.I., Prieto-Samsonov, D.L., de la Riva, G.A., and Selman-Housein, G. (1998) Herbicide-resistant sugarcane (*Saccharum officinarum L.*) plants by *Agrobacterium*-mediated transformation. *Planta* **205**, 20–27.
40. Olhoft, P.M., and Somers, D.A. (2001) L-Cysteine increases *Agrobacterium* mediated T-DNA delivery into soybean cotyledonary-node cells. *Plant Cell Rep.* **20**, 706–711.
41. Vega, J., Yu, W., Kennon, A.M., Chen, X., and Zhang, Z.J. (2008) Improvement of *Agrobacterium*-mediated transformation in Hi-II maize (*Zea mays* L.) using standard binary vectors. *Plant Cell Rep.* **27**, 297–305.
42. Frame, B.R., Shou, H., Chikwamba, R.K., Zhang, Z., Xiang, C.I., Fonger, T.M., Pegg, S.E.K., Li, B., Nettleton, D.S., Pei, D., and Wang, K. (2002) *Agrobacterium tumefaciens*-mediated transformation of maize embryos using a standard binary vector system. *Plant Physiol.* **129**, 13–22.
43. Dan, Y.A. (2004) A novel plant transformation technology-Lipoic acid. *In vitro Cell Dev. Biol. Plant* **42**, 18.
44. Dan, Y., Armstrong, C.L., Dong, J., Feng, X., Fry, J.E., Keithly, G.E., Martinell, B.J., Roberts, G.A., Smith, L.A., Tan, L.J., and

Duncan, D.R. (2009) Lipoic acid – a unique plant transformation enhancer. *In Vitro Cell Dev. Biol. Plant* (DOI: 10.1007/s11627-009-9227-5).

45. Cheng, M., Fry, J.E., Pang, S., Zhou, I., Hironaka, C., Duncan, D.R.I., Conner, T.W.L., and Wang, Y. (1997) Genetic transformation of wheat mediated by *Agrobacterium tumefacien*. *Plant Physiol.* **115**, 971–980.
46. Tang, W. (2003) Additional virulence genes and sonication enhance *Agrobacterium tumefaciens*-mediated loblolly pine transformation. *Plant Cell Rep.* **21**, 555–562.
47. Pena, L., Perez, R.M., Cervera, M., Juarez, J.A., and Navarro, L. (2004) Early events in *Agrobacterium*-mediated genetic transformation of citrus explants. *Ann. Bot. (Lond.)* **94**, 67–74.
48. Arias, R.S., Filichkin, S.A., and Strauss, S.H. (2006) Arias divide and conquer: development and cell cycle genes in plant transformation. *Trends Biotechnol.* **24**, 267–273.
49. Grafi, G. (2004) How cells dedifferentiate: a lesson from plants. *Dev. Biol.* **268**, 1–6.
50. Riou-Khamlichi, C., Huntley, R., Jacqmard, A., and Murray, J.A. (1999) Cytokinin activation of *Arabidopsis* cell division through a D-type cyclin. *Science* 283, 1541–1544.
51. De Veylder, L., Beeckman, T., Beemster, G.T.S., Engler, J.D.A., Ormenese, S., Maes, S., Naudts, M., Der Schueren, E.V., Jacqmard, A., Engler, G., and Inze, D. (2002) Control of proliferation, endoreduplication and differentiation by the *Arabidopsis* E2Fa–DPa transcription factor. *EMBO J.* **21**, 1360–1368.
52. Gordon-Kamm, W., Dilkes, B.P., Lowe, K., Hoerster, G., Sun, X., Ross, M., Church, L., Bunde, C., Farrell, J., Hill, P., Maddock, S., Snyder, J., Sykes, L., Li, Z., Woo, Y.-M., Bidney, D., and Larkins, B.A. (2002) Stimulation of the cell cycle and maize transformation by disruption of the plant retinoblastoma pathway. *Proc. Natl. Acad. Sci. USA* **99**, 11975–11980.
53. Mysore, K.S., Nam, J., and Gelvin, S. (2000) An *Arabidopsis* histone H2A mutant is deficient in *Agrobacterium* T-DNA integration. *Proc. Natl. Acad. Sci. USA* **97**, 948–953.
54. Vlieghe, K., Vuylsteke, M., Florquin, K., Rombauts, S., Maes, S., Ormenese, S., Hummelen, P.V., de Peer, Y.V., Inze, D., and De Veylder, L. (2003) Microarray analysis of E2Fa–DPa-overexpressing plants uncovers a cross-talking genetic network between DNA replication and nitrogen assimilation. *J. Cell Sci.* **116**, 4249–4259.
55. Zambre, M., Geerts, P., Maquet, A., Montagu, M.V., Dillen, W., and Angenon, G. (2001) Regeneration of fertile plants from callus in *Phaseolus polyanthus* Greenman (year bean). *Ann. Bot. (Lond.)* **88**, 371–377.
56. Taoka, K.-I., Yanagimoto, Y., Daimon, Y., Hibara, K.-I., Aida, M., and Tasaka, M. (2004) The NAC domain mediates functional specificity of CUP-SHAPED COTYLEDON proteins. *Plant J.* **40**, 462–473.
57. Slightom, J.L., Durand-Tardif, M., Jouanin, L., and Tepfer, D. (1986) Nucleotide sequence analysis of TL-DNA of *Agrobacterium rhizogenes* agropine type plasmid. Identification of open reading frames. *J. Biol. Chem.* **261**, 108–121.
58. Collier, R., Fuchs, B., Walter, N., Kevin, L.W., and Taylor, C.G. (2005) Ex vitro composite plants: an inexpensive, rapid method for root biology *Plant J.* **43**, 449–457.
59. Giri, A., and Narasu, M.L. (2000) Transgenic hairy roots: recent trends and applications. *Biotechnol. Adv.* **18**, 1–22.
60. Georgiev, M.I., Pavlov, A.I., and Bley, T. (2007) Hairy root type plant in vitro systems as sources of bioactive substances. *Appl. Microbiol. Biotechnol.* **74**, 1175–1185.
61. Seki, H., Nishizawa, T., Tanaka, N., Niwa, Y., Yoshida, S., and Muranaka, T. (2005) Hairy root-activation tagging: a high-throughput system for activation tagging in transformed hairy roots. *Plant Mol. Biol.* **59**, 793–807.
62. Narayanan, R.A., Atz, R., Denny, R., Young, N.D., and Somers, D.A. (1999) Expression of soybean cyst nematode resistance in transgenic hairy roots of soybean. *Crop Sci.* **39**, 1680–1686.
63. Skarjinskaia, M., Karl, J., Araujo, A., Ruby, K., Rabindran, S., Streatfield, S.J., and Yusibov, V. (2008) Production of recombinant proteins in clonal root cultures using episomal expression vectors. *Biotechnol. Bioeng.* **100**, 814–819.
64. Kuster, H., Vieweg, M.F., Manthey, K., Baier, M.C., Hohnjec, N., and Perlick, A.M. (2007) Identification and expression regulation of symbiotically activated legume genes. *Phytochemistry* **68**, 8–18.
65. Breitler, J.C., Labeyrie, A., Meynard, D., Legavre, T., and Guiderdoni, E. (2002) Efficient microprojectile bombardment-mediated transformation of rice using gene cassettes. *Theor. Appl. Genet.* **104**, 709–719.
66. Christou, P. (1995) Strategies for variety-independent genetic transformation of important cereals, legumes and woody species

utilising particle bombardment. *Euphyica* **85**, 13–27.

67. Campbell, B.T., Baeziger, P.S., Mitra, A., Sato, S., and Clemente, T. (2000) Inheritance of multiple genes in wheat. *Crop Sci.* **40**, 1133–1141.
68. Schmidt, M.A., Lafayette, P.R., Artelt, B.A., and Parrott, W.A. (2008) A comparison of strategies for transformation with multiple genes via microprojectile-mediated bombardment. *In Vitro Cell Dev. Biol. Plant* **44**, 162–168.
69. Agrawal, P.K., Kohli, A., Twyman, R.M., and Christou, P. (2005) Transformation of plants with multiple cassettes generates simple transgene integration patterns and high expression levels. *Mol. Breed.* **16**, 247–260.
70. Schmidt, M.A., Tucker, D.M., Cahoon, E.B., and Parrott, W.A. (2005) Towards normalization of soybean somatic embryo maturation. *Plant Cell Rep.* **24**, 383–391.
71. Pawlowski, W.P., and Somers, D.A. (1996) Transgene inheritance in plants genetically engineered by microprojectile bombardment. *Mol. Biotechnol.* **6**, 17–30.
72. Kohli, A., Gahakwa, D., Vain, P., Laurie, D.A., and Christou, P. (1999) Transgene expression in rice engineered through particle bombardment: molecular factors controlling stable expression and transgene silencing. *Planta* **208**, 88–97.
73. Lowe, B.A., Prakash, N.S., Melissa, W., Mann, M.T., Spencer, T.M., and Boddupalli, R.S. (2009) Enhanced single copy integration events in corn via particle bombardment using low quantities of DNA. *Transgenic Res.* **18**, 831–840 (DOI: 10.1007/s11248-009-9265-0).
74. Shimamoto, K., Teralda, R., Izawa, T., and Fujimoto, H. (1989) Fertile transgenic rice plants regenerated from transformed protoplasts. *Nature* **338**, 274–276.
75. Salmenkallio-Marttila, M., Aspegren, K., Kerman, S., Kurt, U., Mannonen, L., Ritala, A., Teeriz, T.H., and Kauppinen, V. (1995) Transgenic barley (*Hordeum vulgare* L.) by electroporation of protoplasts *Plant Cell Rep.* **15**, 301–304.
76. Daveya, M.R., Anthonya, P., Powera, J.B., and Loweb, K.C. (2005) Plant protoplasts: status and biotechnological perspectives. *Biotechnol. Adv.* **23**, 131–171.
77. Sawahel, W.A. (2001) Stable genetic transformation of cotton plants using polybrene spermidine treatment. *Plant Mol. Biol. Rep.* **19**, 377a–377f.
78. Gad, A.E., Rosenberg, N., and Altman, A. (1990) Liposome-mediated gene delivery into plant cells. *Physiol. Plant.* **79**, 177–183.
79. Wordragen, M.V., Roshani, S., Ruud, V., Regis, P., Abvan, K., and Pim, Z. (1997) Liposome-mediated transfer of YAC DNA to tobacco cells. *Plant Mol. Biol. Rep.* **15**, 170–178.
80. Kaeppler, H., Somers, D.A., Rines, H.W., and Cockburn, A.F. (1992) Silicon carbide fiber-mediated stable transformation of plant cells. *Theor. Appl. Genet.* **84**, 560–566.
81. Frame, B.R., Drayton, P.R., Bagnall, S.V., Lewnau, C.J., Bullock, W.P., Wilson, H.M., Dunwell, J.M., Thompson, J.A., and Wang, K. (1994) Production of fertile transgenic maize plants by silicon carbide whisker-mediated transformation. *Plant J.* **6**, 941–948.
82. Singh, N., and Chawla, H.S. (1999) Use of silicon carbide fibers for *Agrobacterium*-mediated transformation in wheat. *Curr. Sci.* **76**, 1483–1485.
83. Nagatani, N., Honda, H., Shimada, T., and Kobayashi, T. (1997) DNA delivery into rice cells and transformation using silicon carbide whiskers. *Biotechnol. Tech.* **11**, 781–786.
84. Torney, F., Trewyn, B.G., Lin, V.S.Y., and Wang, K. (2007) Mesoporous silica nanoparticles deliver DNA and chemicals into plants. *Nature Nanotech.* **2**, 295–300.
85. Nandiyanto, A.B.D., Kim, S.G., Iskandar, F., and Okuyama, K. (2009) Synthesis of silica nanoparticles with nanometer-size controllable mesopores and outer diameters. *Microporous Mesoporous Mater.* **120**, 447–453.
86. Crossway, A., Oakes, J.W., Irvine, J.M., Ward, B., Knauf, V.C., and Shewmaker, C.K. (1986) Integration of foreign DNA following microinjection of tobacco mesophyll protoplasts. *Mol. Gen. Genet.* **202**, 179–185.
87. Jones-Villeneuve, E., Huang, B., Prudhome, I., Bird, S., Kemble, R., Hattori, J., and Miki, B. (1995) Assessment of microinjection for introducing DNA into uninuclear microspores of rapeseed. *Plant Cell Tiss. Organ Cult.* **40**, 97–100.
88. Holm, P.B., Olsen, O., Schnorf, M., Brinch-Pederse, H., and Knudsen, S. (2000) Transformation of barley by microinjection into isolated zygote protoplasts. *Transgenic Res.* **9**, 21–32.
89. Lilly, J.W., Havey, M.J., Jackson, S.A., and Jiang, J. (2001) Cytogenomic analysis reveal the structural plasticity of the chloroplast genome in higher plants. *Plant Cell* **13**, 245–254.

90. Maliga, P. (2004) Plastid transformation in higher plants. *Annu. Rev. Plant Biol.* **55**, 289–313.
91. Verma, D., and Daniell, H. (2007) Chloroplast vector systems for biotechnology applications. *Plant Physiol.* **145**, 1129–1143.
92. Boynton, J.E., Gillham, N.W., Harris, E.H., Hosler, J.P., Johnson, A.M., and Jones, A.R. (1988) Chloroplast transformation in *Chlamydomonas* with high velocity microprojectiles. *Science* **240**, 1534–1538.
93. Lee, S.M., Kang, K., Chung, H., Yoo, S.H., Xu, X.M., Lee, S.B., Cheong, J.J., Daniell, H., and Kim, M. (2006) Plastid transformation in the monocotyledonous cereal crop, rice (*Oryza sativa*) and transmission of transgenes to their progeny. *Mol. Cells* **21**, 401–410.
94. Kumar, S., Dhingra, A., and Daniell, H. (2004) Stable transformation of the cotton plastid genome and maternal inheritance of transgenes. *Plant Mol. Biol.* **56**, 203–216.
95. Daniell, H., Chebolu, S., Kumar, S., Singleton, M., and Falconer, R. (2005) Chloroplast derived vaccine antigens and other therapeutic proteins. *Vaccine* **23**, 1779–1783.
96. Rommens, C.M. (2004) All-native DNA transformation: a new approach to plant genetic engineering. *Trends Plant Sci.* **9**, 457–464.
97. Bernasconi, P. et al. (1995) A naturally occurring point mutation confers broad range tolerance to herbicides that target acetolactate synthase. *J. Biol. Chem.* **270**, 17381–17385.
98. Baerson, S.R., Rodriguez, D.J., Tran, M., Feng, M., Biest, N.A., and Dill, G.M. (2002) Glyphosate-resistant goosegrass identification of a mutation in the target enzyme 5-enolpyruvylshikimate-3-phosphate synthase. *Plant Physiol.* **129**, 1265–1275.
99. Song, J., Bradeen, J.M., Naess, S.K., Raasch, J.A.,Wielgus, S.M., Haberlach, G.T., Liu, J., Kuang, H., Austin-Phillips, S., Buell, C.R., Helgeson, J.P., and Jiang, J. (2003) Gene RB cloned from *Solanum bulbocastanum* confers broad spectrum resistance to potato late blight. *Proc. Natl. Acad. Sci. USA* **100**, 9128–9133.
100. Pechan, T., Cohen, A., Williams, W.P., and Luthe, D.S. (2002) Insect feeding mobilizes a unique plant defense protease that disrupts the peritrophic matrix of caterpillars. *Proc. Natl. Acad. Sci. USA* **99**, 13319–13323.
101. Wang, E., Wang, R., DeParasis, J., Loughrin, J.H., Gan, S., and Wagner, G.J. (2001) Suppression of a P450 hydroxylase gene in plant trichome glands enhances natural-product-based aphid resistance. *Nat. Biotechnol.* **19**, 371–374.
102. Kasuga, M., Liu, Q., Miura, S., Yamaguchi-Shinozaki, K., and Shinozaki, K. (1999) Improving plant drought, salt, and freezing tolerance by gene transfer of a single stress-inducible transcription factor. *Nat. Biotechnol.* **17**, 287–291.
103. Miki, B., and Mc Hugh, S. (2004) Selectable marker genes in transgenic plants: applications, alternatives and biosafety. *J. Biotechnol.* **107**, 193–232.
104. Daley, M., Knauf, V.C., Summerfelt, K.R., and Turner, J.C. (1998) Co-transformation with one *Agrobacterium tumefaciens* strain containing two binary plasmids as a method for producing marker-free transgenic plants. *Plant Cell Rep.* **17**, 489–496.
105. Komari, T., Hiei, Y., Saito, Y., Murai, N., and Kumashiro, T. (1996) Vectors carrying two separate T-DNAs for co-transformation of higher plants mediated by *Agrobacterium tumefaciens* and segregation of transformants free from selection markers. *Plant J.* **10**, 165–174.
106. Hua, Y., and Rommens, C.M. (2007) Transposition-based plant transformation. *Plant Physiol.* **143**, 570–578.
107. Cotsaftis, O., Sallaud, C., Breitler, J.C., Meynard, D., Greco, R., Pereira, A., and Guiderdoni, E. (2002) Transposon-mediated generation of T-DNA- and marker-free rice plants expressing a *Bt* endotoxin gene. *Mol. Breed.* **10**, 165–180.
108. Saelim, L., Phansiri, S., Suksangpanomrung, M., Netrphan, S., and Narangajavana, J. (2009) Evaluation of a morphological marker selection and excision system to generate marker-free transgenic cassava plants. *Plant Cell Rep.* **28**, 445–455.
109. Ebinuma, H., and Komamine, A. (2001) MAT (Multi-Auto-Transformation) vector system. The oncogenes of *Agrobacterium* as positive markers for regeneration and selection of marker-free transgenic plants. *In Vitro Cell Dev. Biol. Plant* **37**, 103–113.
110. Kuraya, Y., Ohta, S., Fukuda, M., Hiei, Y., Murai, N., Hamada, K., Ueki, T., Imaseki, H., and Komari, T. (2004) Suppression of transfer of non-T-DNA "vector backbone" sequences by multiple left border repeats in vectors for transformation of higher plants mediated by *Agrobacterium tumefaciens*. *Mol. Breed.* **14**, 309–320.
111. Yang, A., Su, Q., and An, L. (2009) Ovary-drip transformation: a simple method for

directly generating vector- and marker-free transgenic maize (*Zea mays* L.) with a linear GFP cassette transformation. *Planta* **229**, 793–801.

112. Yang, A., Su, Q., An, L., Liu, J., Wu, W., and Qiu, Z. (2009) Detection of vector- and selectable marker-free transgenic maize with a linear *GFP* cassette transformation via the pollen-tube pathway. *J. Biotechnol.* **139**, 1–5.
113. Ye, X., Williams, E.J., Shen, J., Esser, J.A., Nichols, A.M., Petersen, M.W., and Gilbertson, L.A. (2008) Plant development inhibitory genes in binary vector backbone improve quality event efficiency in soybean transformation. *Transgenic Res.* **17**, 827–838.
114. Lu, H.J., Zhou, X.R., Gong, Z.X., and Upadhyaya, N.M. (2001) Generation of selectable marker-free transgenic rice using double right-border (DRB) binary vectors. *Aust. J. Plant Physiol.* **28**, 241–248.
115. Hajdukiewicz, P., Svab, Z., and Maliga, P. (1994) The small, versatile *pPZP* family of *Agrobacterium* binary vectors for plant transformation. *Plant Mol. Biol.* **25**, 989–994.
116. Fu, X., Duc, L.T., Fontana, S., Bong, B.B., Tinjuangjun, P., Sudhakar, D., Twyman, R.M., Christou, P., and Kohli, A. (2000) Linear transgene constructs lacking vector backbone sequences generate low-copy-number transgenic plants with simple integration patterns. *Transgenic Res.* **9**, 11–19.
117. Dale, E., and Ow, D. (1991) Gene transfer with subsequent removal of the selection gene from the host genome. *Proc. Natl. Acad. Sci. USA* **88**, 10558–10562.
118. Wang, Y., Chen, B., Hu, Y., Li, J., and Lin, Z. (2005) Inducible excision of selectable marker gene from transgenic plants by the Cre/lox site-specific recombination system. *Transgenic Res.* **14**, 605–614.
119. Srivastava, V., and Ow, D.W. (2001) Single-copy primary transformants of maize obtained through the co-introduction of a recombinase-expressing construct. *Plant Mol. Biol.* **46**, 561–566.
120. Zuo, J., Niu, Q.-W., Moller, S.G., and Chua, N.-H. (2001) Chemical-regulated, site-specific DNA excision in transgenic plants. *Nat. Biotechnol.* **19**, 157–161.
121. Bai, X., Qiuyun, W., and Chu, C. (2008) Excision of a selective marker in transgenic rice using a novel Cre/loxP system controlled by a floral specific promoter. *Transgenic Res.* **17**, 1035–1043.
122. McCormac, A.C., Fowler, M.R., Chen, D.F., and Elliott, M.C. (2001) Efficient co-transformation of *Nicotiana tabacum* by two independent T-DNAs, the effect of T-DNA size and implications for genetic separation. *Transgenic Res.* **10**, 143–155.
123. Miller, M., Tagliani, L., Wang, N., Berka, B., Bidney, D., and Zhao, Z.Y. (2002) High-efficiency transgene segregation in co-transformed maize plants using an *Agrobacterium tumefaciens* 2T-DNA binary system. *Transgenic Res.* **11**, 381–396.
124. Matthews, P.R., Waterhouse, P.M., Thornton, S., Fieg, S.J., Gubler, F., and Jacobsen, J.V. (2001) Marker gene elimination from transgenic barley, using co-transformation with adjacent "twin T-DNAs" on a standard *Agrobacterium* transformation vector. *Mol. Breed.* 7, 195–202.
125. Karimi, M., Inze, D., and Depicker, A. (2002) GATEWAY vectors for *Agrobacterium*-mediated plant transformation. *Trends Plant Sci.* 7, 193–195.
126. Karimi, M., Bjorn, D.M., and Hilson, P. (2005) Modular cloning in plant cells. *Trends Plant Sci.* **10**, 103–105.
127. Karimi, M., Depicker, A., and Hilson, P. (2007) Recombinational cloning with plant gateway vectors. *Plant Physiol.* **145**, 1144–1154.
128. Chen, Q.-J., Zhou, H.-M., Chen, J., and Wang, X.C. (2006) A gateway-based platform for multigene plant transformation. *Plant Mol. Biol.* **62**, 927–936.
129. Helliwell, C., and Waterhouse, P. (2003) Constructs and methods for high-throughput gene silencing in plants. *Methods* **30**, 289–295.
130. Daxinger, L., Hunter, B., Sheikh, M., Jauvion, V., Gasciolli, V., Vaucheret, H., Matzke, M., and Furner, I. (2008) Unexpected silencing effects from T-DNA tags in *Arabidopsis*. *Trends Plant Sci.* **13**, 4–6.
131. Heilersig, B.H.J.B., Loonen, A.E.H.M., Wolters, A.-M.A., and Visser, R.G.F. (2006) Presence of an intron in inverted repeat constructs does not necessarily have an effect on efficiency of post-transcriptional gene silencing. *Mol. Breed.* **17**, 307–316.
132. Earley, K.W., Haag, J.R., Pontes, O., Opper, K., Juehne, T., Song, K., and Pikaard, C.S. (2006) Gateway-compatible vectors for plant functional genomics and proteomics. *Plant J.* **45**, 616–629.
133. Nakagawa, T., Kurose, T., Hino, T., Tanaka, K., Kawamukai, M., Niwa, Y., Toyooka, K., Matsuoka, K., Jinbo, T., and Kimura, T. (2007) Development of series of gateway binary vectors, pGWBs, for realizing efficient

construction of fusion genes for plant transformation. *J. Biosci. Bioeng.* **104**, 34–41.

134. Nakagawa, T., Nakamura, S., Tanaka, K., Kawamukai, M., Suzuki, T., Nakamura, K., Kimura, T., and Ishiguro, S. (2008) Development of R4 gateway binary vectors (R4pGWB) enabling high-throughput promoter swapping for plant research. *Biosci. Biotechnol. Biochem.* **72**, 624–629.
135. Curtis, M.D., and Grossniklaus, U. (2003) A gateway cloning vector set for high-throughput functional analysis of genes in planta. *Plant Physiol.* **133**, 462–469.
136. Tzfira, T., Tian, G.-W., Lacroix, B., Vyas, S., Li, J., Leitner-Dagan, Y., Krichevsky, A., Taylor, T., Vainstein, A., and Citovsky, V. (2005) pSAT vectors: a modular series of plasmids for autofluorescent protein tagging and expression of multiple genes in plants. *Plant Mol. Biol.* **57**, 503–516.
137. Zhong, S., Lin, Z., Fray, R.G., and Grierson, D. (2008) Improved plant transformation vectors for fluorescent protein tagging. *Transgenic Res.* **17**, 985–989.
138. Martin, K., Kopperud, K., Chakrabarty, R., Banerjee, R., Brooks, R., and Goodin, M.M. (2009) Transient expression in *Nicotiana benthamiana* fluorescent marker lines provides enhanced definition of protein localization, movement and interactions in planta. *Plant J.* **59**, 150–162.
139. Copeland, N.G., Jenkins, N.A., and Court, D.L. (2001) Recombineering: a powerful new tool for mouse functional genomics. *Nat. Rev. Genet.* **2**, 769–779.
140. Rozwadowski, K., Yang, W., and Kagale, S. (2008) Homologous recombination-mediated cloning and manipulation of genomic DNA regions using gateway and recombineering systems. *BMC Biotechnol.* **8**, 88.
141. Cai, C.Q., Doyon, Y.W., Ainley, M., Miller, J.C., et al. (2009) Targeted transgene integration in plant cells using designed zinc finger nucleases. *Plant Mol. Biol.* **69**, 699–709.
142. Durai, S., Mani, M., Kandavelou, K., Wu, J., Porteus, M.H., and Chandrasegaran, S. (2005) Zinc finger nucleases: custom-designed molecular scissors for genome engineering of plant and mammalian cells. *Nucleic Acids Res.* **33**, 5978–5990.
143. Lloyd, A., Plaisier, C.L., Carroll, D., and Drews, G.N. (2005) Targeted mutagenesis using zinc-finger nucleases in *Arabidopsis*. *Proc. Natl. Acad. Sci. USA* **102**, 2232–2237.
144. Ordiz, M.I., Barbas, C.F., and Beachy, R.N. (2002) Regulation of transgene expression in plants with polydactyl zinc finger transcription factors *Proc. Natl. Acad. Sci. USA* **99**, 13290–13295.
145. Szxzepek, M., Brondani, V., Buchd, I., Serrano, L., Segal, D., and Cathomen, T. (2007) Structure-based redesign of the dimerization interface reduces the toxicity of zinc-finger nucleases. *Nat. Biotechnol.* **25**, 786–793.
146. Weichang, Y., Fangpu, H., Gao, Z., Vega, J.M., and Birchler, J.A. (2007) Construction and behavior of engineered minichromosomes in maize. *Proc. Natl. Acad. Sci. USA* **104**, 8924–8929.
147. Yu, W., Han, F., and Birchler, J.A. (2007) Engineered minichromosomes in plants. *Curr. Opin. Biotechnol.* **18**, 425–431.
148. Kato, A., Zheng, Y.Z., Auger, D.L., Phelps-Durr, T., Bauer, M.J., Lamb, J.C., and Birchler, J.A. (2005) Minichromosomes derived from the B chromosome of maize. *Cytogenet. Genome. Res.* **109**, 156–165.
149. Veena (2008) Engineering plants for future: tools and options *Physiol. Mol. Biol. Plants* **14**, 131–135.
150. Vain, P. (2005) Plant transgenic science knowledge. *Nat. Biotechnol.* **23**, 1348–1349.
151. Chapotin, S.M., and Wolt, J.D. (2007) Genetically modified crops for the bioeconomy: meeting public and regulatory expectations. *Transgenic Res.* **16**, 675–688.
152. Kim, E.H., Suh, S.C., Park, B.S., Shin, K.S., Kweon, S.J., Han, E.J., Park, S.-H., Kim, Y.-S., and Kim, J.-K. (2009) Chloroplast-targeted expression of synthetic *cry1Ac* in transgenic rice as an alternative strategy for increased pest protection. *Planta* **230**, 397–405.
153. Kim, T.-G., Baek, M.-Y., Lee, E.-K., Kwon, T.-H., and Yang, M.-S. (2008) Expression of human growth hormone in transgenic rice cell suspension culture. *Plant Cell Rep.* **27**, 885–891.
154. De Padua, V.L.M., Ferreira, R.P., Meneses, L., Uchoa, L., Marcia, M.-P., and Mansur, E. (2001) Transformation of Brazilian elite *Indica-Type* Rice (*Oryza Sativa* L.) by electroporation of shoot apex explants. *Plant Mol. Biol. Rep.* **19**, 55–64.
155. Cho, M.-J., Choi, H.W., Okamoto, D., Zhang, S., and Lemaux, P.G. (2003) Expression of green fluorescent protein and its inheritance in transgenic oat plants generated from shoot meristematic cultures. *Plant Cell Rep.* **21**, 467–474.
156. Um, M.K., Park, T.I., Kim, Y.J., Seo, H.Y., Kim J.G., Kwon, S.Y., Kwak, S.-S., Yun, D.-J., and Yun, S.J. (2007) Particle

bombardment mediated transformation of barley with an Arabidopsis NDPK2 cDNA. *Plant Biotechnol. Rep.* **1**, 71–77.

157. Manickavasagam, M., Ganapathi, A., Anbazhagan, V.R., Sudhakar, B., Selvaraj, N., Vasudevan, A., and Kasthurirengan, S. (2004) *Agrobacterium*-mediated genetic transformation and development of herbicide-resistant sugarcane (*Saccharum* species hybrids) using axillary buds. *Plant Cell Rep.* **23**, 134–143.

158. Seema, G., Pande, H.P., Lal, J., and Madan, V.K. (2001) Plantlet regeneration of sugarcane varieties and transient GUS expression in calli by electroporation. *Sugar Tech.* **3**, 27–33.

159. Gurel, S., Gurel, E., Kaur, R., Wong, J., Meng, L., Tan, H.-Q., and Lemaux P.G. (2009) Efficient, reproducible *Agrobacterium*-mediated transformation of sorghum using heat treatment of immature embryos. *Plant Cell Rep.* **28**, 429–444.

160. Wu, H., Doherty, A., and Jones, H.D. (2008) Efficient and rapid *Agrobacterium*-mediated genetic transformation of durum wheat (*Triticum turgidum* L. var. durum) using additional virulence genes. *Transgenic Res.* **17**, 425–436.

161. Buck, S.D., Podevin, N., Nolf, J., Jacobs, A., and Depicker, A. (2009) The T-DNA integration pattern in *Arabidopsis* transformants is highly determined by the transformed target cell. *Plant J.* **60**, 134–145 (DOI: 10.1111/j.1365-313X.2009.03942).

162. Ueki, S., Lacroix, B., Krichevsky, A., Lazarowitz, S.G., and Citovsky, V. (2009) Functional transient genetic transformation of *Arabidopsis* leaves by biolistic bombardment. *Nat. Prot.* **4**, 71–77.

163. Anuradha, T.S., Jami, S.K., Datla R.S., and Kirti, P.B. (2006) Genetic transformation of peanut (*Arachis hypogaea* L.) using cotyledonary node as explant and a promoterless *gus::npt*II fusion gene based vector. *J. Biosci.* **31**, 235–246.

164. Athmaram, T.N., Bali, G., and Devaiah, K.M. (2006) Integration and expression of Bluetongue VP2 gene in somatic embryos of peanut through particle bombardment method. *Vaccine* **24**, 2994–3000.

165. De Paadua, V.L.M., Pestana, M.C., Margis-Pinheiro, M., De Oliviera, D.E., and Mansur, E. (2000) Electroporation of intact embryonic leaflets of peanut: gene transfer and stimulation of regeneration capacity. *In Vitro Cell Dev. Biol. Plant* **36**, 374–378.

166. Liu, C.-W., Lin, C.-C., Yiu, J.-C., Chen, J.J.W., and Tseng, M.-J. (2008) Expression of a *Bacillus thuringiensis* toxin (*cry1Ab*) gene in cabbage (*Brassica oleracea* L. var. *capitata* L.) chloroplasts confers high insecticidal eYcacy against *Plutella xylostella*. *Theor. Appl. Genet.* **117**, 75–88.

167. Satyavathi, V.V., Prasad, V., Khandelwal, A., Shaila, M.S., and Lakshmi Sita, G. (2003) Expression of hemagglutinin protein of Rinderpest virus in transgenic pigeon pea [*Cajanus cajan* (L.) Millsp.] plants. *Plant Cell Rep.* **21**, 651–658.

168. Spokevicius, A.V., Beveren, K.V., Leitch, M.A., and Bossinger, G. (2005) *Agrobacterium*-mediated *in vitro* transformation of wood-producing stem segments in eucalypts. *Plant Cell Rep.* **23**, 617–624.

169. Zeng, P., Vadnais, D., Zhang, Z., and Polacco, J. (2004) Refined glufosinate selection in *Agrobacterium*-mediated transformation of soybean [*Glycine max* (L.) Merr.]. *Plant Cell Rep.* **22**, 478–482.

170. Tougou, M., Yamagishi, N., Furutani, N., Kaku, K., Shimizu, T., Takahata, Y., Sakai, J.-I., Kanematsu, S., and Hidaka, S. (2009) The application of the mutated acetolactate synthase gene from rice as the selectable marker gene in the production of transgenic soybeans. *Plant Cell Rep.* **28**, 769–776.

171. Gao, X.R., Wang, G.K., Su, Q., Wang, Y., and An, L.J. (2007) Phytase expression in transgenic soybeans: stable transformation with a vector-less construct. *Biotechnol. Lett.* **29**, 1781–1787.

172. Asad, S., Mukhtar, Z., Nazir, F., Hashmi, J.A., Mansoor, S., Zafar, Y., and Arshad, M. (2008) Silicon carbide whisker-mediated embryogenic callus transformation of cotton (*Gossypium hirsutum* L.) and regeneration of salt tolerant plants. *Mol. Biotechnol.* **40**, 161–169.

173. Li, H., Flachowsky, H., Fischer, T.C., Hanke, M.-V., Forkmann, G., Treutter, D., Schwab, W., Vmann, T.H., and Szankowski, I. (2007) Maize *Lc* transcription factor enhances biosynthesis of anthocyanins, distinct proanthocyanidins and phenylpropanoids in apple (*Malus domestica* Borkh.). *Planta* **226**, 1243–1254.

174. Charity, J.A., Holland, L., Grace, L.J., and Walter, C. (2005) Consistent and stable expression of the nptII, uidA and bar genes in transgenic *Pinus radiata* after *Agrobacterium tumefaciens*-mediated transformation using nurse cultures. *Plant Cell Rep.* **23**, 606–616.

175. DeBlock, M., De Brower, D., and Tenning P. (1989) Transformation of *Brassica napus* and *Brassica oleracea* using *Agrobacterium tumefaciens* and the expression of the *bar*

and *neo* genes in the transgenic plants. *Plant Physiol.* **91**, 694–701.

176. Daniell, H., Muthukumar, B., and Lee, S.B. (2001) Marker free transgenic plants: engineering the chloroplast genome without the use of antibiotic selection. *Curr. Genet.* **39**, 109–116.

177. Freyssinet, G., Pelissier, B., Freyssinet, M., and Delon, R. (1996) Crops resistant to oxynils: from the laboratory to the market. *Field Crops Res.* **45**, 125–133.

178. DeBlock, M., Herrera-Estrella, L., Van Montagu, M., Schell, J., and Zambryski, P. (1984) Expression of foreign genes in regenerated plants and in their progeny. *EMBO J.* **3**, 1681–1689.

179. Irdani, T., Bogani, P., Mengoni, A., Mastromei, G., and Buiatti, M. (1998) Construction of a new vector conferring methotrexate resistance in *Nicotiana tabacum* plants. *Plant Mol. Biol.* **37**, 1079–1084.

180. Howe, A.R., Gasser, C.S., Brown, S.M., Padgette, S.R., Hart, J., Parker, G.B., Fromm, M.E., and Armstrong, C.L. (2002) Glyphosate as a selective agent for the production of fertile transgenic maize (*Zea mays* L.) plants. *Mol. Breed.* **10**, 153–164.

181. Barry, G., Kishore, G., Padgette, S., Talor, M., Kolacz, K., Weldon, M., Re, D., Eichholtz, D., Fincher, K., and Hallas, L. (1992) Inhibitors of amino acid biosynthesis: strategies for imparting glyphosate tolerance to plants. In: Singh, B.K., Flores, H.E., Shannon, J.C. (Eds.), Biosynthesis and Molecular Regulation of Amino Acids in Plants. American Society of Plant Physiology, Rockville, MD, pp. 139–145.

182. Waldron, C., Murphy, E.B., Roberts, J.L., Gustafson, G.D., Armour, S.L., and Malcolm, S.K. (1985) Resistance to hygromycin B. *Plant Mol. Biol.* **5**, 103–108.

183. Joersbo, M., Donaldson, I., Kreiberg, J., Petersen, S.G., and Brunstedt, J. (1998) Analysis of mannose selection used for transformation of sugar beet. *Mol. Breed.* **4**, 111–117.

184. Fraley, R.T., Rogers, S.G., Horsch, R.B., Sanders, P.R., Flick, J.S., Adams, S.P., Bittner, M.L., Brand, L.A., Fink, C.L., Fry, J.S., Gallupi, G.R., Goldberg, S.B., Hoffman, N.L., and Woo, S.C. (1983) Expression of bacterial genes in plant cells. *Proc. Natl. Acad. Sci. USA* **80**, 4803–4807.

185. Haldrup, A., Petersen, S.G., and Okkels, F.T. (1998) Positive selection: a plant selection principle based on xylose isomerase, an enzyme used in the food industry. *Plant Cell Rep.* **18**, 76–81.

186. Jefferson, R.A. (1987) Assaying chimeric genes in plants: the GUS gene fusion system. *Plant Mol. Biol. Rep.* **5**, 387–404.

187. Ahlandsberg, S., Sathish, P., Sun, C., and Jansson, C. (1999) Green fluorescent protein as a reporter system in the transformation of barley cultivars. *Physiol. Plant.* **107**, 194–200.

188. Helmer, G., Casadaban, M., Bevan, M., Kayes, L., and Chilton, M.-L. (1984) A new chimeric gene as a marker for plant transformation: the expression of *Escherichia coli* β-galactosidase in sunflower and tobacco cells. *Biotechnology* **2**, 520–527.

189. Ow, D.W., Wood, K.V., DeLuca, M., De Wet, J.R., Helinski, D.R., and Howell, S.H. (1986) Transient and stable expression of the firefly luciferase gene in plant cells and transgenic plants. *Science* **234**, 856–859.

190. Simmonds, J., Cass, L., Routly, E., Hubbard, K., Donaldson, P., Bancroft, B., Davidson, A., Hubbard, S., and Simmonds, D. (2004) Oxalate oxidase: a novel reporter gene for monocot and dicot transformations. *Mol. Breed.* **13**, 79–91.

Chapter 2

Engineering the Plastid Genome of *Nicotiana sylvestris*, a Diploid Model Species for Plastid Genetics

Pal Maliga and Zora Svab

Abstract

The plastids of higher plants have their own ~120–160-kb genome that is present in 1,000–10,000 copies per cell. Engineering of the plastid genome (ptDNA) is based on homologous recombination between the plastid genome and cloned ptDNA sequences in the vector. A uniform population of engineered ptDNA is obtained by selection for marker genes encoded in the vectors. Manipulations of ptDNA include (1) insertion of transgenes in intergenic regions; (2) posttransformation excision of marker genes to obtain marker-free plants; (3) gene knockouts and gene knockdowns, and (4) cotransformation with multiple plasmids to introduce nonselected genes without physical linkage to marker genes. Most experiments on plastome engineering have been carried out in the allotetraploid *Nicotiana tabacum*. We report here for the first time plastid transformation in *Nicotiana sylvestris*, a diploid ornamental species. We demonstrate that the protocols and vectors developed for plastid transformation in *N. tabacum* are directly applicable to *N. sylvestris* with the advantage that the *N. sylvestris* transplastomic lines are suitable for mutant screens.

Key words: Plastid transformation, *Nicotiana sylvestris*, Tobacco, *aadA*, Spectinomycin resistance, Streptomycin resistance

1. Introduction

The plastid genome (ptDNA) of higher plants is highly polyploid. The number of plastids per cell and the number of ptDNA per plastid is dependent on the species and the cell type. For example, an *Arabidopsis thaliana* leaf cell contains about 120 chloroplasts, the green plastid type differentiated for photosynthesis, and these harbor 1,000-1,700 copies of the 154,478-bp plastid genome (1). In comparison, *Nicotiana tabacum* leaf cells contain about 100 chloroplasts harboring ~10,000 copies of the 155,939-bp ptDNA (2). Transformation of the plastid genome was first accomplished

James A. Birchler (ed.), *Plant Chromosome Engineering: Methods and Protocols*, Methods in Molecular Biology, vol. 701, DOI 10.1007/978-1-61737-957-4_2, © Springer Science+Business Media, LLC 2011

in *Chlamydomonas reinhardtii*, a unicellular alga (3), followed by plastid transformation in this laboratory in *N. tabacum*, a multicellular flowering plant (4). Plastid transformation since has been extended to *Porphyridium*, a unicellular red algal species (5) and the mosses *Physcomitrella patens* (6) and *Marchantia polymorpha* (7). In higher plants, plastid transformation is reproducibly performed in *N. tabacum* (8), tomato (9), soybean (10), lettuce (11, 12), and cabbage (13). Monocots as a group are still recalcitrant to plastid transformation.

Plastid transformation in each of these systems is based on homologous recombination. Plastid transformation vectors are *Escherichia coli* plasmid derivatives that do not replicate in plastids, so that the marker genes encoded in the vectors are expressed only if they integrate in the plastid genome via cloned ptDNA fragments flanking the marker gene (1–2 kb each side). In some of the plastid transformation vectors the restriction sites have been removed so that the plastid vectors can be used for cloning in *E. coli* (9, 14–16). The transforming DNA is introduced into plastids by the biolistic process (4, 8) or by polyethylene glycol treatment (17, 18), followed by selection for the marker gene. The marker genes employed for selective amplification of transformed ptDNA are plastid small rRNA genes with point mutations conferring spectinomycin resistance (4, 17, 18); or chimeric genes that confer resistance to spectinomycin and streptomycin (*aadA*)(8), kanamycin (*neo* or *aph(3′)IIa*) (19, 20) (21), or the amino acid analogues 4-methylindole (4MI) and 7-methyl-DL-tryptophan (7MT) (*ASA2*) (22). If the transforming DNA is introduced into protoplasts, the DNA delivery is by PEG treatment (17, 18). DNA delivery into leaf chloroplasts (4, 8) and the plastids of tissue culture cells (10, 23) is by the biolistic approach.

Engineering of the plastid genome includes four types of manipulations. The first type of manipulation is insertion of the marker gene and the gene of interest in intergenic regions. If mutations are included in the plastid-targeting region, the mutation may or may not be introduced into the plastid gene dependent on the site of recombination as described (24–26). The second type of manipulation is posttransformation marker gene excision. When we contemplate posttransformation excision of the marker genes, we flank the marker gene by recombinase target sites in the transformation vector. Excision of the marker gene is accomplished by a plastid-targeted recombinase expressed from a nuclear gene introduced by Agrobacterium-mediated transformation or by crossing. The example we show for the insertion of a transgene in Fig. 1 is transformation of the *N. sylvestris* plastid genome with the pCK2 plastid transformation vector that carries a selectable spectinomycin resistance (*aadA*) marker gene and the aurea *bar*^{au} gene of interest. Recombination between the pCK2 vector and ptDNA (dashed lines) yielded Ns-pCK2 ptDNA

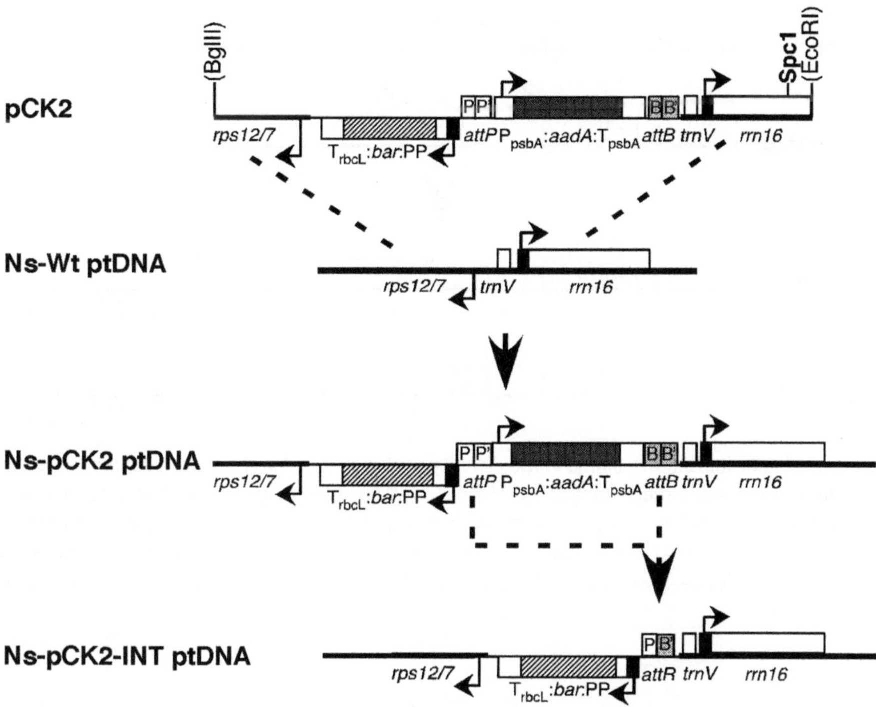

Fig. 1. Construction of transplastomic *Nicotiana sylvestris* plants carrying the aurea *bar*^{au} gene in their plastid genome. Recombination between the pCK2 vector and wild-type ptDNA (*dashed lines*) yields the Ns-pCK2 transplastome carrying the *bar*^{au} gene and the selectable spectinomycin resistance (*aadA*) gene. Note that *aadA* is flanked with the phiC31 phage recombinase *attB* (BB') and *attP* (PP') target sites. Excision of *aadA* by Int leaves behind a recombinant PB′ target site and the *bar*^{au} gene in the Ns-pCK2-INT ptDNA. For further details see reference on the construction of transplastomic *bar*^{au} *N. tabacum* plants (41).

carrying both transgenes. Excision of *aadA* by Int, the phiC31 phage site-specific recombinase, via the *attP* and *attB* target sites left behind only the aurea *bar*^{au} gene and a recombinant target site in the Ns-pCK2-INT ptDNA.

The third type of plastid genome engineering is knocking out plastid genes to query their function. The knockout lines are obtained by replacement of endogenous genes with marker genes by homologous recombination. The first examples of plastid gene knockouts were lacking the plastid *rbcL* (27) and *rpoB* (28) genes. By now at least 25 plastid gene knockout plants have been described (29). The example we show in Fig. 2 is a knockdown variant, in which introduction of an editing site in the *psbF* gene yielded an editing-dependent, functionally impaired version of the *psbF* gene. Transformation of the *N. sylvestris* plastid genome with vector pRB8, dependent on the site of recombination, yielded two types of transplastomic plants carrying *aadA*: editing-dependent, slow-growing plants with reduced chlorophyll content (T1-ptDNA) and plants with a wild-type phenotype (T2-ptDNA).

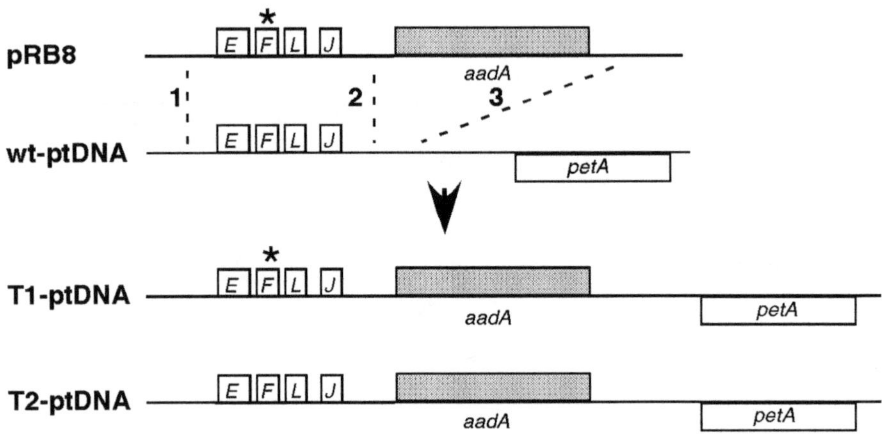

Fig. 2. Construction of *psbF* knockdown plants in *Nicotiana sylvestris* by transformation with the pRB8 plastid vector. The pRB8 plasmid carries a selectable spectinomycin resistance (*aadA*) gene targeting the insertion of *aadA* gene in the *psbJ-petA* intergenic region. The targeting region contains the tobacco *psbE* operon (*psbE*, *psbF**, *psbL*, and *psbJ* genes) with the engineered spinach *psbF* editing site marked by an asterisk (24). Recombination (*dashed lines*) between sites 1 and 3 yielded pigment-deficient knockdown plants (T1 ptDNA), whereas recombination between sites 2 and 3 yielded normal green plants (T2 ptDNA).

The fourth type of manipulation is cotransformation carried out with a mix of two plastid transformation vectors. First, the transplastomic clones are selected by the marker in vector one; then these transplastomic clones are screened for sequences from the second vector that does not carry a selectable marker. Cotransformation by the two plasmids is relatively frequent: plastid genomes recovered after transformation with the mixed vector carry the second, nonselected modification in ~20% of the clones (30, 31). Independent transformation of ptDNA copies with vectors targeted to alternative sites and subsequent sorting could also be used to obtain marker-free transplastomic plants carrying the nonselected genes (32). No cotransformation experiments have been carried out in *N. sylvestris* yet. In Fig. 3 we schematically show introduction of a nonselected marker by cotransformation in *N. tabacum* (30). For reviews on the applications of plastome engineering in basic science and biotechnology, see refs. (14, 29, 33–37).

Most experiments on plastid genome engineering have been carried out in *N. tabacum*, an allotetraploid species. Because of our interest in the identification of nuclear genes that control plastid inheritance, we were looking for a diploid *Nicotiana* species in which a mutant screen can be carried out. We report here protocols for plastid transformation in *Nicotiana sylvestris*. *N. sylvestris* shares all the advantages of *N. tabacum*: facile tissue culture regeneration, transformation of nuclear and plastid genomes, numerous progeny (greater than million) and short generation time (4 months) with the additional advantage that it has a

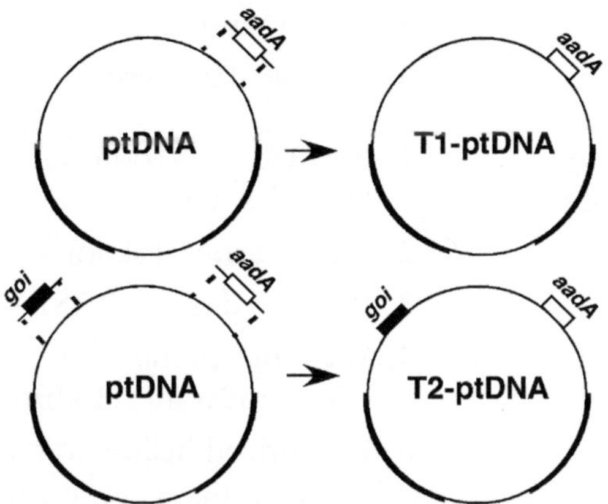

Fig. 3. Cotransformation with mixed plasmids to insert nonselected sequences in the plastid genome without physical linkage to the selectable marker. The nonselected sequence may be a gene (30) or a protein tag (31). Plastids contain multiple copies of the plastid genome (ptDNA). The selectable spectinomycin resistance (*aadA*) gene (*open box*) and nonselected gene-of-interest (*goi*; *filled box*) are targeted to different regions of ptDNA. Clones carrying *aadA* genes (T1-ptDNA and T2-ptDNA) are selected by spectinomycin resistance and distinguished from spontaneous mutants by streptomycin/spectinomycin resistance (see text). Approximately 20% of the transplastomic clones carries both *aadA* and the nonselected *goi* (T2-ptDNA). Based on ref. (30, 31).

diploid nuclear genome ($n = 12$; 2,600 Mb; (38)). Because the diploid *N. sylvestris* is the maternal progenitor of the allotetraploid *N. tabacum*, the plastid genomes of the two species are very similar (39). We report here that vectors pRB8 (24), pMSK56 (40), and pCK2 (41) originally developed for *N. tabacum* can be used for plastid transformation in *N. sylvestris*. We currently use the transplastomic *N. sylvestris* lines in a genetic screen for mutations in plastid inheritance.

2. Materials

2.1. Cleaning of Gold Particles

1. 0.6 μm gold microcarrier available from Bio-Rad.
2. Ice-cold 70% ethanol.
3. Ice-cold sterile distilled water.
4. 50% glycerol, sterilized by autoclaving.

2.2. Coating of Gold Particles with Plasmid DNA

1. Plasmid DNA prepared using the QIAGEN Maxi Kit (Qiagen Inc., Valencia, CA).
2. 2.5 M $CaCl_2$ solution.
3. 0.1 M spermidine free base (Sigma, St Louis, MO).

4. 70% ethanol.
5. Ice-cold 100% ethanol.

2.3. DNA Delivery into Chloroplasts by the Biolistic Process

1. Bio-Rad PDS1000/He biolistic gun (Bio-Rad Laboratories, Hercules, CA).
2. Macrocarrier disk for PDS1000/He biolistic gun (Bio-Rad).
3. Rupture disk for PDS1000/He biolistic gun (Bio-Rad).
4. Vacuum pump for the gun ThermoSavant VLP285 (ThermoSavant, Holbrook, NY).
5. Pressurized helium in tank, 99.999% pure, moisture free.
6. Laminar flow hood for bombardment and tissue culture.
7. Whatman No. 4 filter paper disks sterilized by autoclaving in glass Petri dish.
8. Sterile *Nicotiana sylvestris* TW137 plants in Magenta boxes (see Note 1) grown on RM plant maintenance medium (see Note 2).
9. Petri plate (100 × 15 mm) containing solid RMOP medium (20 ml) for leaf bombardment (see Note 3).
10. Deep (100 × 20 mm) Petri plates with 50 ml selective RMOP containing 500 mg/L Spectinomycin dihydrochloride pentahydrate (Sigma, St Louis, MO) (see Note 4).
11. 70% ethanol.
12. Plastic wrap (strips) that is permeable to gas exchange, for example, Glad ClingWrap.

2.4. Identification of Transplastomic Plants

1. Deep (100 × 20 mm) Petri dishes containing RM plant maintenance medium (see Note 2).
2. Deep (100 × 20 mm) Petri dishes containing a selective spectinomycin (500 mg/L; Sigma, St Louis, MO) RMOP plant regeneration medium (see Note 3).
3. Deep (100 × 20 mm) Petri dishes containing 500 mg/L each of spectinomycin (Sigma, St Louis, MO; Catalog No. S9007) and streptomycin (Sigma, St Louis, MO) in the RMOP plant regeneration medium (see Note 4).
4. Magenta boxes containing selective spectinomycin (500 mg/L; Sigma, St Louis, MO) RM plant maintenance medium (see Note 2).
5. Deep (100 × 20 mm) Petri dishes containing RM plant maintenance medium (see Note 2).
6. Captan (Bonide Products Inc., Oriskany, NY), a fungicide; active ingredient *N*-Trichloromethylthio-4-cyclohexane-1, 2-dicarboximide.
7. Plastic trays (10″ × 20″) with holes on bottom with garden soil and transparent plastic domes.

3. Methods

The protocol we describe here was used to obtain transplastomic *N. sylvestris* after bombardment of tissue culture grown leaves with the pRB8 (24), pMSK56 (40) and pCK2 (41) plastid transformation vectors. The Ns-pRB8 and Ns-pMSK56 plants carry a selectable spectinomycin resistance (*aadA*) gene. The *aadA* gene in the Ns-pMSK56 plasmids is translationally fused with the green fluorescent protein. The Ns-pCK2 plants carry an aurea bar^{au} gene linked to *aadA*. In the Ns-pCK2-INT derivatives the *aadA* gene has been excised with Int, the phiC31 phage site-specific integrase. The protocol we describe here for *N. sylvestris* complements two other protocols published on plastid transformation in *N. tabacum* from this laboratory: one describing plastid transformation and marker gene excision by the CRE-*loxP* site-specific recombination system (42); the second on the applications of plastid transformation to study mRNA editing (43).

3.1. Cleaning of Gold Particles

This protocol was modified from Bio-Rad Bulletin 9075.

1. Weigh out 30 mg 0.6 μm gold microcarrier in a 1.5 ml Eppendorf tube and add 1 ml ice-cold 70% ethanol. Place tube in a Vortex microtube holder and shake vigorously for 5 min. Let the particles settle in the tube for 15 min at room temperature (20–25°C).
2. Spin in microcentrifuge at 3,000 rpm (600 × *g*) for 1 min to compact gold. Remove the ethanol with a pipette and add 1 ml ice-cold sterile distilled water. Vortex the tube to suspended the particles. Allow the gold particles to settle at room temperature for 10 min.
3. Sediment the gold particles by spinning in a microcentrifuge at 3,000 rpm (600 × *g*) for 1 min. Remove water with pipette and add 1 ml ice-cold sterile distilled water.
4. Repeat washing the gold particles with water by repeating steps 2 and 3.
5. Suspend the gold particles by vortexing and store the tube at room temperature for 10 min to allow the particles to settle.
6. Spin the tube in a microcentrifuge at 5,000 rpm (1,700 × *g*) for 15 s, then remove water completely. Add 500 μl 50% glycerol and vortex for 1 min to resuspended particles. Gold concentration will be 60 mg/ml. The clean gold can be stored for 2 weeks at room temperature.

3.2. Coating of Gold Particles with Plasmid DNA

This protocol was modified from Bio-Rad Bulletin 9075.

1. Place Eppendorf tube containing gold in a Vortex microtube holder and shake at setting 3. While tube is shaking, remove

50 μl aliquots of gold and pipette into ten 1.5 ml Eppendorf tubes in a rack (see Note 5). While tubes are shaking, add 5 μl plasmid DNA (1 μg/μl), 50 μl 2.5 M $CaCl_2$, and 20 μl 0.1 M spermidine free base. Make sure to add the components in this order and that the contents are thoroughly mixed before adding the next component.

2. Shake tubes on Vortex at setting 3 for 5 min.
3. Sediment gold by spinning in microcentrifuge at 3,000 rpm (600 × *g*) for 1 min. Remove the supernatant and add 140 μl 70% ethanol to each tube. Tap tube lightly until the pellet just starts to come into solution to make sure pellet is not tightly packed. If gold does not go into solution by gently tapping the tube, break up pellet by pipetting up and down.
4. Sediment gold by spinning in a microcentrifuge at 3,000 rpm for 1 min. Remove supernatant and add 140 μl ice-cold 100% ethanol to each tube. Lightly tap tube until the pellet just starts to come into solution.
5. Sediment gold by spinning in a microcentrifuge at 5,000 rpm (1,700 × *g*) for 15 s. Resuspend coated gold pellet in 50 μl 100% ethanol by gently tapping tube. Pellet should easily enter solution. Shake tubes at setting 3 while waiting to use them for bombardment. If tubes are sitting for a long period of time before bombardment, replace ethanol in tube with fresh 100% ethanol.

3.3. DNA Delivery into Chloroplasts by the Biolistic Process

The protocol we describe here is for biolistic transformation of *N. sylvestris* leaves collected from plants grown in sterile culture. Bombardment of 20–30 leaves with the Bio-Rad PDS1000/He gun yields 10–40 transplastomic clones. The Hepta-adaptor version of the gun (which is simultaneously using seven macrocarriers) is more efficient; bombardment of five leaves is sufficient to obtain 5–25 transplastomic clones.

1. Place leaves for biolistic transformation abaxial side up on two sterile Whatman No. 4 filter papers on top of solid RMOP medium (20 ml) in a 100 × 15 mm Petri plate. Use more than one leaf if necessary to cover the central area of the Petri plate (see Note 6).
2. Set up the biolistic gun in a sterile laminar flow hood. Before bombardment, wipe off main chamber, rupture disk retaining cap, microcarrier launch assembly, and the target shelf of the gun with a cloth soaked in 70% ethanol.
3. Sterilize rupture disks (1,100 psi), macrocarriers, macrocarrier holders, and stopping screens by soaking in 100% ethanol (5 min) then air dry them in tissue culture hood in open container.

4. Turn on helium tank and set pressure in regulator (distal to tank) for 1,300 psi (200–300 psi above rupture disk value).
5. Turn on vacuum pump and gene gun. Set the vacuum rate on the gene gun to 7 and the vent rate to 2.
6. Prepare DNA-coated gold particles as described above. Pipette 10 μl of DNA-coated gold onto macrocarrier or "flying disk" in holder and let the disk air dry for 5 min. Five samples may be made up at one time if bombardment is carried out with one macrocarrier (or seven, if Hepta-adaptor is used).
7. Place rupture disk into retaining cap and screw in place tightly.
8. Put stopping screen and flying disk (face down) in microcarrier launch assembly and place in chamber just below rupture disk. For description see Bio-Rad Bulletin 9075.
9. Place leaf on thin RMOP plate (100 × 15 mm) into chamber 9 cm (fourth shelf from top) below the microcarrier launch assembly and close the door (see Note 7).
10. Press vacuum button to open valve. When vacuum reaches 28 inches of mercury (Hg) hold down fire button until the pop from the gas breaking the rupture disk is heard (see Note 8).
11. Immediately release the vacuum and remove the leaf sample.
12. Repeat steps 6–11 until all leaf samples are bombarded. When finished, turn off helium tank, and release pressure by holding down the fire button while vacuum is on. Turn off vacuum pump.
13. Place clear plastic sleeves over plates containing bombarded leaf samples and incubate in culture room. Incubation allows time for marker gene expression before selection is started.
14. After 2 days, cut bombarded leaves into small (1 cm square) pieces and place abaxial side up in deep plates (100 × 20 mm) containing a selective RMOP medium with 500 mg/L spectinomycin. Place only seven pieces per plate as the leaf pieces will grow and expand (see Note 9).
15. Seal each plate with plastic wrap that is permeable to gas exchange and incubate plates in culture room for 4–12 weeks.

3.4. Identification of Transplastomic Plants

1. The bombarded green leaf sections turn pale on the selective spectinomycin medium and expand. Cells carrying transformed spectinomycin-resistant plastids turn green and overcome the inhibition of shoot regeneration by spectinomycin. As a result, transplastomic cells form green shoots on the pale leaf sections. However, mutations in the 16S rRNA also confer spectinomycin resistance and yield green shoots (8).

Dissect green shoots that appear 4–12 weeks after bombardment onto RM plates for rooting and classification as a transplastomic clone or a spontaneous mutant. Each shoot at a distinct location derives from an independent event; therefore, they should be treated as independently derived clones. We mark the independent clones by a serial number.

2. To distinguish transplastomic clones from spontaneous mutants, transfer small (1 cm^2) leaf sections of the regenerated shoots (Subheading 3.4, step 1) onto (a) selective streptomycin–spectinomycin RMOP medium and (b) spectinomycin RMOP medium. Transgenic clones carrying an *aadA* gene are resistant to both spectinomycin and streptomycin, whereas spontaneous spectinomycin-resistant mutants are resistant only to spectinomycin (8). Resistance is manifested as formation of green calli with regenerating shoots; sensitivity is indicated by the bleaching of leaf sections and the absence of shoot regeneration. Classify mutants on the streptomycin–spectinomycin medium; pick shoots of the transplastomic clones from spectinomycin and transfer them onto RM plant maintenance medium (see Note 10). Carry out DNA gel blot analyses in leaves regenerated on spectinomycin medium to confirm plastid transformation and to test the uniformity of ptDNA population (42). At this stage we identify the transplastomic clones by the initials of the plant species and the transforming plasmid name, for example, Ns-pCK2-6A for *N. sylvestris* transformed with plasmid pCK2, serial number 6 and one or more letters of the alphabet to indicate the number of shoot regeneration cycles on selective medium. Always test plants from three to four independently transformed clones because ~10% of plants regenerated in tissue culture are sterile due to somaclonal variation.
3. Because the plastid genome is highly polyploid, the plant may be chimeric even if it looks homoplastomic on the gel blots. Therefore, we repeat plant regeneration on the selective medium and verify the uniformity of ptDNA by gel blot analyses. Plants regenerated 2× on selective spectinomycin medium are normally homoplastomic and are marked by the addition of two letters of the alphabet, for example, Ns-pCK2-6AB.
4. Transfer rooted transplastomic shoots to greenhouse to obtain seed. Gently break up the agar; wash roots to remove the agar-solidified RM medium; drench the soil after planting with Captan to fend off fungal infection; and cover plantlets with a plastic dome to prevent desiccation. Grow plants in shade for about a week, and then remove plastic dome and expose the plants to full sunlight.

5. Collect mature seedpods from transplastomic plants and let the seed dry on the lab bench at room temperature for 1 week. Remove moisture for long-term storage by a 2-week storage in seed drier.
6. Germinate surface-sterilized seeds on RM-spectinomycin (500 mg/L) medium (see Note 2). Seedlings carrying transformed ptDNA will be dark green, whereas sensitive seedlings will be white. One hundred percent green seedlings confirm uniform population of transformed ptDNA in the transplastomic plants.

4. Notes

1. *Nicotiana sylvestris* TW137 plants in Magenta boxes. The *Nicotiana sylvestris* ecotype TW137 was obtained from the USDA seed collection at NC State University. The fertile TW137 was converted to a cytoplasmic male sterile (CMS) form by repeated pollination of the *N. tabacum* CMS92 line with *N. sylvestris* TW137 pollen. We obtained the CMS92 *N. tabacum* line with the *Nicotiana undulata* cytoplasm from Ezra Galun (44). The seeds are vapor sterilized by storing them for 3 h in an open 1.5 ml Eppendorf tube in a closed desiccator over the mix of 100 ml Clorox bleach and 3 ml concentrated (37.8%) HCl. The seed is germinated in deep (100 × 20 mm) Petri dishes on RM plant maintenance medium (60 ml/dish; see below). The seedlings are grown to plants in Magenta boxes containing 80 ml RM medium.
2. The RM plant maintenance medium (MS medium, ref. (45)) per liter contains: 100 ml 10× macronutrients, 10 ml 100× micronutrients, 5 ml 1% Fe–EDTA, 30 g sucrose, pH 5.6–5.8 with 1 M KOH; 7 g agar). RM Medium 10× macronutrient solution per liter contains: 19 g KNO_3, 3.7 g $MgSO_4 \cdot 7H_2O$, 4.4 g $CaCl_2 \cdot 2H_2O$, 1.7 g KH_2PO_4, 16.5 g $(NH_4)NO_3$. RM Medium 100× micronutrient solution per liter contains: 169 mg $MnSO_4 \cdot H_2O$, 62 mg H_3BO_3, 86 mg $ZnSO_4 \cdot 7H_2O$, 8.3 mg KI, 2.5 ml $Na_2MoO_4 \cdot 2H_2O$ (1 mg/ml), 2.5 ml $CuSO_4 \cdot 5H_2O$ (1 mg/ml), 0.25 ml $CoCl_2 \cdot 6H_2O$ (1 mg/ml). Agar, plant tissue culture tested (Sigma, St Louis, MO).
3. RMOP shoot regeneration medium (46) per liter contains: 100 ml RM medium 10× macronutrients, 10 ml RM medium 100× micronutrients, 5 ml 1% Fe–EDTA, 1 ml thiamine (1 mg/ml), 0.1 ml alpha-naphthaleneacetic acid (NAA 1 mg/ml in 0.1 M NaOH), 1 ml 6-benzylaminopurine (BAP 1 mg/ml in 0.1 M HCl), 0.1 g myo-inositol, 30 g sucrose, pH 5.8 with 1 M KOH, 7 g agar.

4. The selective RMOP medium contains filter-sterilized 500 mg/L spectinomycin dihydrochloride (Sigma, St Louis, MO), or 500 mg/L streptomycin sulfate (Sigma, St Louis, MO), or both. The composition of RMOP medium is described above.
5. With one DNA construct 20–30 single leaves are bombarded. Each tube (30 mg) of gold is sufficient for 50 bombardments (two DNA constructs). One may use freshly prepared gold particles, or stored gold. If using stored gold, vortex tube for 5 min before coating with DNA.
6. If you are using greenhouse leaves, you need to surface sterilize the leaves and cut out a segment that covers the plate. Greenhouse leaves may be surface sterilized by rinsing the leaves in water containing a drop of surfactant (Tween 80 or liquid soap) to remove dirt; dipping the leaves in 70% ethanol; and placing them in diluted commercial bleach (tenfold diluted Clorox; final sodium hypochlorite concentration is ~0.6%) for 3 min. The bleach should be thoroughly removed by rinsing the leaves in sterile distilled water 5×.
7. Leaves should be prepared in advance of transformation and may be stored on top of the filter paper in a "ready-to-shoot" state for at least a half day.
8. If gun is fired at a lower vacuum pressure, the DNA-coated particles will lack the momentum to penetrate cells and no transplastomic lines will be obtained. If you have no experience with biolistic transformation, we recommend that you test particle coating and DNA delivery using transient expression of a nuclear *uidA* gene that encodes the enzyme β-glucuronidase, the activity of which can be readily detected by histochemical staining (47, 48).
9. Leaf pieces from 1 leaf fit on 3–5 plates. If the leaf sections are too large, there will be insufficient nutrient in the medium to support growth for up to 12 weeks, the time frame within which transplastomic clones appear. Diagnostic sign of overcrowding is absence of spontaneous spectinomycin-resistant mutants and transplastomic clones. Overcrowding may be also caused by less than the desired 50 ml culture medium in a deep plate.
10. We prefer to work with shoots regenerated on spectinomycin, because streptomycin is a mutagen.

Acknowledgments

This work was supported by grants from the USDA Biotechnology Risk Assessment Research Grant Program Award No. 2005-33120-16524 and 2008-03012.

References

1. Zoschke, R., Liere, K., and Borner, T. (2007) From cotyledon to mature plant: Arabidopsis plastidial genome copy number, RNA accumulation and transcription are differentially regulated during leaf development, *Plant J* **50**, 710–722.
2. Shaver, J. M., Oldenburg, D. J., and Bendich, A. J. (2006) Changes in chloroplast DNA during development in tobacco, *Medicago truncatula*, pea, and maize, *Planta* **224**, 72–82.
3. Boynton, J. E., Gillham, N. W., Harris, E. H., Hosler, J. P., Johnson, A. M., Jones, A. R., Randolph-Anderson, B. L., Robertson, D., Klein, T. M., Shark, K. B., and Sanford, J. C. (1988) Chloroplast transformation in *Chlamydomonas* with high velocity microprojectiles, *Science* **240**, 1534–1538.
4. Svab, Z., Hajdukiewicz, P., and Maliga, P. (1990) Stable transformation of plastids in higher plants, *Proc Natl Acad Sci USA* **87**, 8526–8530.
5. Lapidot, M., Raveh, D., Sivan, A., Arad, S. M., and Shapira, M. (2002) Stable chloroplast transformation of the unicellular red alga Porphyridium species, *Plant Physiol* **129**, 7–12.
6. Sugiura, C., and Sugita, M. (2004) Plastid transformation reveals that moss trnR-CCG is not essential for plastid function, *Plant J* **40**, 31–321.
7. Chiyoda, S., Linley, P. J., Yamato, K. T., Fukuzawa, H., Yokota, A., and Kohchi, T. (2007) Simple and efficient plastid transformation system for the liverwort *Marchantia polymorpha* L. suspension-culture cells, *Transgenic Res* **16**, 41–49.
8. Svab, Z., and Maliga, P. (1993) High-frequency plastid transformation in tobacco by selection for a chimeric *aadA* gene, *Proc Natl Acad Sci USA* **90**, 913–917.
9. Ruf, S., Hermann, M., Berger, I. J., Carrer, H., and Bock, R. (2001) Stable genetic transformation of tomato plastids: foreign protein expression in fruit, *Nat Biotechnol* **19**, 870–875.
10. Dufourmantel, N., Pelissier, B., Garcon, F., Peltier, G., Ferullo, J. M., and Tissot, G. (2004) Generation of fertile transplastomic soybean, *Plant Mol Biol* **55**, 479–489.
11. Lelivelt, C., McCabe, M., Newell, C., de Snoo, B., Van Dunn, K., Birch-Machin, I., Gray, J. C., Mills, K., and Nugent, J. M. (2005) Plastid transformation in lettuce (*Lactuca sativa* L), *Plant Mol Biol* **58**, 763–774.
12. Kanamoto, H., Yamashita, A., Asao, H., Okumura, S., Takase, H., Hattori, M., Yokota, A., and Tomizawa, K. (2006) Efficient and stable transformation of *Lactuca sativa* L. cv. Cisco (lettuce) plastids, *Transgenic Res* **15**, 205–217.
13. Liu, C. W., Lin, C. C., Chen, J. J., and Tseng, M. J. (2007) Stable chloroplast transformation in cabbage (*Brassica oleracea* L. var. *capitata* L.) by particle bombardment, *Plant Cell Rep* **26**, 1733–1744.
14. Lutz, K. A., Azhagiri, A. K., Tungsuchat-Huang, T., and Maliga, P. (2007) A guide to choosing vectors for transformation of the plastid genome of higher plants, *Plant Physiol* **145**, 1201–1210.
15. Sinagawa-Garcia, S. R., Tungsuchat-Huang, T., Paredes-Lopez, O., and Maliga, P. (2009) Next generation synthetic vectors for transformation of the plastid genome of higher plants, *Plant Mol Biol* **70**, 487–498.
16. Zhou, F., Karcher, D., and Bock, R. (2007) Identification of a plastid intercistronic expression element (IEE) facilitating the expression of translatable monocistronic mRNAs from operons, *Plant J* **52**, 961–972.
17. Golds, T., Maliga, P., and Koop, H. U. (1993) Stable plastid transformation in PEG-treated protoplasts of *Nicotiana tabacum*, *Biotechnology* **11**, 95–97.
18. O'Neill, C., Horvath, G. V., Horvath, E., Dix, P. J., and Medgyesy, P. (1993) Chloroplast transformation in plants: polyethylene glycol (PEG) treatment of protoplasts is an alternative to biolistic delivery systems, *Plant J* **3**, 729–738.
19. Carrer, H., Hockenberry, T. N., Svab, Z., and Maliga, P. (1993) Kanamycin resistance as a selectable marker for plastid transformation in tobacco, *Mol Gen Genet* **241**, 49–56.
20. Huang, F. C., Klaus, S. M. J., Herz, S., Zuo, Z., Koop, H. U., and Golds, T. J. (2002) Efficient plastid transformation in tobacco using the *aphA-6* gene and kanamycin selection, *Mol Genet Genomics* **268**, 19–27.
21. Lutz, K., Corneille, S., Azhagiri, A. K., Svab, Z., and Maliga, P. (2004) A novel approach to plastid transformation utilizes the phiC31 phage integrase, *Plant J* **37**, 906–913.
22. Barone, P., Zhang, X. H., and Widholm, J. M. (2009) Tobacco plastid transformation using the feedback-insensitive anthranilate synthase [alpha]-subunit of tobacco (ASA2) as a new selectable marker, *J Exp Bot* **60**, 3195–3202.
23. Langbecker, C. L., Ye, G. N., Broyles, D. L., Duggan, L. L., Xu, C. W., Hajdukiewicz, P. T., Armstrong, C. L., and Staub, J. M. (2004)

High-frequency transformation of undeveloped plastids in tobacco suspension cells, *Plant Physiol* **135**, 39–46.

24. Bock, R., Kössel, H., and Maliga, P. (1994) Introduction of a heterologous editing site into the tobacco plastid genome: the lack of RNA editing leads to a mutant phenotype, *EMBO J* **13**, 4623–4628.
25. Whitney, S. M., and Andrews, T. J. (2001) Plastome-encoded bacterial ribulose-1,5-bisphosphate carboxylase/oxygenase (Rubisco) supports photosynthesis and growth of tobacco, *Proc Natl Acad Sci USA* **98**, 14738–14743.
26. Sharwood, R. E., von Caemmerer, S., Maliga, P., and Whitney, S. M. (2008) The catalytic properties of hybrid rubisco comprising tobacco small and sunflower large subunits mirror the kinetically equivalent source Rubiscos and can support tobacco growth, *Plant Physiol* **146**, 83–96.
27. Kanevski, I., and Maliga, P. (1994) Relocation of the plastid *rbcL* gene to the nucleus yields functional ribulose-1,5-bisphosphate carboxylase in tobacco chloroplasts, *Proc Natl Acad Sci USA* **91**, 1969–1973.
28. Allison, L. A., Simon, L. D., and Maliga, P. (1996) Deletion of *rpoB* reveals a second distinct transcription system in plastids of higher plants, *EMBO J* **15**, 2802–2809.
29. Maliga, P. (2004) Plastid transformation in higher plants, *Ann Rev Plant Biol* **55**, 289–313.
30. Carrer, H., and Maliga, P. (1995) Targeted insertion of foreign genes into the tobacco plastid genome without physical linkage to the selectable marker gene, *Biotechnology* **13**, 791–794.
31. Rumeau, D., Becuwe-Linka, N., Beyly, A., Carrier, P., Cuine, S., Genty, B., Medgyesy, P., Horvath, E., and Peltier, G. (2004) Increased zinc content in transplastomic tobacco plants expressing a polyhistidine-tagged Rubisco large subunit, *Plant Biotechnol J* **2**, 389–399.
32. Ye, G. N., Colburn, S., Xu, C. W., Hajdukiewicz, P. T. J., and Staub, J. M. (2003) Persistance of unselected transgenic DNA during a plastid transformation and segregation approach to herbicide resistance, *Plant Physiol* **133**, 402–410.
33. Lutz, K. A., and Maliga, P. (2007) Construction of marker-free transplastomic plants, *Curr Opin Biotechnol* **18**, 107–114.
34. Bock, R. (2001) Transgenic plastids in basic research and plant biotechnology, *J Mol Biol* **312**, 425–438.
35. Bock, R. (2007) Plastid biotechnology: prospects for herbicide and insect resistance, metabolic engineering and molecular farming, *Curr Opin Biotechnol* **18**, 100–106.
36. Daniell, H., Chebolu, S., Kumar, S., Singleton, M., and Falconer, R. (2005) Chloroplast-derived vaccine antigens and other therapeutic proteins, *Vaccine* **23**, 1779–1783.
37. Bock, R., and Warzecha, H. (2010) Solar-powered factories for new vaccines and antibiotics, *Trends Biotechnol* **28**(5), 246–252.
38. Leitch, I. J., Hanson, L., Lim, K. Y., Kovarik, A., Chase, M. W., Clarkson, J. J., and Leitch, A. R. (2008) The ups and downs of genome size evolution in polyploid species of *Nicotiana* (Solanaceae), *Ann Bot* **101**, 805–814.
39. Yukawa, M., Tsudzuki, T., and Sugiura, M. (2006) The chloroplast genome of *Nicotiana sylvestris* and *Nicotiana tomentosiformis*: complete sequencing confirms that the *Nicotiana sylvestris* progenitor is the maternal genome donor of *Nicotiana tabacum*, *Mol Genet Genomics* **275**, 367–373.
40. Khan, M. S., and Maliga, P. (1999) Fluorescent antibiotic resistance marker to track plastid transformation in higher plants, *Nat Biotechnol* **17**, 910–915.
41. Kittiwongwattana, C., Lutz, K. A., Clark, M., and Maliga, P. (2007) Plastid marker gene excision by the phiC31 phage site-specific recombinase, *Plant Mol Biol* **64**, 137–143.
42. Lutz, K. A., Svab, Z., and Maliga, P. (2006) Construction of marker-free transplastomic tobacco using the Cre-*loxP* site-specific recombination system, *Nat Protocols* **1**, 900–910.
43. Lutz, K. A., and Maliga, P. (2007) Transformation of the plastid genome to study RNA editing, *Methods Enzymol* **424**, 501–518.
44. Galun, E., Arze-Gonen, P., Fluhr, R., Edelman, M., and Aviv, D. (1982) Cytoplasmic hybridization in *Nicotiana*: mitochondrial DNA analysis in progenies resulting from fusion between protoplasts having different organelle constitutions, *Mol Gen Gent* **186**, 50–56.
45. Murashige, T., and Skoog, F. (1962) A revised medium for the growth and bioassay with tobacco tissue culture, *Physiol Plant* **15**, 473–497.
46. Sidorov, V., Menczel, L., and Maliga, P. (1981) Isoleucine-requiring *Nicotiana* plant deficient in threonine deaminase, *Nature* **294**, 87–88.
47. Jefferson, R. A., Kavanagh, T. A., and Bevan, M. W. (1987) GUS fusions: beta-glucuronidase as a sensitive and versatile gene fusion marker in higher plants, *EMBO J* **6**, 3901–3907.
48. Gallagher, S. R., (Ed.) (1992) *GUS protocols: using the GUS gene as a reporter of gene expression*, Academic Press, San Diego.

Chapter 3

Homologous Recombination in Plants: An Antireview

Michal Lieberman-Lazarovich and Avraham A. Levy

Abstract

Homologous recombination (HR) is a central cellular process involved in many aspects of genome maintenance such as DNA repair, replication, telomere maintenance, and meiotic chromosomal segregation. HR is highly conserved among eukaryotes, contributing to genome stability as well as to the generation of genetic diversity. It has been intensively studied, for almost a century, in plants and in other organisms. In this antireview, rather than reviewing existing knowledge, we wish to underline the many open questions in plant HR. We will discuss the following issues: how do we define homology and how the degree of homology affects HR? Are there any plant-specific HR qualities, how extensive is functional conservation and did HR proteins acquire new functions? How efficient is HR in plants and what are the *cis* and the *trans* factors that regulate it? Finally, we will give the prospects for enhancing the rates of gene targeting and meiotic HR for plant breeding purposes.

Key words: Homologous recombination, Meiotic recombination, Zinc-finger nuclease, Gene targeting, Chromatin remodeling, Plant breeding

1. Introduction

A number of recent reviews have covered what is known on various aspects of homologous recombination in plants (1–8). Here, we wish to write an "antireview," i.e., to focus on what is not known. We will provide a brief update on the current knowledge in the plant HR field in order to point out what we view as the existing huge gaps in our understanding of this process.

James A. Birchler (ed.), *Plant Chromosome Engineering: Methods and Protocols*, Methods in Molecular Biology, vol. 701, DOI 10.1007/978-1-61737-957-4_3,

2. What is Meant by Homologous?

Homologous recombination (HR) is a complex process whereby DNA segments that share significant sequence homology are exchanged. This definition, however, does not state what is considered to be extensive length of homology that can be used as a substrate for the HR machinery. In yeast, it is thought that as little as 30 bp are sufficient (9). In plants, we know that a few hundred base pairs can engage in HR (10), but we do not know whether there is a lower limit, nor what is the dependence on the type of partners. Sequence microhomology is often found at the borders of nonhomologous end-joining (NHEJ) events in plants (11). Such microhomologies seem to serve as sticky ends that stabilize the ligation of nonhomologous ends without involving HR proteins. Another level of complexity that will be discussed further is the degree of homology. How divergent can partners be to still be considered as homologous by the HR machinery?

3. What is Meant by Homologous Recombination?

There are many ways to catalyze strand exchange between homologous sequences and the outcomes can be quite diverse. In all HR cases there is a need for a homology search. Remarkably, the basic question of homology search, namely, how do homologous partners find each other? How do the various DNA molecules move within the nucleus until partners come into the physical proximity that is necessary to enable strand invasion? Is all the genome scanned to find a proper alignment? Does the scanning process proceed in a random or organized manner? Is strand invasion involved in the identification of homology or is there another, yet undiscovered, scanning process? This topic is very poorly understood in plants, as well as in bacterial (12) and yeast models (13), and remains one of the most puzzling and basic open questions in the field. Out of the many alternative pathways that can be used to recombine, we still understand very little on how the cell "decides" in which path to proceed. The choice can be between various mechanisms involving Holliday junctions, an interference-independent pathway, single-strand annealing (SSA), synthesis-dependent strand annealing (SDSA) (see review (14)) or some other yet undiscovered pathways.

4. Anything Special About HR in Plants?

Prior to asking whether there is something special about HR in plants, one should think whether there is anything special in plants' life cycle that might, through natural selection, affect HR (15). One aspect is that the lack of a germline and the vegetative propagation of several species, may suggest that somatic HR plays a greater role in plants than in other species, which rely solely on sexual reproduction from a germline that is differentiated early on in development. Another potentially relevant fact is that most plants are polyploids (16). Whole genome duplication confers buffering of deleterious mutations and thus might enable the polyploid plant genome to be more tolerant to genetic rearrangements than the diploid one. For example, there might be a "release" of selection or repression of HR between direct or inverted repeats. Another challenge of polyploidy is to maintain fertility, which might be reduced through HR between homoeologs (in allopolyploids) and through multivalent pairing during meiosis. Note that we do not know how polyploidy per se, might affect HR rates and patterns in an autopolyploid or allopolyploid genome.

Finally, due to their sessile nature, plants are constantly exposed to DNA-damaging agents, such as UV and, in some areas, heavy metals (17). Did HR evolve differently in plants than in other kingdoms due to this chronic exposure? Would such a difference be due to quantitative (regulation, level of expression) or qualitative (the machinery itself) dissimilarities? Does HR play an important role in the adaptation of plants to different habitats (e.g., high altitude, high UV, or contaminated soils)? Plants provide a wonderful system to address such questions due to their sessility and the wide ecological distribution of model species such as Arabidopsis. Interestingly, plants grown in the Chernobyl area were found to exhibit higher mutation and HR rates than plants grown in noncontaminated areas (18, 19). We might assume that the HR machinery is necessary for survival under such conditions, although this has not been proven. Testing such hypothesis is not an easy task. It would require performing "evolution" experiments, namely, growing HR mutants and WT in genotoxic and in normal habitat for a number of years and comparing their fitness. In bacteria that grow under chronic exposure to irradiation, e.g., *Deinococcus radiodurans*, HR plays an important role in DNA repair (20). However, it is still debated whether this is the main cause for radiation resistance or whether other pathways (e.g., NHEJ) are responsible for this phenomenon (12).

The HR machinery is, together with other DNA maintenance functions such as replication and repair, one of the most conserved in the cell. Protein homologies run as deep as between prokaryotic proteins, e.g., RecA, vs. their eukaryotic homologs Rad51 (21). So is there anything special about the plant HR machinery? There are undoubtedly many differences that have been pointed out (see above reviews). These differences concern mostly the different number of homologs, or the lack thereof for certain key genes when comparing yeast and plants or plants and mammals. Are these differences the result of random drift between species, or were these differences selected to serve some plant-specific HR needs?

There are plant-specific processes, not related to HR per se, that have "adopted" HR proteins, assigning them with new functions. The Rad51D protein, for instance, is an AtRad51 paralog which plays a role in HR (1, 22). It was found, in a genetic screen in Arabidopsis, that Rad51D is a repressor of Sni1, by itself a repressor of pathogenesis-related proteins, linking Rad51D with the plant's defense response (23). Hence, it seems that Rad51D has acquired a new, plant-specific function following gene duplication (neofunctionalization). Interestingly, a mutation in *RAD51D* was also shown to enhance the *tebichi* (*teb-1*) developmental phenotype, using a *teb-1 rad51d* double mutant (24). Mutation in another *AtRAD51* paralog, *XRCC2*, also enhances the *tebichi* phenotype (24). These data point to novel plant-specific functions of DNA repair genes related to the pathogenesis response or to development.

Similarly, in a genetic screen for hyper-recombinogenic plants, Molinier et al. found that *CENTRIN2* (*CEN2*) modulates both HR and nucleotide excision repair in Arabidopsis (25). Later on, interactions were found between CEN2 and the CUL4-DDB1A-DDB2 E3 ligase, suggesting that these proteins are involved in the same pathway (26). The tomato *DDB1* homolog was previously characterized as the gene responsible for the *high pigment-1* mutant phenotype (27). Altogether, these data are linking DNA repair, HR, protein degradation, photomorphogenesis and secondary metabolites production, again pointing to neofunctionalization of HR genes in new plant-specific traits.

Other examples where HR is linked with plant-specific cellular mechanisms include the *abo4* mutant, which was isolated based on its sensitivity to abscisic acid (ABA) and shows a 60-fold increase in somatic HR (28) or the chromatin-remodeling gene, *AtBRM,* that controls shoot and flower development (29) and is involved in intrachromosomal recombination (ICR) (30).

What is the "physiological" basis – if any – for these diverse types of neofunctionalization? Has nature simply tinkered with HR proteins, borrowing domains such as helicases, DNA binding, ATPases, etc., for non-HR functions? Or is there a more rational explanation, such as a connection to DNA replication, or – who knows – programmed DNA damage may serve as a signal for certain types of biological processes not related to HR.

5. How Efficient is HR in Plants?

It is often mentioned that HR is not efficient in plants, possibly owing to the repetitive nature of the plant genome. This common view should be refined. Gene targeting (GT) rates are indeed very low in higher plants, with reports ranging from 10^{-3} to 10^{-6} (31), but other types of HR are not fundamentally different than in other species. Moreover, the correlation between the repetitiveness of the plant genome and HR is not clear-cut. GT rates are relatively similarly low in Arabidopsis, rice and tobacco which have a genome with few, average, and abundant amount of repeats, respectively. More significantly, the genome of the moss *Physcomitrella patens* is proficient at gene targeting (32) even though it is approximately threefold larger than that of Arabidopsis and it is rich in repetitive DNA (33). Is this due to peculiar recombination machinery, or owing to an organization and structure of Physcomitrella's chromatin that is different from that of higher plants?

5.1. Meiotic Recombination

Regarding meiotic HR, plants are not different from other eukaryotic species with only a few crossovers per chromosome arm (except for very short arms). There is no linear dependence between chromosome size and the number of crossovers it experiences. Plant chromosomes, as those of most eukaryotic species, must have their "obligatory chiasma" to ensure proper segregation, and are subject to genetic interference. As the size of chromosomes is highly variable in plants, the kb/cM ratio greatly varies, from a few hundred Kbs per cM (e.g., Arabidopsis) to values that are two orders of magnitude higher in plant species with a very large genome. Increasing meiotic crossovers would be of immense value to breeders. Unraveling the pathways that control meiotic crossovers could help achieve this goal. One pathway that controls meiotic crossovers is interference. A MSH4-independent meiotic recombination pathway, that is not sensitive to interference was discovered in plants (34). This pathway, however, seems to promote only up to ~15% of the crossovers and it is an open question at this point whether it is possible to boost it and whether there are other major bottlenecks. One may also enhance the frequency of DNA double-strand breaks (DSBs), which are initiating the crossover events. However, there seems to be a great excess of breaks compared to crossovers, with these breaks being repaired by a pathway (probably SDSA) that leads to gene conversion and does not promote crossovers (35). Therefore, there is no guarantee that such an approach would be successful.

5.2. Somatic Recombination

The early studies on the molecular control of HR were done using plasmids transformed into plant cells. Extrachromosomal recombination (intra- or interplasmids) was then monitored using reporter genes (36). In these experiments, the rates that were reported were relatively high compared to intramolecular HR (see below), suggesting that the structure of the DNA in a plasmid is more prone to HR than that of a chromosome, which is packaged in chromatin. Alternatively, it is plausible that the large amount of DNA molecules introduced into the cell (which was not controlled in these early experiments) has enhanced the chances for HR. These experiments may suggest that the cell machinery is not strongly limiting, provided the substrate has the proper quantity or quality.

Another type of HR that has often been used to quantify the effect of various mutants, as well as biotic and abiotic factors, is the assay of somatic recombination between repeats, in direct or inverse orientation, namely ICR. With these assays, recombination can be scored either through selectable markers (36) or through reporter genes (37). ICR between such repeats in *cis* is in the 10^{-3} to 10^{-5} range. However, it can be enhanced by 2–3 orders of magnitude upon DSB induction (38). In an experiment that enabled the determination of the outcome of I-SceI-induced DSB repair, namely homologous vs. nonhomologous, it turned out that up to one-third of the DSBs were repaired via ICR (39). Similarly, when repeats are present from both sides of a break induced by a transposon excision, as in the P locus of maize, a large proportion of the repair events occurs through HR rather than via the typical end-joining and formation of transposon excision footprints (40). This suggests that while spontaneous HR between nearby repeats in *cis* is not very frequent, ICR is an efficient DSB-repair pathway. How is ICR frequency affected by the distance between the repeats is not well established. We can assume that the frequency decreases with an increase in the inter-repeat distance, maybe as a function of the extent of DSB end resection that enables annealing of homologous segments, but according to what function? Do repeats found on the same chromosome, i.e., physically linked, always have a higher chance to engage in HR than repeats located on different chromosomes (ectopic HR)? Indeed spontaneous HR between ectopic repeats is extremely low (below the 10^{-7} range). It can also be induced by a DSB, but even then remains in the 10^{-5} range (41, 42).

6. The Regulation of Homologous Recombination

In a recent review, Li et al. (2007) conveniently distinguished between the *cis* and *trans* factors that affect HR (8). The *cis* factors are related to the DNA itself or to the chromatin and the *trans* factors correspond to the machinery that regulates or catalyzes the HR reaction.

6.1. The cis Regulators

The *cis* factors related to the DNA include the degree of sequence similarity and the modification of the DNA such as DNA methylation.

6.1.1. Sequence Divergence

It is now well established that even minor sequence divergence between HR partners can cause a strong reduction in HR, somatic or meiotic (43). The antirecombination effect of sequence divergence can be abolished in some mismatch repair mutants (44, 45). In plants, most reported studies do not separate all the products of a HR event, as can be done through tetrad analysis. It is thus difficult to estimate whether HR is restricted to purely identical sequence segments, or whether recombination intermediates, such as double Holliday junctions (DHJs) that contain mismatches are processed into crossovers. All that can be said is that crossovers occurred between two polymorphic markers. We cannot say whether these markers are included in the intermediate. In maize, the intergenic region, which is usually repetitive and divergent, is a cold (or even frozen) spot of recombination (46). Genes are highly conserved in sequence between maize varieties. Therefore, this supports the possibility that crossovers are restricted to regions of very high, and possibly full, identity. In some allopolyploid organisms, mechanisms have evolved to prevent pairing and crossover between homoeologous chromosomes. Such pairing might indeed promote the formation of multivalents and affect fertility. In wheat, the *Ph1* locus, which prevents homoeologous pairing, has been the subject of intensive studies. It co-segregates with a cluster of repeats of cyclin-dependant kinase genes (47). Mismatch repair genes, which were candidates for *Ph*-like activity did not map to the *Ph1* locus (48); however, one mismatch repair member mapped to *Ph2*, another suppressor of homoeologous pairing (49). In *Brassica napus*, *PrBn*, a gene that controls the pairing between homoeologs has also been characterized (50). Here, as well as in the above cases, the mode of action on homoeologous pairing suppression remains unclear.

6.1.2. Chromatin Structure and Methylation

Intuitively, it is expected that accessibility to the DNA would facilitate the physical interaction between homologous partners and thus promote HR. Indeed, as mentioned above, plasmids introduced into plant cells seem to recombine with each other at relatively high rates. In addition, plant mutants in CAF1, a chromatin assembly factor, have a loose chromatin structure and show an increase in ICR by 2 orders of magnitude (51, 52). Along the same line, the upregulation of *RAD54*, a chromatin-remodeling gene, enhances the rate of GT (53), while the downregulation of other *SWI2/SNF2* chromatin remodelers *At2g46020* and *At5g44800* (30) reduces the rate of ICR. Similarly, altering the activity of *MIM* (54) or *BRU1* (55) through overexpression or mutation affects the rates of ICR. Finally, regions of heterochromatin, like knobs, or regions around centromeres are cold spots of recombination. Altogether, this points to a tight connection between DNA packaging and HR. Interestingly, heterochromatic regions tend also to be hypermethylated. It would be of interest to test the effect of cytosine methylation on HR. Small RNAs, which also affect chromatin structure and methylation could thus have a direct or indirect effect on HR. Can we manipulate chromatin structure to enhance GT and meiotic HR, which are both of interest to the breeder? This might be tricky, as the *caf1* mutant, which shows the most dramatic increase in HR, has a very stunted phenotype. Strategies for transient and localized remodeling of chromatin are therefore attractive directions for future efforts. Another holy grail in the field of plant HR is to understand the basis for high GT rates in *Physcomitrella*. Chromatin packaging may be a key to this puzzle.

6.1.3. Hot Spots

As discussed above, the HR coldness of heterochromatin regions has probably a simple physical cause, namely, accessibility. Understanding the nature of hotspots is more challenging. Mapping of recombination breakpoints on chromosome 4 of Arabidopsis has confirmed the lack of homogeneity in the rate of HR along the chromosomes, with nearby segments varying in HR frequency by orders of magnitude. The causes for this variability could not be predicted by simple sequence features (6). Interestingly, the number of Rad51/Dmc1 foci exceed by far the amount of crossovers (35) and are more homogenous in distribution than crossovers (56). This suggests that what determines the variability in crossover events along the chromosome is not the occurrence of DSBs, but rather the way these DSBs resolve, as crossover or noncrossover events. This choice of pathway (COs vs non-COs) is poorly understood. It might determine both hotspots localization and genetic interference.

6.2. The trans Regulators

In addition to the *cis* factors, related to the DNA sequence itself, *trans*-acting factors, namely, the plant proteins that are part of the HR machinery, have been extensively studied and recent reviews have covered these proteins (see above reviews). Here, we chose to focus on the core components of the HR machinery, namely, the homologs of the yeast Rad51, Rad52, and Rad54 proteins (14). Although HR is well conserved among eukaryotes, the differences that were observed raise many questions. The yeast Rad52 protein is central in HR (57) and its expression in human cells increases GT rates (58). The plant homolog is still elusive. It has been speculated that there is no Rad52 homolog in plants. Alternatively, a distant homolog remains undiscovered, and/or the Rad52 function, namely Rad51 loading on single-stranded DNA, is fulfilled by a different protein such as BRCA2, as suggested for mammalian cells (57). A *BRCA2* ortholog was found in plants (59) and was suggested to have a similar function. Rad51 foci are formed in plants (35), suggesting that loading of Rad51 on nuclear filaments takes place. It is still unclear which proteins carry out this loading function in plants.

Considering the physical interaction between Rad52 and Rad51, as shown in yeast (60) and in human (61, 62), it would be interesting to screen the plant proteome for Rad51 "interactors" in search of new players of HR, including the elusive plant Rad52 (or a functional equivalent). It can be done by screening a yeast-two-hybrid library, or by a pull down assay and analysis of the whole complex, as was done recently for the mammalian Rad52 (63). Similarly, it might be instructive to search for new "interactors" of the plant Rad54, as well.

In Arabidopsis, like in vertebrates, five *RAD51* paralogs were identified, yet their exact role is not clear. As reviewed in Bleuyard et al. 2006, the five paralogs were shown to be involved in HR and DNA repair, as well as in meiotic HR. The *RAD51* paralogs form the same complexes as in vertebrates, implying functional conservation of these proteins. However, while in vertebrates (and also in *Drosophila* and *Caenorhabditis elegans*) inactivation of *RAD51* leads to embryonic lethality, *atrad51* Arabidopsis mutant plants are viable (64). The molecular basis for this significant difference is still unknown.

The SWI2/SNF2 chromatin-remodeling factor Rad54 enables, by nucleosome positioning, the core HR events: strand invasion and branch migration (polymerization). Thus, it can be considered a major regulator of HR. As such a central player, it might not be so surprising that the yeast Rad54 can facilitate GT in Arabidopsis (53) – a cross-kingdoms function. Still, many questions arise; how does Rad54 function in Arabidopsis without its yeast counterparts? Does it function alone, as suggested by its in vitro activities (65), or through interaction with its plant counterparts? Indeed, it has been shown that Rad54 interacts with AtRad51 (and AtRad54 with Rad51) in a yeast-two-hybrid assay (66). Despite similarities in function with its yeast ortholog, AtRad54 complements the yeast mutant phenotype only partially (66). This maybe underlies the fact that the plant protein is less active than its yeast ortholog in promoting GT, possibly as a result of natural selection.

7. Gene Targeting

The homologous integration of an extrachromosomal DNA molecule into a homologous chromosomal target site is a powerful tool for the precise engineering of plant genomes. Therefore, it is not surprising that it has been a major goal of the plant community since the first report on GT in plants (67). After years of frustrating attempts to develop efficient gene targeting in plants (31) there may be light at the end of the tunnel. We discuss below the challenges of understanding the mechanism of GT and the future prospects. In fact, it is easy to understand why GT is inefficient in plants: A small vector must scan a huge genome, which is nicely packaged in chromatin, identify the target, get into physical contact with the target, invade it, and engage in strand exchange. This seems like mission impossible, and indeed, two recent reviews (12, 13) describe the biophysical constraints of such homology search and open many questions on how it can work at all? Therefore, by addressing the question of why it does work in certain systems, such as yeast or *Physcomitrella*, one might get insight into the true bottlenecks of the system. So far, most biologists interested in GT have not tried to understand how it really works, but rather have tried to make it work. This has been achieved in mouse as in plants, by the development of clever selection systems, such as positive–negative selection (68) or by artificially creating a recombination hotspot through DSB induction. The first work in plants showed that transformed DNA could be captured at a DSB site induced via cleavage by the I-SceI meganuclease at an I-SceI recognition site (69). Then, a new generation of chimeric meganucleases, zinc-finger nucleases (ZFNs), was developed (70), in which a modular DNA-binding domain fused to a nuclease can be designed to bind any genomic target at relatively high specificity. Recently, this new technology was implemented for the targeting of endogenous plant genes (71, 72). At the mechanistic level, the broken chromosome is probably an accessible entry point that can be recognized by the vector during its scan of the genome. Indeed, the broken chromosome is probably virtually immobile (73) compared to the small vector that can more freely roam the nucleus. So, it makes more sense that the vector goes to the break rather than the break to the vector. Once the physical contact is established, it is still unclear how gene replacement occurs, namely, through an SDSA-like mechanism (74, 75) or through the classical DSB-repair pathway with formation of DHJs (76). Meganuclease technologies have brought new hopes to the field of GT and gene therapy. Nevertheless, there are still topics that must be perfected to turn it into a routine application. ZF design must be improved to really achieve the binding of any sequence. Possibly, it might be necessary to predict or test chromatin structure in the target area for optimal design of the target. Indeed, the findings that chromatin remodeling affects GT in plants (53) supports the need to include the chromatin component in any GT strategy. Ideally, one would combine both efficient design algorithms and evolution methods to develop the perfect nuclease – specific and efficient – an "intelligent evolution" approach. Finding methods to estimate the frequency of off-site breaks induction is critical for gene therapy applications. In plants, off-site mutations might be less critical as backcrosses can be made to get rid of unplanned mutations. However, an excess of ZFNs-induced breaks can also be toxic to the cell and create large deletions and chromosomal rearrangements that may cause gamete sterility, if not cell death. Preferably, the DSB-causing agent should be expressed both temporally and spatially in such a way that induces the break specifically at the desired time and place. Transient expression of highly sequence-specific ZFNs should be used for this purpose as was done in maize and tobacco (71, 72).

8. Conclusions

Plants have contributed their share to the field of HR and, hopefully, will continue to do so in the future. We have tried to point to the many open questions that must be addressed in order to better understand the mechanisms of HR, the control of genome dynamics and stability, and to enhance the rates of HR in plants. The latter is of importance for plant breeding. Increasing meiotic crossover rates is necessary for accelerating breeding processes and for breaking undesirable linkages. It will probably require a better understanding of bottlenecks such as DSB induction and genetic interference. It is also important to better understand HR between homoeologous chromosomes in order to transfer genes between related species. Broadening the gene pool that can be used to improve our crops is essential to face the challenge of plant yield improvement in a changing environment and to find new sources of resistance to biotic and abiotic stresses. Finally, there are some reasons for optimism on the GT front, and we might be at the beginning of a new exciting journey that might make GT routine in plant biology. Nevertheless, careful optimization of the method will be required for each plant species and possibly for each cell type where GT experiments will be carried out. Finally, the emergence of new technologies of genotyping, sequencing, protein design and imaging are opening opportunities for making new discoveries on the HR mechanism and for building new HR-based tools.

Acknowledgments

We would like to thank Dr. Even-Faitelson and Dr. Samach for useful comments and the EU-FP7 Recbreeb and US-Israel BARD grants for financial support.

References

1. Bleuyard, J. Y., Gallego, M. E., and White, C. I. (2006) Recent advances in understanding of the DNA double-strand break repair machinery of plants. *DNA Repair (Amst)* **5**, 1–12.
2. Wijnker, E. and de Jong, H. (2008) Managing meiotic recombination in plant breeding. *Trends Plant Sci* **13**, 640–6.
3. Schuermann, D., Molinier, J., Fritsch, O., and Hohn, B. (2005) The dual nature of homologous recombination in plants. *Trends Genet* **21**, 172–81.
4. Muyt, A. D., Mercier, R., Mezard, C., and Grelon, M. (2009) Meiotic recombination and crossovers in plants. *Genome Dyn* **5**, 14–25.
5. Mercier, R. and Grelon, M. (2008) Meiosis in plants: ten years of gene discovery. *Cytogenet Genome Res* **120**, 281–90.
6. Mezard, C., Vignard, J., Drouaud, J., and Mercier, R. (2007) The road to crossovers: plants have their say. *Trends Genet* **23**, 91–9.
7. Kumar, S., Allen, G. C., and Thompson, W. F. (2006) Gene targeting in plants: fingers on the move. *Trends Plant Sci* **11**, 159–61.
8. Li, J., Hsia, A. P., and Schnable, P. S. (2007) Recent advances in plant recombination. *Curr Opin Plant Biol* **10**, 131–5.
9. Haber, J. E. (2000) Partners and pathways repairing a double-strand break. *Trends Genet* **16**, 259–64.
10. Puchta, H. and Hohn, B. (1991) A transient assay in plant cells reveals a positive correlation between extrachromosomal recombination rates and length of homologous overlap. *Nucleic Acids Res* **19**, 2693–700.
11. Gorbunova, V. and Levy, A. A. (1999) How plants make ends meet: DNA double-strand break repair. *Trends Plant Sci* **4**, 263–69.
12. Weiner, A., Zauberman, N., and Minsky, A. (2009) Recombinational DNA repair in a cellular context: a search for the homology search. *Nat Rev Microbiol* **7**, 748–55.
13. Barzel, A. and Kupiec, M. (2008) Finding a match: how do homologous sequences get together for recombination? *Nat Rev Genet* **9**, 27–37.
14. Li, X. and Heyer, W. D. (2008) Homologous recombination in DNA repair and DNA damage tolerance. *Cell Res* **18**, 99–113.
15. Walbot, V. (1985) On the life strategies of plants and animals. *Trends Genet* **1**, 165–9.
16. Doyle, J. J., Flagel, L. E., Paterson, A. H., Rapp, R. A., Soltis, D. E., Soltis, P. S., and Wendel, J. F. (2008) Evolutionary genetics of genome merger and doubling in plants. *Annu Rev Genet* **42**, 443–61.
17. Britt, A. B. (1999) Molecular genetics of DNA repair in higher plants. *Trends Plant Sci* **4**, 20–25.
18. Kovalchuk, O., Arkhipov, A., Barylyak, I., Karachov, I., Titov, V., Hohn, B., and Kovalchuk, I. (2000) Plants experiencing chronic internal exposure to ionizing radiation exhibit higher frequency of homologous recombination than acutely irradiated plants. *Mutat Res* **449**, 47–56.
19. Kovalchuk, I., Kovalchuk, O., Arkhipov, A., and Hohn, B. (1998) Transgenic plants are sensitive bioindicators of nuclear pollution caused by the Chernobyl accident. *Nat Biotechnol* **16**, 1054–9.
20. Slade, D., Lindner, A. B., Paul, G., and Radman, M. (2009) Recombination and replication in DNA repair of heavily irradiated *Deinococcus radiodurans*. *Cell* **136**, 1044–55.
21. Sung, P. (1994) Catalysis of ATP-dependent homologous DNA pairing and strand exchange by yeast RAD51 protein. *Science* **265**, 1241–3.
22. Bleuyard, J. Y., Gallego, M. E., Savigny, F., and White, C. I. (2005) Differing requirements for the Arabidopsis Rad51 paralogs in meiosis and DNA repair. *Plant J* **41**, 533–45.
23. Durrant, W. E., Wang, S., and Dong, X. (2007) Arabidopsis SNI1 and RAD51D regulate both gene transcription and DNA recombination during the defense response. *Proc Natl Acad Sci USA* **104**, 4223–7.
24. Inagaki, S., Nakamura, K., and Morikami, A. (2009) A link among DNA replication, recombination, and gene expression revealed by genetic and genomic analysis of TEBICHI gene of *Arabidopsis thaliana*. *PLoS Genet* **5**, e1000613.
25. Molinier, J., Ramos, C., Fritsch, O., and Hohn, B. (2004) CENTRIN2 modulates homologous recombination and nucleotide excision repair in Arabidopsis. *Plant Cell* **16**, 1633–43.
26. Molinier, J., Lechner, E., Dumbliauskas, E., and Genschik, P. (2008) Regulation and role of Arabidopsis CUL4-DDB1A-DDB2 in maintaining genome integrity upon UV stress. *PLoS Genet* **4**, e1000093.
27. Lieberman, M., Segev, O., Gilboa, N., Lalazar, A., and Levin, I. (2004) The tomato homolog of the gene encoding UV-damaged DNA binding protein 1 (DDB1) underlined as the gene that causes the high pigment-1 mutant phenotype. *Theor Appl Genet* **108**, 1574–81.

28. Yin, H., Zhang, X., Liu, J., Wang, Y., He, J., Yang, T., Hong, X., Yang, Q., and Gong, Z. (2009) Epigenetic regulation, somatic homologous recombination, and abscisic acid signaling are influenced by DNA polymerase epsilon mutation in Arabidopsis. *Plant Cell* **21**, 386–402.
29. Farrona, S., Hurtado, L., Bowman, J. L., and Reyes, J. C. (2004) The *Arabidopsis thaliana* SNF2 homolog AtBRM controls shoot development and flowering. *Development* **131**, 4965–75.
30. Shaked, H., Avivi-Ragolsky, N., and Levy, A. A. (2006) Involvement of the Arabidopsis SWI2/SNF2 chromatin remodeling gene family in DNA damage response and recombination. *Genetics* **173**, 985–94.
31. Mengiste, T. and Paszkowski, J. (1999) Prospects for the precise engineering of plant genomes by homologous recombination. *Biol Chem* **380**, 749–58.
32. Schaefer, D. G. and Zryd, J. P. (1997) Efficient gene targeting in the moss *Physcomitrella patens*. *Plant J* **11**, 1195–206.
33. Rensing, S. A., Lang, D., Zimmer, A. D., Terry, A., Salamov, A., Shapiro, H., Nishiyama, T., Perroud, P. F., Lindquist, E. A., Kamisugi, Y., Tanahashi, T., Sakakibara, K., Fujita, T., Oishi, K., Shin, I. T., Kuroki, Y., Toyoda, A., Suzuki, Y., Hashimoto, S., Yamaguchi, K., Sugano, S., Kohara, Y., Fujiyama, A., Anterola, A., Aoki, S., Ashton, N., Barbazuk, W. B., Barker, E., Bennetzen, J. L., Blankenship, R., Cho, S. H., Dutcher, S. K., Estelle, M., Fawcett, J. A., Gundlach, H., Hanada, K., Heyl, A., Hicks, K. A., Hughes, J., Lohr, M., Mayer, K., Melkozernov, A., Murata, T., Nelson, D. R., Pils, B., Prigge, M., Reiss, B., Renner, T., Rombauts, S., Rushton, P. J., Sanderfoot, A., Schween, G., Shiu, S. H., Stueber, K., Theodoulou, F. L., Tu, H., Van de Peer, Y., Verrier, P. J., Waters, E., Wood, A., Yang, L., Cove, D., Cuming, A. C., Hasebe, M., Lucas, S., Mishler, B. D., Reski, R., Grigoriev, I. V., Quatrano, R. S., and Boore, J. L. (2008) The Physcomitrella genome reveals evolutionary insights into the conquest of land by plants. *Science* **319**, 64–9.
34. Higgins, J. D., Armstrong, S. J., Franklin, F. C., and Jones, G. H. (2004) The Arabidopsis MutS homolog AtMSH4 functions at an early step in recombination: evidence for two classes of recombination in Arabidopsis. *Genes Dev* **18**, 2557–70.
35. Franklin, A. E., McElver, J., Sunjevaric, I., Rothstein, R., Bowen, B., and Cande, W. Z. (1999) Three-dimensional microscopy of the Rad51 recombination protein during meiotic prophase. *Plant Cell* **11**, 809–24.
36. Puchta, H. and Hohn, B. (1991) The mechanism of extrachromosomal homologous DNA recombination in plant cells. *Mol Gen Genet* **230**, 1–7.
37. Swoboda, P., Gal, S., Hohn, B., and Puchta, H. (1994) Intrachromosomal homologous recombination in whole plants. *EMBO J* **13**, 484–9.
38. Puchta, H. (1999) Use of I-Sce I to induce DNA double-strand breaks in Nicotiana. *Methods Mol Biol* **113**, 447–51.
39. Siebert, R. and Puchta, H. (2002) Efficient repair of genomic double-strand breaks by homologous recombination between directly repeated sequences in the plant genome. *Plant Cell* **14**, 1121–31.
40. Athma, P. and Peterson, T. (1991) *Ac* induces homologous recombination at the maize P locus. *Genetics* **128**, 163–73.
41. Shalev, G. and Levy, A. A. (1997) The maize transposable element *Ac* induces recombination between the donor site and an homologous ectopic sequence. *Genetics* **146**, 1143–51.
42. Puchta, H. (1999) Double-strand break-induced recombination between ectopic homologous sequences in somatic plant cells. *Genetics* **152**, 1173–81.
43. Opperman, R., Emmanuel, E., and Levy, A. A. (2004) The effect of sequence divergence on recombination between direct repeats in Arabidopsis. *Genetics* **168**, 2207–15.
44. Emmanuel, E., Yehuda, E., Melamed-Bessudo, C., Avivi-Ragolsky, N., and Levy, A. A. (2006) The role of AtMSH2 in homologous recombination in *Arabidopsis thaliana*. *EMBO Rep* **7**, 100–5.
45. Li, L., Jean, M., and Belzile, F. (2006) The impact of sequence divergence and DNA mismatch repair on homeologous recombination in Arabidopsis. *Plant J* **45**, 908–16.
46. Dooner, H. K. and He, L. (2008) Maize genome structure variation: interplay between retrotransposon polymorphisms and genic recombination. *Plant Cell* **20**, 249–58.
47. Al-Kaff, N., Knight, E., Bertin, I., Foote, T., Hart, N., Griffiths, S., and Moore, G. (2008) Detailed dissection of the chromosomal region containing the Ph1 locus in wheat *Triticum aestivum*: with deletion mutants and expression profiling. *Ann Bot* **101**, 863–72.
48. Korzun, V., Borner, A., Siebert, R., Malyshev, S., Hilpert, M., Kunze, R., and Puchta, H. (1999) Chromosomal location and genetic mapping of the mismatch repair gene homologs MSH2, MSH3, and MSH6 in rye and wheat. *Genome* **42**, 1255–7.

49. Dong, C., Whifford, R., and Langridge, P. (2002) A DNA mismatch repair gene links to the Ph2 locus in wheat. *Genome* **45**, 116–24.
50. Nicolas, S. D., Leflon, M., Monod, H., Eber, F., Coriton, O., Huteau, V., Chevre, A. M., and Jenczewski, E. (2009) Genetic regulation of meiotic cross-overs between related genomes in *Brassica napus* haploids and hybrids. *Plant Cell* **21**, 373–85.
51. Endo, M., Ishikawa, Y., Osakabe, K., Nakayama, S., Kaya, H., Araki, T., Shibahara, K. I., Abe, K., Ichikawa, H., Valentine, L., Hohn, B., and Toki, S. (2006) Increased frequency of homologous recombination and T-DNA integration in Arabidopsis CAF-1 mutants. *EMBO J* **25**, 5579–90.
52. Kirik, A., Pecinka, A., Wendeler, E., and Reiss, B. (2006) The chromatin assembly factor subunit FASCIATA1 is involved in homologous recombination in plants. *Plant Cell* **18**, 2431–42.
53. Shaked, H., Melamed-Bessudo, C., and Levy, A. A. (2005) High-frequency gene targeting in Arabidopsis plants expressing the yeast RAD54 gene. *Proc Natl Acad Sci USA* **102**, 12265–9.
54. Hanin, M., Mengiste, T., Bogucki, A., and Paszkowski, J. (2000) Elevated levels of intrachromosomal homologous recombination in Arabidopsis overexpressing the MIM gene. *Plant J* **24**, 183–9.
55. Takeda, S., Tadele, Z., Hofmann, I., Probst, A. V., Angelis, K. J., Kaya, H., Araki, T., Mengiste, T., Scheid, O. M., Shibahara, K., Scheel, D., and Paszkowski, J. (2004) BRU1, a novel link between responses to DNA damage and epigenetic gene silencing in Arabidopsis. *Genes Dev* **18**, 782–93.
56. Anderson, L. K. and Stack, S. M. (2005) Recombination nodules in plants. *Cytogenet Genome Res* **109**, 198–204.
57. Mortensen, U. H., Lisby, M., and Rothstein, R. (2009) Rad52. *Curr Biol* **19**, R676–7.
58. Di Primio, C., Galli, A., Cervelli, T., Zoppe, M., and Rainaldi, G. (2005) Potentiation of gene targeting in human cells by expression of *Saccharomyces cerevisiae* Rad52. *Nucleic Acids Res* **33**, 4639–48.
59. Siaud, N., Dray, E., Gy, I., Gerard, E., Takvorian, N., and Doutriaux, M. P. (2004) Brca2 is involved in meiosis in *Arabidopsis thaliana* as suggested by its interaction with Dmc1. *EMBO J* **23**, 1392–401.
60. Tsutsui, Y., Khasanov, F. K., Shinagawa, H., Iwasaki, H., and Bashkirov, V. I. (2001) Multiple interactions among the components of the recombinational DNA repair system in *Schizosaccharomyces pombe*. *Genetics* **159**, 91–105.
61. Milne, G. T. and Weaver, D. T. (1993) Dominant negative alleles of RAD52 reveal a DNA repair/recombination complex including Rad51 and Rad52. *Genes Dev* **7**, 1755–65.
62. Shen, Z., Cloud, K. G., Chen, D. J., and Park, M. S. (1996) Specific interactions between the human RAD51 and RAD52 proteins. *J Biol Chem* **271**, 148–52.
63. Du, Y., Zhou, J., Fan, J., Shen, Z., and Chen, X. (2009) Streamline proteomic approach for characterizing protein-protein interaction network in a RAD52 protein complex. *J Proteome Res* **8**, 2211–7.
64. Li, W., Chen, C., Markmann-Mulisch, U., Timofejeva, L., Schmelzer, E., Ma, H., and Reiss, B. (2004) The Arabidopsis AtRAD51 gene is dispensable for vegetative development but required for meiosis. *Proc Natl Acad Sci USA* **101**, 10596–601.
65. Alexeev, A., Mazin, A., and Kowalczykowski, S. C. (2003) Rad54 protein possesses chromatin-remodeling activity stimulated by the Rad51-ssDNA nucleoprotein filament. *Nat Struct Biol* **10**, 182–6.
66. Klutstein, M., Shaked, H., Sherman, A., Avivi-Ragolsky, N., Shema, E., Zenvirth, D., Levy, A. A., and Simchen, G. (2008) Functional conservation of the yeast and Arabidopsis RAD54-like genes. *Genetics* **178**, 2389–97.
67. Paszkowski, J., Baur, M., Bogucki, A., and Potrykus, I. (1988) Gene targeting in plants. *EMBO J* **7**, 4021–6.
68. Terada, R., Urawa, H., Inagaki, Y., Tsugane, K., and Iida, S. (2002) Efficient gene targeting by homologous recombination in rice. *Nat Biotechnol* **20**, 1030–4.
69. Salomon, S. and Puchta, H. (1998) Capture of genomic and T-DNA sequences during double-strand break repair in somatic plant cells. *EMBO J* **17**, 6086–95.
70. Bibikova, M., Carroll, D., Segal, D. J., Trautman, J. K., Smith, J., Kim, Y. G., and Chandrasegaran, S. (2001) Stimulation of homologous recombination through targeted cleavage by chimeric nucleases. *Mol Cell Biol* **21**, 289–97.
71. Shukla, V. K., Doyon, Y., Miller, J. C., DeKelver, R. C., Moehle, E. A., Worden, S. E., Mitchell, J. C., Arnold, N. L., Gopalan, S., Meng, X., Choi, V. M., Rock, J. M., Wu, Y. Y., Katibah, G. E., Zhifang, G., McCaskill, D., Simpson, M. A., Blakeslee, B., Greenwalt, S. A., Butler, H. J., Hinkley, S. J., Zhang, L., Rebar, E. J., Gregory, P. D., and Urnov, F. D. (2009) Precise genome modification in the

crop species Zea mays using zinc-finger nucleases. *Nature* **459**, 437–41.

72. Townsend, J. A., Wright, D. A., Winfrey, R. J., Fu, F., Maeder, M. L., Joung, J. K., and Voytas, D. F. (2009) High-frequency modification of plant genes using engineered zinc-finger nucleases. *Nature* **459**, 442–5.
73. Soutoglou, E., Dorn, J. F., Sengupta, K., Jasin, M., Nussenzweig, A., Ried, T., Danuser, G., and Misteli, T. (2007) Positional stability of single double-strand breaks in mammalian cells. *Nat Cell Biol* **9**, 675–82.
74. Gorbunova, V. and Levy, A. A. (1997) Non-homologous DNA end joining in plant cells is associated with deletions and filler DNA insertions. *Nucleic Acids Res* **25**, 4650–7.
75. Rubin, E. and Levy, A. A. (1997) Abortive gap repair: the underlying mechanism for *Ds* elements formation. *Mol Cell Biol* **17**, 6294–302.
76. Szostak, J. W., Orr-Weaver, T. L., Rothstein, R. J., and Stahl, F. W. (1983) The double-strand break repair model of recombination. *Cell* **33**, 25–35.

Chapter 4

Chromosome Painting for Plant Biotechnology

Akio Kato, Jonathan C. Lamb, Patrice S. Albert, Tatiana Danilova, Fangpu Han, Zhi Gao, Seth Findley, and James A. Birchler

Abstract

Fluorescence in situ hybridization (FISH) is an invaluable tool for chromosome analysis and engineering. The ability to visually localize endogenous genes, transposable elements, transgenes, naturally occurring organellar DNA insertions – essentially any unique sequence larger than 2 kb – greatly facilitates progress. This chapter details the labeling procedures and chromosome preparation techniques used to produce high-quality FISH signals on somatic metaphase and meiotic pachytene spreads.

Key words: FISH, Chromosomes, Retrotransposons, Metaphase spreads

1. Introduction

Many aspects of chromosome engineering are facilitated by the ability to recognize specific chromosomes in somatic preparations primarily from root tip tissues. In this chapter, we summarize labeling procedures and chromosome preparation techniques and their modifications that have proven useful in analyzing maize chromosomes. Many variations work equally well, and the protocols can be individualized to other circumstances. These techniques, however, have now been extended to a variety of other plant species, and we include a section on the approach that researchers might take to further their application to even more species. We assume that the reader has a working knowledge of basic molecular biological techniques and access to a high-quality fluorescence microscope.

The set of protocols outlined in this chapter has been used to localize single genes and other small unique sequences (1,2), repetitive DNA sequence clusters (3–5), transposable elements (6),

James A. Birchler (ed.), *Plant Chromosome Engineering: Methods and Protocols*, Methods in Molecular Biology, vol. 701,
DOI 10.1007/978-1-61737-957-4_4,

and nuclear-inserted mitochondrial DNA (7). The techniques have also been used to study allopolyploids and interspecific hybrids via retroelement genome painting (8, 9) and to karyotype F1 hybrids (5). See also the review by Kato et al. (10).

2. Materials

2.1. Preparation of PCR Product for Labeling

1. Primers (see Note 1).
2. Template DNA (see Note 2).
3. PCR hot start reaction mix, e.g., JumpStart RedTaq ReadyMix (Sigma-Aldrich, St. Louis MO).
4. Thermal cycler.
5. PCR product clean up kit, e.g., Wizard SV Gel and PCR Clean Up System (Promega, Madison, WI).
6. Nano-volume spectrophotometer (e.g., NanoDrop® Thermo Fisher Scientific, Pittsburgh, PA).

2.2. Nick Translation Reaction: Direct Labeling of SMALL Chromosomal Targets (<30 kb)

1. DNA (see Note 3). Store at –20°C.
2. Nick translation buffer: 500 mM Tris–HCl pH 7.8 (with HCl), 50 mM $MgCl_2$, 100 mM 2-mercaptoethanol (see Notes 4 and 5). Optional addition, 100 µg/ml bovine serum albumin fraction V (Sigma-Aldrich, St. Louis, MO). Store at –20°C.
3. Unlabeled dNTP mix: A solution made from individual stocks of the three dNTPs that will *not* be labeled during the reaction (see Note 6). The concentration of each dNTP in the solution is 2 mM (in nuclease-free water). Store at –20°C.
4. Labeled dNTP (see Note 7): Light-sensitive. Store at –20°C or according to manufacturer's recommendations.
5. DNA polymerase I, 10 U/µl. (Invitrogen, Carlsbad, CA). Store at –20°C.
6. DNase I (see Note 8). Store at –20°C.
7. Stop buffer: 0.5 M EDTA, pH to 8.0 with NaOH (see Note 9). Autoclave. Store at room temperature to avoid precipitation.

2.3. Nick Translation Reaction: Direct Labeling of LARGE Chromosomal Targets (>30 kb)

1. Same components as listed for small target labeling.
2. Labeled dNTP (see Note 10).

2.4. Purification and Precipitation of Red or Far-Red Probes

1. 1× TE buffer: Make 10× stock [100 mM Tris–HCl pH 7.6 (with HCl), 10 mM EDTA]. Pass through a 0.45-µm filter and autoclave. Prepare working solution of 1× TE by diluting

with water. Also needed is a 1× working stock at pH 7.8 for preparation of salmon sperm DNA stock (see item 8 below). Store at room temperature.

2. Bio-Gel® P-60 Gel, Medium Grade (Bio-Rad, Hercules, CA) equilibrated in sterile 1× TE pH 7.6: Make in advance to allow time for gel to hydrate (overnight at 4°C). Each column will use a little more than 2 ml of hydrated gel (~11 ml hydrated/g powder). Gel expands considerably. Some 1× TE should remain above the hydrated gel. Store at 4°C. Good for ~1 year; sodium azide not added. Gel is not degassed prior to use.
3. Two pairs of tweezers for handling glass wool.
4. 1.6- or 1.7-ml microcentrifuge tubes – three per each 2-μg probe; four per each 5-μg probe.
5. Silane-treated glass wool (Supelco, Bellefonte, PA).
6. Glass Pasteur pipets, 5 3/4″ (Thermo Fisher Scientific, Pittsburgh, PA), one per probe.
7. Metal rod, ~3 mm diameter, to tamp glass wool into pipet (old aluminum inoculating loop handle works well).
8. Salmon sperm DNA: Purchase lyophilized powder (Sigma-Aldrich, St. Louis, MO) and make stock (see Note 11) or purchase 300–2,000 bp salmon sperm DNA (Stratagene/ Agilent Technologies, Cedar Creek, TX). Store at −20°C.
9. 3 M sodium acetate: Adjust to pH 5.2 with glacial acetic acid. Autoclave and store at room temperature (see Note 12).
10. Ethanol: Need 100% and a 70% aqueous solution.
11. 2× SSC, 1× TE: Make by diluting 20× stock solution of SSC [3 M sodium chloride, 0.3 M sodium citrate, pH 7.0 (see Note 13), pass through a 0.45-μm filter (first wet the filter with small volume of water), autoclave, and store at room temperature] and 10× stock solution of TE pH 7.6, see Subheading 2.4, item 1).
12. Alternative to Bio-Gel® purification for some probes is to use PCR clean-up kit spin columns (see Note 14).

2.5. Precipitation of Blue or Green Probes

1. Salmon sperm DNA: 10 μg/μl (see Subheading 2.4, item 8).
2. 3 M sodium acetate pH 5.2 (see Subheading 2.4, item 9).
3. 5× TAE buffer with salmon sperm DNA (optional, see Note 15): Make 10× TAE stock [100 mM Tris base, 10 mM EDTA, adjusted to pH 5.2 with acetic acid]. Pass through a 0.45-μm filter. Autoclave and store at room temperature. Working solution is 5× TAE with salmon sperm DNA added to a final concentration of 140 ng/μl. Store at room temperature.
4. Ethanol: Need 100% and a 70% aqueous solution.

2.6. Testing Probes and Cocktail Mixes

1. Individual probes. Store at –20°C in the dark.
2. 2× SSC, 1× TE (see Subheading 2.4, item 11).
3. Blocking DNA: 140 ng/μl autoclaved salmon sperm DNA in 2× SSC, 1× TE. Store at room temperature.

2.7. Preparation of Metaphase Spreads from Root Tissue

1. 1× Citric Buffer: Make a 5× working stock [50 mM sodium citrate, 50 mM EDTA, adjusted to pH 5.5 by adding citric acid]. Autoclave and store at room temperature. Dilute to 1× with water. Use chilled on ice.
2. Digestive Enzyme Solution: For ~10 ml, mix thoroughly in tube or weigh boat on ice 0.1 g (1% w/w) of Pectolyase Y-23 (Karlan Research Products, Cottonwood, AZ), 0.2 g (2% w/w) of Cellulase Onozuka R-10 (Karlan Research Products, Cottonwood, AZ), and 9.7 g of 1× citric buffer (see Note 16). Quickly dispense as 20-μl aliquots into thin-walled 0.5-ml PCR tubes (make sure caps tightly closed), quick freeze on dry ice, and store at –20°C. Good for ~1 year.
3. Nitrous oxide gas (induces metaphase arrest in mitotic cells). Purchase from local medical gas supply company.
4. Pressure chamber for nitrous oxide treatment. Custom-manufactured (see Note 17).
5. Acetic acid: Use glacial acetic acid to make a 90% aqueous solution for fixing root tissue; store at 4°C; 100% glacial or 90% glacial – 10% methanol will also be needed (see Subheading 3.7, step 11).
6. Ethanol: 70% aqueous solution. Store at room temperature but use chilled on ice.
7. 37°C water or dry bath for tissue digestion.
8. Humid chamber: high humidity environment for spreading of cells (see Note 18).
9. UV Cross-linker that can deliver a total energy of 120–125 mJ/cm^2 at “Optimum” cross-link setting.
10. Miscellaneous supplies: filter paper (Whatman 1 or thicker), razor blade or scalpel, tweezers, dissecting needle with pointed end ground off, glass microscope slides (1″ × 3″).
11. Root tips: obtain from newly germinated seeds, from young plants, or from new growth on root-pruned plants.

2.8. Denaturation, Hybridization, and Slide Processing

1. Have probe cocktail(s) and cross-linked slides ready and labeled.
2. Boiling water bath: Two components – hot plate and baking pan that holds boiling water (see Note 19).
3. Denaturation pan assembly: A smaller baking pan (aluminum, if possible) that can float freely in the water bath and a plastic

pipet tip box lid (or equivalent) to cover the slides/probes in the pan. The bottom of the baking pan is covered with 2–3 layers of Kimwipes that are sprayed with purified water just before use. The water bath and floating pan are covered with aluminum foil to retain heat and block light during the denaturation step.

4. Optional: A pair of 3/8″ square × 9–12″ long wooden sticks (cut from 3′ stock; dowel rod section of hardware store) (see Note 20).
5. Ice bucket filled to the top with crushed ice and a small beaker filled with slushy ice.
6. Metal plate(s) on which to quick chill several slides simultaneously (e.g., two at 8 × 15 × 0.6 mm thick). One may use stainless-steel slide tray, turned upside down, for this purpose.
7. 2× SSC, 1× TE containing 140 ng/μl salmon sperm DNA: Make by diluting 20× SSC and 10× TE (see Subheading 2.4, item 11), and 10 μg/μl autoclaved salmon sperm DNA (see Subheading 2.4, item 8).
8. Cover slips: 22 × 22 mm plastic (Thermo Fisher Scientific, Pittsburgh, PA) and 24 × 50 mm glass, No. 1 (Corning, from Thermo Fisher).
9. Tweezers for handling cover slips and either a large serrated pair of tweezers or tongs for removing denaturation pan from boiling water bath.
10. Mini microcentrifuge (2,000 rcf) for quick spin of chilled, denatured probes.
11. Humidified storage container for hybridization: reusable airtight "food" container, bottom lined with Kimwipes moistened with water.
12. Dedicated incubator/oven set to 55°C for hybridization.
13. 2× SSC: Make by diluting 20× stock (see Subheading 2.4, step 11).
14. Two Coplin jars – one at room temperature and one preheated to 55°C into which ~45 ml of preheated 2× SSC (not the buffer containing TE) will be added.
15. Antifade mounting medium: Vectashield® and Vectashield® containing 1.5 μg/ml DAPI (Vector Laboratories, Burlingame, CA) (see Note 21).
16. Fluorescence microscope with 20× or 40× oil lens for scanning slides, 100× oil lens for image acquisition (flat field and color corrected best), filter sets for fluorochromes to be used (see Note 22), cooled CCD (charge-coupled device) camera, and acquisition software.

3. Methods

The success of fluorescence in situ hybridization (FISH) depends on producing high-quality metaphase spreads and probes with a high signal-to-noise ratio. The techniques that best accomplish this for most plants include nitrous oxide treatment to increase the number of metaphase spreads and enzymatic maceration followed by dropping of the cell suspension onto the slide to provide quality chromosome spreads with little cytoplasmic background (modified from Kato (11, 12)). Probes are made by a direct labeling method and do not use antibodies (3, 13).

The FISH procedure comprises several distinct steps – preparation of the DNA that will be fluorescently labeled, the nick translation reaction that directly incorporates a fluorochrome-labeled dNTP into the newly synthesized probe (see Note 23), probe purification and precipitation, making probe cocktail mixes, metaphase spread preparation, and the actual denaturation, hybridization, and slide processing steps.

In addition to the basic procedure are special topics sections, which include strategies for single gene/small target detection (see Note 56), more color options for FISH probes (see Note 57), 5′ end-labeled oligos as FISH probes (see Note 58), a streamlined FISH protocol for large targets (see Note 59), meiotic FISH techniques (see Note 60), retroelement genome painting (see Note 61), and guidelines for developing a karyotyping cocktail for other plant species (see Note 62).

3.1. Preparation of PCR Product for Labeling

1. A relatively large amount of DNA is necessary for each slide (20–200 ng). Therefore, 10–100 μg of the DNA insert should be amplified. Use a hot-start PCR protocol with conditions appropriate for the sequence being amplified (see Note 24). Total reaction volumes of 300–400 μl will produce ample amounts of purified product of at least 200 ng/μl, the concentration on which the probe-labeling nick translation protocol is based (see Note 25).
2. Verify product size by running some of the reaction mix on a gel.
3. Purify the PCR product; process ~400 μl through one spin column to increase concentration.
4. If possible, use a nano-volume spectrophotometer to determine the concentration and quality of the product.

3.2. Nick Translation Reaction: Direct Labeling of SMALL Chromosomal Targets (<30 kb)

1. Use for lower copy number or large single-gene targets, including maize subtelomere clones 1.1 and 4-12-1. Assemble the following on ice:

DNA: 2 μg (200 ng/μl)	10.0 μl (see Note 26)
10× Nick translation buffer	2.0 μl
Nonlabeled dNTP mix (2 mM each)	2.0 μl
Labeled-dNTP (1 mM)	0.5 μl (see Notes 7 and 27)

2. Add the following enzymes and thoroughly mix by pipetting. Do not vortex.

DNA polymerase I (10 U/μl)	8.0 μl
DNase (100 mU/μl)	0.4 μl
Total volume	22.9 μl

3. Incubate at 15°C for 2 h (see Note 28).
4. Optional stopping point: Completed reactions may be stored at –20°C in the dark.
5. To continue, add 2 μl of stop buffer to thawed reaction (optional) and proceed to Subheadings 3.4 or 3.5.

3.3. Nick Translation Reaction: Direct Labeling of LARGE Chromosomal Targets (>30 kb)

1. Use for higher copy number targets (see Note 29). Assemble the following on ice:

DNA: 5 μg (200 ng/μl)	25.0 μl (see Note 30)
10× Nick translation buffer	5.0 μl
Nonlabeled dNTP mix (2 mM each)	5.0 μl
Labeled-dNTP (1 mM)	1.0 μl (see Note 10)
Sterile ultrapure or nuclease-free water	6.75 μl (optional)

2. Add the following enzymes and thoroughly mix by pipetting. Do not vortex.

DNA polymerase I (10 U/μl)	6.25 μl
DNase (100 mU/μl)	1.0 μl
Total volume	43.25 or 50 μl

3. Incubate at 15°C for 2 h (see Note 28).
4. Optional stopping point. Completed reactions may be stored at –20°C in the dark. To continue, add 5 μl of stop buffer (optional). Proceed to Subheadings 3.4 or 3.5.

3.4. Purification and Precipitation of Red or Far-Red Probes

1. Probes labeled with either red or far-red fluorochromes need to be purified through a Bio-Gel® P-60 column prior to the precipitation step (see Note 14). Blue and green probes do not benefit from this procedure. Prepare column (see Note 31) and label collection tubes sequentially. Stand the prepared column in a 1.6-ml microcentrifuge tube placed in a tube rack.
2. Slowly add nick translation reaction to the column. Cover with a cardboard box if in lighted area and wait a minute to allow probe to enter gel matrix.
3. Add 50 μl of 1× TE. Elute. Discard the eluate.

   Add 350 μl of 1× TE. Elute. Discard the eluate. Move column to the second 1.6-ml tube (see Note 32.).
4. For a 2-μg probe,

   Add 400 μl of 1× TE. Elute. *Save* eluate. Move column to the third tube.

   Add 400 μl of 1× TE. Elute. *Save* eluate. [If purifying >2 μg of DNA, another elution may be needed to ensure collection of all labeled probe.]
5. For a 5-μg probe,

   Add 350 μl of 1× TE. Elute. *Save* eluate. Move column to the third tube.

   Add 350 μl of 1× TE. Elute. *Save* eluate. Move column to the fourth tube.

   Add 350 μl of 1× TE. Elute. *Save* eluate.
6. To each elution tube, add autoclaved salmon sperm DNA (3 μl of 10 μg/μl for a 2-μg probe; 5 μl for a 5-μg probe).
7. Add 0.1 sample volume of 3 M sodium acetate pH 5.2 and 2.5 sample volumes of 100% ethanol (see Note 33). Vortex.
8. Store the tube at –20°C for 2 h minimum (or overnight).
9. Centrifuge at 16,000 rcf for 30 min (see Note 34). Remove the supernatant.
10. Wash the DNA pellets with 70% ethanol. Follow with a quick 100% ethanol wash. Remove as much of the ethanol as possible (see Note 35).
11. Air-dry the probe pellets in the dark for 5–30 min (humidity-dependent). Check pellets frequently; do not over dry (makes them harder to dissolve).
12. For a 2-μg probe, dissolve the pellets in a *combined volume* of 10 μl (NOT 10 μl each, see Note 36) 2× SSC, 1× TE buffer (pH 7.6) for a probe concentration designated as 200 ng/μl. Time, brief heat treatment at 65°C, and/or vortexing may be needed to dissolve pellet. For a 5-μg probe, dissolve in a combined volume of 25 μl.

13. Consolidate total volume into the tube with the most incorporated fluorochrome (brightest pellet, often from the first elution). Store at –20°C in the dark.

3.5. Precipitation of Blue or Green Probes

1. Add 1× TE pH 7.6 to a total volume of 350–400 μl and transfer to a 1.6-ml microcentrifuge tube (see Notes 37 and 38).
2. Add autoclaved salmon sperm DNA (30 or 50 μg, based on amount of DNA labeled; see Subheading 3.4, step 6).
3. Add 0.1 sample volume of 3 M sodium acetate pH 5.2 and 2.5 sample volumes of 100% ethanol. Vortex.
4. Precipitate DNA at –20°C for 2 h minimum (or overnight).
5. Centrifuge at 16,000 rcf for 30 min (see Note 34). Remove the supernatant.
6. Wash the DNA pellet with 70% ethanol. Follow with a quick 100% ethanol wash. Remove as much of the ethanol as possible (see Note 35).
7. Air-dry the probe pellet in the dark for 5–30 min (humidity-dependent). Check pellet frequently; do not over dry (harder to dissolve).
8. Dissolve a 2-μg probe in 10 μl and a 5-μg probe in 25 μl 2× SSC, 1× TE buffer (pH 7.6) for a probe concentration designated as 200 ng/μl.

3.6. Testing Probes and Cocktail Mixes

Each new probe must be tested for overall signal strength and signal-to-noise ratio. A good concentration range at which to start is 10–20 ng/μl (i.e., to test a probe of unknown signal properties at 20 ng/μl, use 1 μl of 200 ng/μl probe plus 4 μl 2× SSC, 1× TE and 5 μl of salmon sperm DNA blocking solution; see Note 39). Differently colored probes may be tested on the same slide. Final concentrations of different probes labeled with the same fluorochrome will need to be modified so that each signal is optimal at the same acquisition exposure time. Volume adjustments are made with 2× SSC, 1× TE. For more complex cocktail mixes, it might be necessary to dissolve probe DNA pellets at concentrations >200 ng/μl. The minimum volumes that can be used are listed above and referred to in Subheading 3.8. Although using these volumes is more economical, some areas under the 22×22 mm cover slip will not be fully bathed in probe. If your application requires complete probe coverage, use 8–10 μl each of blocking solution and probe cocktail.

3.7. Preparation of Metaphase Spreads from Root Tissue

1. Germinate kernels in moist vermiculite for 2–3 days at 30°C (maize), or harvest young roots from plants. Choose roots 1–5 cm in length (2–4 cm better). Keep them moist.
2. Cut 1–1.5 cm of root tip and treat with nitrous oxide at 10 atm for 1–3 h (see Note 40).

3. Immediately fix root tips in ice-cold 90% acetic acid for 10 min (not longer than 1 h).
4. Optional stopping point: Store in new tube containing 70% ethanol at −20°C. For larger target applications, roots keep well for a year or more.
5. To continue: Wash root tips in ice-cold 1× citric buffer for 10 min.
6. Remove the sticky substance from the root cap by rolling on dry filter paper or by holding the cut end and wiping in the direction of the cap (moistening filter paper with citric buffer, optional).
7. Optional step: Removal of the root cap (see Note 41).
8. Cut off about 0.7–1 mm of the distal section containing the actively dividing meristem region (more opaque, ~1–2 mm long) and transfer to a tube containing 20 μl of ice-cold enzyme solution.
9. Incubate at 37°C for 30–60 min, average ~47 min, (see Note 42). Plunge tube into ice.
10. Fill tube with 70% ethanol, remove. Rinse twice again with 70% ethanol (see Notes **43** and 44).
11. Replace ethanol with 30 μl of 100% acetic acid or 90% acetic acid – 10% methanol.
12. Carefully break the root section with a rounded off dissecting needle. Stir or tap the tube with your finger several times to suspend the cells. Keep cells on ice.
13. Place labeled microscope slides in a humid chamber (see Note 45).
14. Drop 5.5–9 μl of the cell suspension on each slide. Volume used is dependent on cell density (see Note 46). (Optional: The bottom of the slide can be etched to indicate the location of cells.)
15. When dry, view through the 10× and 40× objective lenses of a compound microscope to select slides with the best spreads (quality and number) for hybridization (see Note 47).
16. Cross-link chromatin to slides by exposure to UV light. Set cross-linker to "Optimum" (see Note 48).

3.8. Denaturation, Hybridization, and Slide Processing

1. Preheat water bath to a rolling boil; do not heat the denaturation pan assembly (see Note 49).
2. Place metal plate in full bucket of ice; insure maximum contact.
3. Place slides on top of a pair of square sticks so that the cells are located between the sticks.
4. Pipet 5 μl of salmon sperm DNA (140 ng/μl 2× SSC, 1× TE) onto the center of each cell spread. Using tweezers, apply a

22×22 mm plastic cover slip; one of the corners needs to hang over the edge of the slide just a little (facilitates removal prior to addition of probe).

5. Denature root tip DNA (on slide) and probe DNA cocktail (in 200-μl thin-wall PCR tube) by floating in the boiling water bath for 5 min (100°C) (see Notes 50 and 51).
6. Quickly cool the probe in the ice slush and the slides on the prechilled metal plate for 1–2 min. Keep probe in the dark.
7. Centrifuge probes briefly to return liquid to the bottom of the tube. Return to ice.
8. With tweezers, lift plastic cover slip (or lift off, flip over, and set down). Pipet 5 μl of probe onto the cells and replace cover slip in same orientation. (Cells cannot be allowed to dry.)
9. Place slides in a humid storage container lined with moistened Kimwipes for 4 h to overnight at 55°C.
10. Place slides in a Coplin jar (see Note 52) containing room-temperature 2× SSC briefly (quick dip to 5 min) to remove the cover slip and excess probe.
11. Using tweezers, transfer slides to a Coplin jar containing 55°C 2× SSC and wash for 20 min at 55°C. The temperature and 2× SSC concentration can be varied for different stringency requirements (see Note 53).
12. Remove slides and blot excess 2× SSC from bottom and edges of slides. Do not allow tops to dry (see Note 54).
13. Apply one drop of Vectashield® mounting medium (with or without DAPI) and carefully apply a 24×50 mm glass cover slip.
14. Acquire images.
15. Slides may be stored at 4°C in the dark (see Note 55).

4. Notes

1. Most of the common plasmids (pUC18, pBluescript, etc.) contain standard primer sequences (T3, T7, M13F, M13R, SP6, etc.) that can be used for PCR amplification. These primers can be ordered as custom primers [e.g., 25 nmol synthesis from Integrated DNA Technologies (http://www.idtdna.com/order/order.aspx) or from Sigma (http://www.sigmaaldrich.com/life-science/custom-oligos.html)]. It is also possible to design primers to amplify specific regions. One example of software that can be used to determine optimal primers from a DNA sequence is at http://frodo.wi.mit.edu/primer3/.

2. The DNA to be labeled is usually amplified from a plasmid or cosmid, but can be a noncloned PCR product. Plasmid DNA can be purified from a bacterial culture using either standard molecular biology protocols or a kit. Note that reamplification of PCR-product DNA will, over time, introduce and amplify errors.
3. Although PCR products are commonly labeled for FISH studies, plasmid DNA and BAC clone DNA can also be labeled.
4. The water used to prepare this and most other solutions had a resistivity of >16 MΩ-cm and flowed through a 0.2-μm filter prior to use.
5. 2-Mercaptoethanol is toxic; handle accordingly. Prepare only a small volume of the buffer (5–10 ml), aliquot, and store at −20°C. The nick translation reaction still works after the characteristic odor of the 2-mercaptoethanol is depleted, but probably not at optimum levels.
6. Make ~1 ml and aliquot in volumes based on usage. Store at −20°C. It is helpful to color-code the tubes to match the color of the dNTP to be labeled to avoid inadvertently using the wrong mix.
7. Labeled dNTPs commonly used for small target detection include Texas Red®-5-dCTP (Perkin Elmer, Boston, MA), Alexa Fluor® 488-5-dUTP (Invitrogen Molecular Probes, Carlsbad, CA), and Fluorescein-12-dUTP (available from either supplier). Texas Red® is intrinsically the brightest and should be used to label the smallest targets.
8. DNase I can be purchased ready-made (Invitrogen, Carlsbad, CA) or as a powder (Recombinant Grade I, Roche Applied Science, Indianapolis, IN). DNase from different sources has different levels of activity. This also seems true for different lot numbers of the same product. To prepare a working stock solution (100 mU/μl) from the powdered form of the enzyme:
   (a) First prepare 2× DNase buffer and 1× DNase buffer (50% glycerol).

   2× DNase buffer: 100 mM Tris–HCl pH 7.5, 10 mM $MgCl_2$, 0.2 mM phenylmethylsulfonyl-fluoride (optional, hazardous material; if used, this substance is dissolved in ethanol first and then added to the buffer), 2 mM 2-mercaptoethanol (hazardous material), and 200 μg/ml of bovine serum albumin.

   (b) 10 U/μl primary stock: 10,000 U of DNase I are dissolved in 0.5 ml of 2× DNase buffer on ice. Add 0.5 ml of sterile glycerol and mix gently.

(c) 100 mU/μl intermediate stock: Add 20 μl of primary stock to 80 μl of cold 1× DNase buffer. Concentration is 2 U/μl.

(d) Working stock: Add 10 μl of the 2 U/μl DNase solution to 190 μl of cold 1× DNase buffer.

(e) Store all solutions at –20°C. Freezing smaller volume aliquots of the more dilute stocks is recommended. The diluted DNase is relatively unstable and should be checked before use. 10 mU of DNase in 10 μl of DNase buffer digests 1 μg of DNA (1 kb) below 300 bp within 10 min at room temperature.

9. This concentration of EDTA will not go into solution until the pH is almost 8. The use of EDTA as a stop buffer is optional, but it is a required component of other solutions.

10. In addition to the fluorophores listed for small targets (see Note 7), Cyanine 5-dUTP (Perkin Elmer, Boston, MA) and Cascade Blue®-7-dUTP (Invitrogen Molecular Probes, Carlsbad, CA) can be used for the largest targets. There are few options for a blue fluorochrome. Perkin Elmer discontinued production of coumarin-5-dUTP, and Enzo's coumarin-13-dUTP cannot efficiently be incorporated into DNA via nick translation.

11. Dissolve DNA in 1× TE buffer pH 7.8 at a concentration of 10 μg/μl. Place on a rotary shaker (overnight, if necessary). The DNA should be completely dissolved prior to autoclaving for 30 min (shears DNA). Run sample on gel to determine the size of the DNA fragments. We use fragments ~100–400 bp in length (300–2,000 bp as purchased also works). If probe and slide are denatured separately, size is less important. Store at –20°C in 1-ml aliquots. If a white precipitate (probably undissolved DNA) is visible, centrifuge for 5 min at 16,000 rcf and transfer DNA to new tube. Final concentration, determined spectrophotometrically, will not be exactly 10 μg/μl; 7–11 μg/μl works well.

12. Alternatively, for DNA precipitation, a mixture of ethanol:sodium acetate (9:1 v/v) can be used. Make with 100% ethanol and 3 M sodium acetate pH 5.2. Store at room temperature.

13. Optional addition: EDTA to a final concentration of 20 mM.

14. Column purification will result in cleaner probes, but if the DNA used in the nick translation reaction is at least 2.6 kb in length (smaller products not tested; Cyanine 5-labeled probes not tested), a fast way to remove most of the unincorporated nucleotides is to pass 2–8 μg of nick-translated reaction product through a PCR clean-up kit spin column. Promega Wizard

columns work well (others not tried). Elute DNA with nuclease-free water to a concentration of ~100–150 ng/µl if possible. Use enough water to elute all the DNA from the column. Use the aqueous probe as it is or concentrate by vacuum evaporation. Barely dry pellets may be dissolved in 2× SSC, 1× TE. Probes thus purified work best on spreads with little cytoplasmic background.

15. This solution is needed only if following the original precipitation protocol for blue or green probes (see Subheading 3.5, step 1, Note 38).
16. If background cytoplasm is a problem, as is the case with some maize lines, increase the cellulase concentration to 4%.
17. The pressure chamber and coupling components must be able to withstand pressures safely exceeding the160 psi (1,100 p*K*a) working pressure and have a valve through which the nitrous oxide enters and after treatment is vented. See example in Fig. 1.
18. One version of a humid chamber consists of an open-top cardboard box (~3″ high), subdivided into sections that will easily accommodate the width and number of microscope slides to be processed. All surfaces are lined with multiple layers of paper towels (stapled into position), which are moistened with purified water prior to use. Slides placed in the chamber are elevated (e.g., on dry wooden stake or upside down tube rack). Covering the chamber with a large Kimwipe (either damp or dry) is optional. The chamber is reused.

Fig. 1. One example of a nitrous oxide chamber. This unit is steel and has a threaded brass lid. The main chamber dimensions are 2.25″ I.D. × 3″ O.D. × 5″ tall. The mostly recessed *o*-ring at the opening of the chamber provides for a tight seal. Inset shows a close-up of part of the needle valve (right side is toward the chamber).

19. The baking pan in contact with the hot plate should not be made of thin gauge aluminum, which tends to warp. A thermostatically controlled electric skillet with lid may be used in lieu of the water bath.
20. Sticks are not required, but elevating the slides on them makes it easier to apply cover slips.
21. Vectashield® without DAPI is used when a blue fluorescent probe is in the cocktail mix. Alternatively, DAPI-containing Vectashield® diluted 1/20 with DAPI-free Vectashield® can be used to lightly stain chromosomes but still allow detection of strong blue signals. This dilution is also used when the signal from full strength DAPI is too bright. We have not tried the "Hard-Set" version of Vectashield®.
22. If imaging red and far-red signals from the same specimen, the filter sets must be custom made to avoid bleed through of bright red signals into the far-red channel and vice versa. As less overlap is achieved, signal strengths are decreased. Consult a microscopy optics specialist for specifications.
23. The direct labeling procedure utilizes a nick translation reaction in which dNTPs are incorporated into the PCR product of the sequence of interest. Only one of the four dNTPs is tagged with a fluorochrome. Nick translation involves the activities of two enzymes. DNase I randomly nicks the template DNA. The second enzyme, DNA polymerase I, has two activities – an exonuclease activity that removes DNA at the nick and a polymerase activity that fills in the gap.

    The nick translation procedure can be modified to alter the degree of incorporation of conjugated nucleotides into the probe. If sensitivity is not an issue, less DNA polymerase I is sufficient for adequate labeling (see Subheading 3.3). However, if the chromosomal target sequence is small, such as a transgene or single-copy endogenous gene, then a higher incorporation rate of the labeled dNTP is desired. Increasing the amount of DNA polymerase (see Subheading 3.2) up to about 20 times the amount in a "standard" reaction will produce a brighter probe (13).

    For detecting small targets on chromosomes, the signal-to-noise ratio must be maintained as high as possible. In addition to the degree of labeling, the size of the probe also affects signal brightness. Ideal probe length is from approximately 50 to 300 bases. In our experience, fragments much larger than 400 bases cause background fluorescence. If probes from long templates are being produced, the addition of extra DNase to the labeling reaction or an increase in the length of the nick translation reaction might be needed to produce a probe of the desired length. Because antibody-based (indirect labeling) detection methods also increase the background noise due to nonspecific adherence to chromosomes, the use of probes that are directly labeled with a fluorescent molecule is preferred.

All fluorescent probes are light-sensitive. Working in subdued light conditions is recommended. Labeled dNTPs, nick translation reactions, purification columns, probe DNA, and hybridized slides need to be stored in the dark and exposed to bright light as little as possible.

24. PCR product sizes known to produce good probes can range from about 300 bp to 5.5 kb, although most are 1–2 kb. Larger products have not been tested.
25. The reaction volumes listed will provide enough DNA to perform two to five 5-μg nick translation reactions depending on amplification efficiency. Only 2 μg of DNA at 200 ng/μl is required if the probe will have limited use; consequently, smaller reaction volumes can be used. Purified PCR products generally remain stable for a number of years at –20°C.
26. If the DNA concentration is greater than 200 ng/μl, add autoclaved or nuclease-free water to make up the difference (i.e., volume of DNA + water = 10 μl).
27. For labeling targets less than ~3 kb, see Note 58.
28. The incubation may be performed in a PCR machine. Set the machine to hold at 1–4°C after the 2-h incubation for insurance against overdigestion or for "overnight" reactions. Alternatively, 15°C water in a covered, thick-walled styrofoam box works well.
29. Chromosomal targets containing high copy numbers of repeated sequences, such as knob (180-bp and TR-1), CentC, Cent4, and the B chromosome-specific repeat in maize are easily detected with this method.
30. If the DNA concentration is greater than 200 ng/μl, add autoclaved or nuclease-free water to make up the difference (i.e., volume of DNA + nonoptional water = 25 μl).
31. To save time, columns can be made during the last half an hour of the nick translation reaction. Make fresh each time. One column can be used to purify 1–8 μg of probe DNA. The tip of a 5 3/4″ long Pasteur pipet is blocked with silane-treated glass wool (use tweezers to insert small amount in open end of pipet and metal rod to guide glass wool to tip; tamp to form a secure plug ~2–3 mm high). Too much glass wool will impede flow. Load cold 1× TE-saturated Bio-Gel® P-60 into the pipet (sterile plastic transfer pipet works well). Do not introduce voids. The pipet is placed at a slight angle in a 1.6-ml microcentrifuge tube (if vertical, may seal tip against tube), and the slurry is added until the top of the fully

settled gel is ~2 mm above the constriction in the pipet. Remove excess gel if necessary. Add TE to wash the column and ensure proper flow. If the eluate is not clear, the glass wool plug has been breached (make new column). Empty the collection tube as necessary. Wait until all of the TE has run through the column before loading the nick translation reaction. The tiny bubbles that may form in the column do not adversely affect purification or probe quality.

32. It takes approximately 3–5 min for 350–400 μl of 1× TE to flow through a column. Do not transfer the column to the next collection tube until all the added TE has been eluted from the column.
33. Alternatively, add 2.5 volumes of 90% ethanol – 10% sodium acetate (3 M, pH 5.2) to each tube.
34. Before removing the supernatant, briefly check the probe using UV light. If some of the probe is visible on the side of the tube, centrifuge again for 5–10 min with the tube in the opposite orientation. Do not UV illuminate far-red probes; the wavelength emitted by cyanine 5 is not within visible range.
35. The pellet will most likely become dislodged in the 100% ethanol rinse, if not in the 70% rinse. Rotating the tube while pouring out the ethanol is helpful. A brief spin will reposition the pellet near the bottom of the tube.
36. To avoid overdrying the pellets, pipet a portion of the total volume of 2× SSC, 1× TE onto each pellet (more on the one with the most probe). If under brief UV light examination, absolutely no probe is visible in the pellet from the last eluate, you may elect not to keep it but to add an equivalent amount of buffer to the retained probe.
37. Alternatively, do not add the 1× TE. In which case, the amounts of added sodium acetate and ethanol are based on the volumes of the initial nick translation reaction and added salmon sperm DNA.
38. Procedure for the original DNA probe precipitation protocol, which uses 5× TAE with 140 ng/μl autoclaved salmon sperm DNA:
    (a) Add 175 μl (2-μg reaction) or 350 μl (5-μg reaction) of the TAE–salmon sperm solution.
    (b) Vortex and transfer entire volume to a 1.6-ml tube.
    (c) Continue protocol starting with Subheading 3.5, step 3.
39. If both probe signal and background are strong, reducing the probe concentration could yield better results. For information pertaining to commonly used nick translated maize probes, see Kato et al. (3).

40. Root tips to be nitrous oxide treated are placed in 0.6-ml or 1.6-ml tubes, each with a hole punched in the lid to facilitate gas exposure and lightly misted with water to prevent desiccation. Many roots may be treated in the same tube. The root cells continue to divide during the treatment, so longer treatment times generally result in more metaphase spreads and chromosomes that are more condensed (shorter arm lengths). When treatment time has elapsed, quickly depressurize the chamber, transfer the tubes to ice, and add the fixative. A delay will result in the resumption of cell division and separation of the metaphase chromosomes into sister chromatids. If adding fixative to more than ~20 samples, it might be helpful to request assistance.
41. If the cap is relatively large, its removal might improve preparation quality by eliminating unwanted cell types. Do not attempt this on narrow diameter roots or if the root meristem is not clearly visible. Although a two-stage process, the tip alone can be removed (stage 1). Then, perform steps 8 and 9 in the protocol. To remove the remainder of the root cap (stage 2), after the digestion step, fill the tube with ice-cold 1× TE to wash the root. Use a glass pipette and rubber bulb to gently move the TE up and down to dislodge the remainder of the root cap. It looks like a small transparent ring, but is not always visible. Unless enough of the tip is removed, dislodging the rest of the cap will be difficult or impossible. Remove TE and return to step 10 of the protocol.
42. The length of time the root meristematic tissue is digested depends on the number and size of the tissue pieces, and the degree of chromosome spread desired. The fixed tissue can be left intact, cut transversely into several pieces, or cut both longitudinally and transversely. Whatever the size, quickly transfer the pieces into the enzyme solution. Ideal conditions for different lines will vary. For smaller diameter roots, two or three roots per enzyme tube might be needed to produce adequate cell density. If the desired meristem region is greater than 1 mm long (or for large diameter roots), it is better to cut the sample into two or more pieces. A properly digested root should yield slightly to pressure from a rounded off dissecting needle; it should not be hard, nor mushy.
43. These steps may be performed by pipetting (plastic transfer pipet with a 200-μl pipet tip stuck on the end) or by pouring (if you are careful). If the tissue is accidentally drawn into the pipet tip, squirt it back into the tube and, if necessary, centrifuge for a few seconds (mini microcentrifuge, 2,000 rcf) to "pellet" the small pieces and loose cells.
44. The choice of solutions in which the roots are broken and the cells are suspended is up to the individual. The simplest

procedure is listed in the Subheading 3. Two alternatives to breaking roots in 100% glacial acetic acid are to use 90% acetic acid–10% methanol or 70% ethanol. If the latter is used, break the root in 70–100 μl ethanol, pellet the cells (~5 s at 2,000 rcf), carefully pour off the ethanol, invert the tube, and blot the rim on a paper towel, then resuspend the cells in 30 μl of either of the above two acetic acid solutions. Factors to consider when deciding on solutions: (1) Breaking in ethanol results in a slightly cleaner preparation but takes longer and some cells will be lost; (2) 100% glacial acetic acid is easy to use but the preparations cannot be frozen; and (3) the acid–methanol resuspended cell preps may be usefully stored at 4°C for about a month (decreased resolution); this solution should be made fresh. If storage at −20°C is desirable, following the 70% ethanol washes, rinse cells several times in 100% ethanol to remove all water and replace with acetic acid–methanol, 90:10 or 75:25. Storage life is not known (see Note 48).

45. Requirement of the humid chamber for drying cell suspensions is dependent on ambient humidity. Use of the chamber is not necessary during "humid" months. However, if not used during the "drier" times of the year, the solvent drop will not evaporate, and the cells will not spread. Therefore, year-round use of the humid chamber is recommended.

46. The pattern of dropped cells can be a single drop in the center of the slide (average 5.5–6 μl) or a grid of smaller drops to cover the same area (e.g., 3 × 3). Also, two hybridizations can be set up on the same slide by positioning the dropped cells such that they can be covered individually by 22 × 22 mm plastic cover slips for hybridization, yet fit under a 24 × 50 (or 60) mm glass cover slip for viewing.

47. The perfect spread is characterized by a full complement of nonoverlapping chromosomes, with little or no cytoplasmic background and physically arranged such that they fit in a single image frame. Individual chromosomes will appear dark and lie flat on the slide (same focal plane; edges not refractile). Most of the interphase nuclei will be reasonably intact (perhaps slightly overdigested). A severely underdigested specimen is identified based on a high percentage of cells with either intact walls or very dense cytoplasm, whereas a severely overdigested specimen is characterized by interphase nuclei lacking substance, metaphase spreads lacking the full complement of chromosomes, and the presence of numerous isolated chromosomes. A good digest may contain all of these morphologies, but the goal is to adjust tissue size and digest times to maximize the number of good spreads. If the tissue was not actively dividing at the time of harvest, few metaphase

spreads will be observed, even in a properly digested cell preparation.

48. Cross-linked slides can be used immediately for hybridization or be stored at 4°C or at –20°C in a frost-free freezer (the latter good for 4–6 months when hybridizing to multicopy target sequences). They can also be left at room temperature for 1 week and still be good to use for hybridizations. However, by 4 weeks, the chromosomes show visible signs of deterioration when viewed at 40× and should not be used. Intermediate times have not been investigated. For detection of small targets, it is not desirable to store slides. Instead, store the cell suspension at –20°C and prepare fresh slides immediately before use. For very small targets (e.g., single genes), it may be necessary to use slides that are prepared from freshly digested cells that have not been stored. To maximize signal quality, store the root tips in 70% ethanol at –20°C until you are ready to do the FISH procedure.

49. If the hybridization is not to very small targets, salmon sperm DNA, 2× SSC 1× TE, and the probes can be mixed together, applied to the slide, covered with a plastic cover slip, and denatured in situ. This occasionally increases background fluorescence but is a quicker method and is useful when many slides are prepared at the same time. If nick-translated probes account for most of the probe cocktail volume, the additional salmon sperm DNA is not necessary. Use 10–15 μl of probe cocktail per slide, depending on desired coverage. See also Note 59.

50. 100 μl of probe can be denatured in a single tube. Probes can also be denatured by placing the tube at ~95°C in a heat block.

51. Prior to insertion of the denaturation pan into the boiling water bath, wet the layer of Kimwipes (drain off excess) and place slides and probe cocktail tube(s) on Kimwipes. Press down simultaneously near both ends of each slide to ensure good contact with the wet surface (improves heat transfer). Cover slides with the plastic pipet box lid (P1000 lid covers four slides; something larger is better). Cocktail mix does not have to be under the lid. Place the pan in the boiling water and cover everything with foil. Much of the 5-min time period is spent bringing the pan and slides to temperature. The actual time spent at a rolling boil (pan moves up and down under foil) is ~1–1.5 min. Do not overboil. When the time is complete, carefully lift off the foil, use tweezers/tongs to remove the denaturation pan and set it on the benchtop; it must be kept level during this maneuver. With hands, remove lid by lifting straight up, keep level, and quickly move it out of the way. These steps nearly eliminate the possibility of condensation falling on the slides and subsequently diluting the probe.

52. Slides can be placed back-to-back to increase the processing capacity of the jar.

53. By changing the temperature and salt concentration, the stringency can be modified [42°C, 2× SSC = low stringency; 55°C, 2× SSC = medium stringency; 60°C, 2× SSC = high stringency; 65°C, 2× SSC = very high stringency; 50°C, 0.1× SSC = medium stringency; 55°C, 0.1× SSC = high stringency; 60°C, 0.1× SSC = very high stringency (most signals will disappear)]. The addition of 50% formamide further increases the stringency but is seldom necessary. Wash for 10–30 min.

54. Avoid excess 2× SSC on the slide as it will dilute the Vectashield®. FISH using preparations that have dried for a short time still produce good, but perhaps not optimal, signals.

55. Line box with foil and dry Kimwipes to absorb excess mounting medium and later immersion oil. Although strong signals retain most of their fluorescence for several months, it is advisable to acquire images from freshly prepared slides.

56. Single gene/small target detection. The above-described procedures consistently allow detection of chromosomal targets larger than approximately 2–3 kb. Even smaller sequences can be detected, although on a lower percentage of cells in a given preparation. For these smaller targets, we recommend labeling probe DNA with Texas Red®, which usually results in a brighter signal compared to probes labeled with the other fluorochromes. Probes should be hybridized to fresh cell preparations (i.e., they are digested, suspended, and dropped on the same day) with selection of preparations of the best quality (with abundant chromosome spreads, free of cell walls and cytoplasmic remains). Using a higher probe concentration or fewer cells per slide and a long hybridization time (overnight) are also helpful.

    The average size of a maize gene is 4 kb, so many maize genes can be visualized with FISH. However, detection of maize genes smaller than 2–3 kb or containing repetitive elements in their introns requires a modified strategy. To localize such targets, repeat-free "pooled" probes that hybridize to adjacent unique chromosomal regions (clustered genes or regions within one gene) are produced. Repetitive elements can be identified in a sequence with various software programs available online, for example, http://www.repeatmasker.org or http://www.girinst.org. Because these programs can be limited by the completeness of repeat element databases, presumptive "unique" sequences as determined by BLAST analysis are hybridized to confirm that they are not repetitive. The regions within "unique" sequences with homology to mRNAs or cDNAs of other plant species have the most promise to be single copy conserved genic regions and can be expected to be present in different maize lines. These sequences can be used to design PCR primers and can be amplified using genomic DNA or BAC DNA as a template. By separately FISH-testing each PCR product to be included in the final pooled collection, sequences that hybridize throughout the genome, since they contain repetitive element homology, can be identified and eliminated. Finally, to produce a pooled probe, selected PCR products can be labeled with a fluorochrome by nick translation separately or together as an equimolar ratio mix.

Because mature mRNA has no introns, cDNAs can be successfully used as probes for genes containing repetitive elements in their introns. Searching the public databases for large maize cDNAs (greater than 4 kb) can identify candidate genes, and several of these have been successfully used as FISH probes.

Many maize genes are present in the genome as tandem arrays. These gene families make good FISH targets because a single member of the family can be used as a probe, which hybridizes to many locations in the cluster. This approach has been used to label the *rp1* and *rp3* rust resistance gene clusters, as well as two alpha-zein clusters and an expansin gene cluster (1). Further, the selected DNA regions can be cloned and amplified using standard primers (M13 or T7, SP6). In this step, it is recommended to prepare the PCR mix in a laminar hood and purify the PCR product in columns with caps to exclude contamination from clones of highly repetitive elements such as knobs or NOR. Because of the long exposure time used to visualize single-gene probes, even low concentrations of labeled contaminants can give bright signals during FISH.

57. More color options for FISH probes. The probe cocktail described by Kato et al. (3) includes one example of combining colors. Specifically, the maize ribosomal 5S gene is hybridized with two probes that label the same sequence. One of these probes is tagged with Texas Red® and the other with Alexa Fluor® 488, which is green. Because hues of light combine additively, the resultant signal is yellow. This is true, however, only when the colors are mixed in the correct ratios. By varying the concentrations of red and green, one can create a palette that includes a deep red orange, orange, yellow, and lime green. Similarly, one can create cyan (blue plus green), magenta (blue plus red), or pink (red plus white, if a cyanine 5-labeled probe is pseudocolored white). The advantages of multiple colors for separate probes are obvious, but this strategy does require some extra adjustments to obtain

the desired result. In addition to concentration, three factors affect color outcome.

1. The nick translation reaction used in the direct labeling procedure is initiated by random events – the action of the DNase. Consequently, some probes will produce stronger signals than others. Combining several probes prior to determining the concentration needed to produce a specific color is helpful. This problem may be eliminated by using 5′ end-labeled oligonucleotides as probes (see Note 58).
2. The intrinsic intensities of the fluorochromes must be taken into consideration. Of those in common usage, Texas Red® produces the strongest signal, followed by Alexa Fluor® 488 and fluorescein, cyanine 5, then Cascade Blue®. When combining colors, remember that equal concentrations will not contribute equally to the end product and that Cascade Blue® will only label very large targets.
3. Keep in mind that the background fluorescence from all the other probes in the cocktail mix will affect the target color combination, as will DAPI if it is used as a counter stain.

The last option for color correction is to adjust the signal strengths of the individual colors with PhotoShop or a similar image processing software.

Below are colors frequently used and the final concentrations of component probes. Note that the concentrations are provided as a starting point and may not result in the exact color indicated.

Yellow (5S)	16.0 ng/μl Alexa Fluor® 488 + 8.0 ng/μl Texas Red®
Orange (Cent4)	13.3 ng/μl Alexa Fluor® 488 + 12.0 ng/μl Texas Red®
Teal (NOR)	0.8 ng/μl Alexa Fluor® 488 + 18.7 ng/μl Cascade Blue®

58. 5′ end-labeled oligos as FISH probes. For routine screening of tissue samples, it is time-consuming to continually make probes for frequently targeted high-copy sequences. Alternatively, one might need to screen a large number of sequences for FISH probe potential. In either case, it may be more convenient to simply use 5′ end-labeled oligonucleotides from companies such as Integrated DNA Technologies (IDT), Invitrogen Molecular Probes, and Sigma. Kits are also available.

    One will have to provide the company with the target sequence, usually 20–50 bases in length, and specify the fluorochrome. Any oligo sequence will have to be: (1) specific to the target and (2) screened so as to minimize self-hybridization (i.e., dimerization and hairpins). Therefore, it will be necessary to BLAST the candidate sequences against the organism's database or against related species if the genome has not yet been sequenced or if there are many repeat sequences throughout the genome. Useful web sites and probe design strategies can be found in Note 56.

Generally, because oligo probes can produce very strong signals, they may have to be substantially diluted to working concentrations. As with nick-translated probes, the final concentration used is dependent on probe signal strength and target sequence copy number. Whereas the final concentrations of individual nick-translated probes vary from 0.2 to 40 ng/μl, the 5′ end-labeled oligos are used at concentrations more in the range of 0.003–1 ng/μl. Because these oligos are not very stable at the more dilute concentrations, it is advisable to immediately make and freeze small-volume aliquots (10–20 μl) of the primary stock (500 ng/μl) and a few vials of a less concentrated intermediate working stock. Make only enough final working stock for short-term use.

Oligo probes are single-stranded, so they do not require denaturation prior to hybridization. (In fact, some are rendered less effective if exposed to high heat conditions.) If the hybridization will include both nick-translated and 5′ end-labeled probes, add the oligo probes after the denaturation step, when the slides are on ice. Because fluorescently labeled oligos are likely to be initially used at relatively high concentrations during repeat screening procedures, it is critical to minimize potential cross-contamination of samples during both slide setup and wash procedures.

5′oligos for maize repeats:

Target	Fluorochrome	Sequence
CentC	Alexa Fluor® 488	CCT AAA GTA GTG GAT TGG GCA TGT TCG
Knob	Alexa Fluor® 488	TCG AAA ATA GCC ATG AAC GAC CAT T
1-26-2	Texas Red®	$G[TAG]_{18}$

A wide range of fluorochromes is available for conjugation to DNA oligonucleotides, both in terms of emitted wavelength as well as fluorochrome moiety within a given color range. Blue-emitting fluorochromes are generally not an option due to lack of availability, insufficient signal strength, and excessive cost. This fact leaves the following three options:

Green fluors. At present, the least expensive fluors are fluorescein-based (e.g. 6-carboxyfluorescein, 6-FAM™, which is available from IDT). Because fluorescein-tagged oligos do not require postsynthesis purification, they can be synthesized at a small scale. While fluorescein is not as bright as red fluors, or as resistant to photobleaching as Alexa Fluor® 488 (another green option), it is the most economical option (currently ~\$20–30 per 25 nM-scale oligo) for screening purposes. In contrast, a single Alexa Fluor® 488 modified oligo can cost in excess of \$300.

Red fluors. Texas Red®-conjugated fluorophores are generally the brightest option and should be reserved for lower copy number repeats, which are harder to detect. Texas Red® tagged oligos do require a minimum synthesis scale (100 nM) and HPLC purification, making them much more expensive (~\$100). Within a given company, multiple Texas Red® options might be available. (Texas Red® 615, from IDT, works quite well.)

Far-red fluors. Cyanine 5 (Cy5) is a far-red fluorochrome that also works quite well but should be reserved for the highest copy number localized repeats. Cy5-tagged oligos also require a minimum synthesis scale (100 nM) and HPLC purification.

59. Streamlined FISH protocol for large targets. This procedure saves time by modifying several of the postdigestion steps outlined in the main protocol. Most notably, the probe DNA and chromosomal DNA (cells on slide) are denatured together instead of separately. Although this protocol can also be used with smaller targets, the results are not as reproducible.
Upon completion of the digestion step:
    1. Instead of using TE, fill the tube with 70% ethanol (use a squeeze bottle if there are many samples). Close cap and invert to mix. Set tube on ice. When tissue settles, remove ethanol. Repeat.
    2. Add 70–100 μl of 70% ethanol. Use a blunt dissecting needle to "flatten" the tissue. Flick tube or vortex briefly to separate cells. Return to ice.
    3. Prepare a solution of 90% acetic acid–10% methanol (~100 μl for three tubes) by mixing glacial acetic acid and 100% methanol. Make fresh each time (both components are highly volatile).
    4. Spin cells approximately 3–5 s in tabletop mini microcentrifuge (2,000 rcf).
    5. Decant the ethanol, taking care to retain cells (some ethanol remains). Blot the opening of the tube on a paper towel, then add 25–30 μl of the acetic acid–methanol solution (100% acetic acid can be used as an alternative). Gently resuspend cells (tap/flick tube, stir with pipet tip).
    6. As in the original protocol, drop 5.5–9 μl of each sample on a labeled microscope slide in a humid chamber. Allow to dry and then cross-link using "optimal" setting. Select best slides for hybridization.

7. Make probe cocktail mix (need 10–15 μl per slide). The larger volume gives better coverage but does use more probe. If using only 5′ end-labeled oligos, add 0.5–1.0 mg of autoclaved salmon sperm DNA per microliter probe cocktail.
8. Pipet probe mix onto each spread and then cover with a 22×22 mm plastic cover slip (no need for over-hanging edge).
9. As in the original protocol, place slides in humidified aluminum pan, cover, and place in boiling water bath for 5 min and then transfer directly to a humidified plastic container for hybridization (4 h to overnight, 55°C).
10. Dip slides in 2× SSC at room temperature only long enough to pop off the plastic cover slip, then wipe off the backs of the slides and apply Vectashield (with or w/o DAPI). Apply large glass cover slip. Note: The 20-min 55°C rinse in 2× SSC specified in the original protocol is not needed unless target is low copy number or too much background is observed.

60. Meiotic FISH. A detailed description of the stages of maize microsporogenesis with photos can be found in Neuffer et al. (14).

Immature tassels are fixed in freshly made ethanol and acetic acid solution (3:1 v/v) for 24 h at 4°C. The tassels are then rinsed three times with 70% ethanol and stored in 70% ethanol at −20°C.

Each floret on the tassel contains two groups of three anthers. Each of the three anthers is at the same stage of meiosis. By examining one of the three, the stage of the remaining two can be determined. The anther to be staged is placed on a dry slide to which a drop of acetocarmine is added. Iron dissecting probes are used to smash apart the anther, and a glass cover slip is applied. The slide is heated with an alcohol lamp until just before the stain begins to boil. This helps flatten cells and also intensifies the chromosomal staining.

After correctly staged anthers are located, the other two anthers can be kept in 70% ethanol at −20°C until used for preparation. Washes and enzyme digestion procedures are the same as described previously for the root tips. The time in enzyme cocktail can be 20–50 min depending on anther size.

The enzyme solution is rinsed away by filling the tube with cold 1× TE. The TE is removed, and the anthers are gently rinsed in 100% ethanol three times. Add 20–35 μl of 3:1–9:1 acetic acid and methanol solution and break apart the anthers with a blunt dissecting probe. Flick the tube gently to further separate cells. About 8 μl of cell suspension is dropped onto a glass slide in a humid chamber and allowed to dry. Examine dried preparations with a phase contrast microscope to confirm the correct staging and spreading of the meiotic chromosomes. UV cross-link the slides as previously described.

Meiotic chromosome spreading and cytoplasm thickness depends on the maize genotype. In some inbred lines, chromosomes do not separate from their cytoplasm during the spreading procedure. As a result, there tends to be higher background after FISH than for mitotic chromosome spreads. If the cytoplasm appears dark (without staining) through a light microscope, then the background after FISH is usually strong. However, for intense signals (knob, CentC, NOR, etc.), cytoplasm is not a problem. For smaller targets, only slides containing meiotic chromosome spreads with little or no visible cytoplasm should be used for FISH.

Some procedures have described various treatments to reduce or eliminate the cytoplasmic background to better detect small target signals. Immersion of cross-linked slides in a methanol, chloroform, and glacial acetic acid mixture (6:3:1 v/v) for 20–30 s can improve the ratio of target/cytoplasm fluorescence (15). Also, increasing the length of incubation of anthers in the cellulase–pectolyase enzyme cocktail reduces the background, as does increasing the concentration of cellulase (see Note 16). Using the FISH procedure outlined for mitotic spreads, we have been able to detect single gene signals on pachytene preparations that have little cytoplasm.

61. Retroelement genome painting. Because the maize genome contains retrotransposon families that are present in thousands of copies, probes prepared from them can be used to paint maize chromosomes. The retroelement families have differentially expanded in the various *Zea* lineages as well as in the sister genus *Tripsacum* (8). Retrotransposons used as FISH probes label chromosomes from different species with different intensities, allowing the chromatin from different species to be distinguished in interspecific hybrids.

    This approach can be used as an alternative to, and has some advantages over, the commonly used genomic in situ hybridization (GISH) technique. Because the different *Zea* species share the same retrotransposons and other sequences, the chromatin of all species is almost equally labeled. Because the retrotransposons are present in different copy number, using them directly as probes allows the chromosomes to be distinguished based on fluorescence intensity. Additionally, because knob heterochromatin sequences are so abundant, it is not possible to block hybridization to the knob regions. If introgressed segments are to be detected, the knobs will give discrete signals that can appear as introgressed chromatin in advanced backcross generations.

To distinguish maize and *Z. diploperennis*, it is possible to use a number of different retrotransposon probes. The long terminal repeats (LTRs) from Grande or Huck work particularly well for this purpose. To distinguish maize from *Tripsacum*, the most common maize retrotransposons will work. Grande, Huck, Prem1, Prem2, Opie, and others will work well. Cinful1 is present at high copy number in both genera.

To identify new retrotransposons that paint *Tripsacum* chromatin but not maize, a library of random DNA sequences from *Tripsacum* was produced and screened. The library was prepared by nebulizing genomic DNA from *Tripsacum*, repairing the ends and ligating into a standard cloning vector. Twenty-five clones were chosen for FISH screening based on strong hybridization to *Tripsacum* genomic DNA. Four of these clones contained elements that strongly hybridized to *Tripsacum* chromosomes but only weakly to maize chromosomes. Other clones contained a 5S ribosomal gene, knob sequences, or retrotransposons, which hybridized equally well to both maize and *Tripsacum* chromosomes. Because FISH screening is rapid and simple, this approach could be easily adapted to other species with genomes that are enriched with repetitive elements.

62. Developing a new karyotyping cocktail for another plant species. Developing a FISH-based karyotyping cocktail requires a collection of probes that can be used in combination with morphological features to identify each chromosome in a karyotype. Not only must individual probes work consistently (they must be bright, low in background, and relatively insensitive to minor variations in hybridization conditions) but also the collection itself must work together. Karyotyping cocktails have been developed for numerous plant species; they generally utilize some combination of probes developed from BACs, repeat-subtracted gene collections, clustered repetitive elements, and clustered genes. Obviously, the utility of an individual probe type depends on the organization and composition of the genome of interest. In maize, for example, due to the high repeat content/low gene density, BAC probes are not a viable option because of the massive cross-hybridization with elements dispersed throughout the genome. In contrast, in soybean, in which about half of the total repeat content is concentrated in pericentromeric regions, low-repeat content BACs can be used.

Each type of FISH probe has strengths and weaknesses. Because of their large inserts (>100 kb), BAC FISH signals can generate excellent single-locus signals in FISH; yet, also because of their large inserts, BACs with repetitive DNA can generate genome-wide background in FISH. Even with a high level of background, BAC FISH signals are often still apparent and may possibly be useful.

Although single-copy gene sequences as small as 2 kb can be detected in mitotic chromosome spreads using nick-translated probes, the FISH signal is not generally bright enough to be used in the context of a karyotyping cocktail because most or all of the other cocktail components using the same fluorescent label will be much brighter. One option is to develop collections of multiple single-copy genes localized to a particular chromosomal segment. Another option is to develop a single gene probe that targets *clusters* of highly homologous genes. For example, both 18S and 5S ribosomal genes are frequently present in gene clusters containing hundreds of copies of nearly identical genes. Thus, even small (2 kb) nick-translated probes or fluorochrome-labeled oligonucleotides can generate bright FISH signals. Other gene clusters are also landmarks that could be used but will have to be identified by candidate or bioinformatic screening. For example, 1-kb probes developed from disease-resistance loci in soybean can be used because such genes are present in clusters containing 10–30 copies within relatively confined (1 megabase) regions.

Repetitive DNA elements represent an extremely valuable resource for karyotyping probes not only because repeats can be present in high-copy clusters but also because elements can have broad and variable distribution across multiple or all chromosomes. Candidate repeat probes will have to be identified by screening collections of genomic sequences obtained by whole-genome shotgun sequencing, BAC-end sequencing, etc. From an efficiency perspective, the larger the sequence collection, the better. This is the case because the most successful repeat probes will likely be developed from the more abundant repeat classes, which can be best assessed using a large collection of sequences. Two starting points for any repeat screens are (candidate) centromeric repeats and trinucleotide repeats. The latter are a type of simple sequence repeats common to many organisms; all 64 variants can be quickly and inexpensively screened using a set of 12 fluorescein-labeled oligonucleotides. Centromeric repeats are also potentially present in widely variable copy number and as sequence variants. In designing oligonucleotide probes (see Note 58) based on centromeric (or other) repeat types, it is certainly worth exploring sequence variants for distribution variation in the genome by making probes in a second fluorochrome color (red).

Acknowledgments

This work was supported by grants from the National Science Foundation DBI 0423898, DBI 0421671 and DBI 0701297.

Note Added in Proof

Cascade Blue®-7-dUTP is no longer available. Coumarin-5-dUTP can be purchased as a custom synthesis from Perkin Elmer.

References

1. Lamb, J. C., Danilova, T., Bauer, M. J., Meyer, J., Holland, J. J., Jensen, M. D., and Birchler, J. A. (2007) Single gene detection and karyotyping using small target FISH on maize somatic chromosomes. *Genetics* **175**, 1047–1058.
2. Danilova, T. V. and Birchler, J. A. (2008) Integrated cytogenetic map of mitotic metaphase chromosome 9 of maize: resolution, sensitivity and banding paint development. *Chromosoma* **117**, 345–356.
3. Kato, A., Lamb, J. C., and Birchler, J. A. (2004) Chromosome painting using repetitive DNA sequences as probes for somatic chromosome identification in maize. *Proceedings of the National Academy of Sciences of the United States of America* **101**, 13554–13559.
4. Lamb, J. C., Meyer, J. M., Corcoran, B., Kato, A., Han, F., and Birchler, J. A. (2007) Distinct chromosomal distributions of highly repetitive sequences in maize. *Chromosome Research* **15**, 33–49.
5. Birchler, J. A., Albert, P. S., and Gao, Z. (2008) Stability of repeated sequence clusters in hybrids of maize as revealed by FISH. *Tropical Plant Biology* **1**, 34–39.
6. Yu, W., Lamb, J. C., Han, F., and Birchler, J. A. (2007) Cytological visualization of DNA transposons and their transposition pattern in somatic cells of maize. *Genetics* **175**, 31–39.
7. Lough, A. N., Roark, L. M., Kato, A., Ream, T. S., Lamb, J. C., Birchler, J. A., and Newton, K. J. (2008) Mitochondrial DNA transfer to the nucleus generates extensive insertion site variation in maize. *Genetics* **178**, 47–55.
8. Lamb, J. C. and Birchler, J. A. (2006) Retroelement genome painting: cytological visualization of retroelement expansions in the genera Zea and Tripsacum. *Genetics* **173**, 1007–1021.
9. Lamb, J. C., Kato, A., Yu, W., Han, F., Albert, P. S., and Birchler, J. A. (2006) Cytogenetics and chromosome analytical techniques, in *Floriculture, Ornamental and Plant Biotechnology: Advances and Topical Issues, Volume IV* (Teixeira da Silva, J. A., ed.) Global Science Books, London, pp. 245–248.
10. Kato, A., Vega, J. M., Han, F., Lamb, J. C., and Birchler, J. A. (2005) Advances in plant chromosome identification and cytogenetic techniques. *Current Opinion in Plant Biology* **8**, 148–154.
11. Kato, A. (1997) An improved method for chromosome counting in maize. *Biotechnic & Histochemistry* **72**, 249–252.
12. Kato, A. (1999) Air drying method using nitrous oxide for chromosome counting in maize. *Biotechnic & Histochemistry* **74**, 160–166.
13. Kato, A., Albert, P. S., Vega, J. M., and Birchler, J. A. (2006) Sensitive FISH signal detection using directly labeled probes produced by high concentration DNA polymerase nick translation in maize. *Biotechnic & Histochemistry* **81**, 71–78.
14. Neuffer, M. G., Coe, E. H., and Wessler, S. R. (1997) The maize organism: the gametophyte (microsporogenesis) in *Mutants of Maize.* Cold Spring Harbor Laboratory Press, New York, pp. 20–28.
15. Krabchi, K., Gadji, M., Yan, J., and Drouin, R. (2006) Dual-color PRINS for in situ detection of fetal cells in maternal blood, in *PRINS and* In Situ *PCR Protocols,* Second Edition (Pellestor, F., ed.) Humana Press, New Jersey, pp. 142, 146.

Chapter 5

Plant B Chromosomes

Andreas Houben, Shuhei Nasuda, and Takashi R. Endo

Abstract

B chromosomes are dispensable elements of the genome that do not recombine with the A chromosomes of the regular complement and that follow their own evolutionary pathway. Here, we survey current knowledge on the DNA/chromatin composition, origin, and drive mechanisms of B chromosomes and discuss the potential research applications of supernumerary chromosomes.

Key words: B chromosome, Genome evolution, Chromatin, Transcription, Centromere, Nondisjunction

1. Introduction

B chromosomes (Bs) are dispensable components of the genomes of numerous plant, fungi, and animal species. Even some supernumerary marker chromosomes of humans show similarities to B chromosomes (1). They do not pair with any of the standard A chromosomes (As) at meiosis, and have irregular modes of inheritance. As they are dispensable for normal growth, Bs were considered nonfunctional and without any essential genes. As a result, B chromosomes follow their own species-specific evolutionary pathways. Because most Bs do not confer any advantages on the organisms that harbor them, they may be thought of as parasitic elements that persist in populations by making use of the cellular machinery required for the inheritance and maintenance of A chromosomes. In low numbers, Bs show little or no impact on the hosts. However, increased numbers of Bs cause phenotypic differences and reduced fertility (reviewed in refs. 2–14).

The distribution of Bs among plants is not random. Among angiosperms, they are more likely to occur in outcrossing than in inbreeding species, and their presence is also positively correlated

James A. Birchler (ed.), *Plant Chromosome Engineering: Methods and Protocols*, Methods in Molecular Biology, vol. 701, DOI 10.1007/978-1-61737-957-4_5, © Springer Science+Business Media, LLC 2011

with genome size and negatively with chromosome number. They are not found any more frequently in polyploids than in diploids or in allopolyploid species (15, 16). In many species, e.g., *Brachyscome* (*Brachycome*) *dichromosomatica* (17), different morphological types of Bs exist within single species. Bs are largely absent from species with small genomes; however, species with large genomes are studied more frequently than species with small genomes, and Bs are more likely to be reported in well-studied species (18). The relationship between genome size and B frequency may be explained also on the grounds that species with large genomes can tolerate extra chromosomes more easily (19) or that the greater amount of noncoding DNA, which is what largely constitutes large genomes, is itself a trigger for B formation (16). According to a survey of 23,652 angiosperm species (about 9% of the estimated 260,000 species), about 8% of monocots and 3% of eudicots have Bs, and of these, by far the largest numbers of species belong to the Poaceae and Asteraceae. These families are also highly species-rich; however, several other families have a higher proportion of species with Bs. Among orders, the two B-chromosome "hot spots" are the Liliales and Commelinales (15, 16).

2. Diverse Routes of B Chromosome Formation

Several scenarios have been proposed for the origin of Bs. Most probably, they arose in different ways in different organisms (reviewed in refs. 10–12). The most widely accepted view is that they are derived from the A chromosome complement. Some evidence also suggests that Bs can be spontaneously generated in response to the new genomic conditions after interspecific hybridization. The involvement of sex chromosomes has also been argued for their origin in some animals (see ref. 11 for examples). Despite the high number of species with Bs, their de novo formation is probably a rare event, because the occurrence of similar B variants within related species suggests a close relationship between the different variants (e.g., *Secale* (20), *Brachyscome* (21)).

The molecular processes that gave rise to Bs during evolution remain unclear, but the characterization of sequences residing on them sheds some light on their origin and evolution. In maize and *Brachyscome dichromosomatica* for example, the Bs contain sequences that originate from different As, so the Bs could represent an amalgamation of these diverse A-derived sequences (22–25).

The actual process of sequence transfer from As to Bs is not clear, but recent results (26–28) indicate that transposition of mobile elements may have played an important role. Analysis of large DNA insert clones demonstrated that maize Bs are composed of B-specific sequences intermingled with those in common with the As. The 22-kb-long B-specific *StarkB* element, for example, has been subject to frequent insertions by LTR-type retroelements (27), in a fashion similar to the nested insertions seen in some intergenic A chromosome regions (29). It is likely that *StarkB* tandem array formation started with a transposition of a *GrandeB* mobile element into an original *non-GrandeB* sequence, which was then amplified, possibly from sister-strand exchange. Subsequently invasion(s) interwoven with amplification resulted in loss of the original tandem structure of *StarkB* (30). Using the divergence of retroelements interrupting B-specific sequences, Lamb et al. (27) have estimated the minimum age of the maize B to be at least 2 million years. Recently established oat–maize B chromosome addition lines (31) should become an ideal material for the further characterization of the maize B sequence composition because of the low level of sequence similarity between oat and maize.

B chromosomes provide an ideal target for transposition of mobile elements (32), and insertion of such elements may therefore be responsible for the generation of structural variability in Bs (11). Indeed, a B-specific accumulation of Ty3/gypsy retrotransposons has been reported for the fish *Alburnus alburnus* (L.) (33). In the same context, note that Bs contain various types of coding and noncoding repeats similar to those found in circular extrachromosomal DNA of various organisms (34). Extrachromosomal DNA with similarity to tandem repeat sequences shared by A and B chromosomes has recently been identified (35), and integration of extrachromosomal DNA into Bs may be favored because transcriptionally less active Bs are subject to reduced negative selection. However, whether an evolutionary link exists between extrachromosomal DNA and the evolution of Bs remains to be determined.

A preferential contribution of rDNA coding repeats to the evolution of B chromosomes has been proposed, because rDNA loci have been detected on Bs of many species (for review, see refs. 9, 36). In the herb *Plantago lagopus* L., the novel B chromosome is a product of a spontaneous amplification process of 5S ribosomal DNA. The process has initiated as a result of two breaks in one of the chromosomes which harbored 5S and 45S rDNA sequences. This initial segment of DNA has, as a result of unscheduled DNA amplification, led to the formation of a ring chromosome. Subsequently, the ring chromosome underwent breakage–fusion–bridge cycle and got established as a linear chromosome after healing of the broken ends (37). Interestingly, amplification of satellite DNA has been used for the formation of engineered mammalian chromosomes (38). However, in contrast to the situation with animals, the molecular mechanism of sequence amplification in plants is poorly understood. However, except for tobacco (39), no amplification-stimulating DNA elements from plants have been identified thus far.

Alternatively, B chromosomal rDNA sites could be a consequence of the reported mobile nature of rDNA (40); Bs may be the preferred "landing sites" because of B chromosome's inactivity. Increasing evidence also indicates that ribosomal sequences can change position within the genome without corresponding changes in the surrounding sequences (41–43).

The sequence of B-located rRNA genes has been used to study the likely origin of Bs by revealing the relatedness of the internal transcribed spacers (ITS) between the different chromosome types. Sequence analysis of A and B ITS sequences of *C. capillaris* (44) and *B. dichromosomatica* (45, 46), and comparisons with sequences of related species, indicates that the B chromosome rRNA genes are probably derived from those of the A chromosomes of the host species. The reason(s) why B chromosomes' rDNA is not (e.g., *B. dichromosomatica* (45) or weakly e.g., *Crepis capillaris* (44)) transcribed is not yet clear. Differences in posttranslational histone modifications, such as histone methylation or acetylation, between A and B chromosomes, which may be responsible for differences in gene activity, have been demonstrated (28, 47–50). Another possibility is that suppression may occur because of nucleolar dominance, so that the rRNA genes on the A chromosomes are active at the expense of those on the B chromosomes (45). The inactivity of B chromosome rDNA might explain the presence of multiple ITS2 sequences, since homogenization of rDNA spacers is thought to occur only in transcribed regions. Concerted evolution is a feature of rDNA repeats (51), but mechanisms that control it may not include nontranscribed rDNA sequences (52, 53). Since no sequence homogenization occurs between A and B chromosomes' rDNAs, and since Bs are not constitutively active, one might expect further sequence drift of B-located rDNA sequences. An increase in the number of different members of the already heterogeneous population of ITS sequences could be the evolutionary consequence, as has been postulated for the B chromosome-like paternal sex ratio chromosome of the wasp *Trichogramma kaykai* (54).

Some findings imply that B chromosomes arise spontaneously in response to the genomic stress after interspecific hybridization, e.g., in *Coix aquatica* Roxb. and *C. gigantea* J. König ex Roxb. (55). After fertilization, the two different parental genomes are combined within a single nucleus, which in most cases is embedded within the maternal cytoplasm. Such a novel genomic constitution may result in conflicts, and as a consequence, a genomic and epigenetic reorganization of the genomes can occur (56). An incomplete loss of one parental genome during hybrid embryogenesis might play a role in the hybrid origin of Bs. Evidence exists that, during the uniparental chromosome elimination process, the centromeres of parental chromosomes undergoing elimination are the last to be lost (57). If such a centric fragment is retained, rather than being eliminated, a subsequent spontaneous doubling could provide an ideal prerequisite for the de novo formation of a supernumerary chromosome. Indeed, a centric fragment was generated during the introgression of a chromosome region from the wasp *Nasonia giraulti* Darling into *N. vitripennis* (Walker). This neo B showed a lower-than-normal Mendelian segregation ratio in meiosis and some mitotic instability, but the transmission rate and mitotic stability then increased over successive generations (58).

A novel mechanism for new chromosome evolution based on recombination of nonhomologous chromosomes during the DNA double-strand repair process at S-phase has been postulated for the formation of the "zebra" chromosome, which is composed of *Elymus trachycaulus/Triticum aestivum* structurally rearranged chromosome fragments (59). Although this restructured chromosome does not represent a B chromosome, it is imaginable that a similar mechanism could also result in the formation of a neo B chromosome.

On the basis of new insights into the mechanisms of chromosome evolution (60–62), we are tempted to ask, as did Patton (63), whether the "by-product" of a Robertsonian translocation between two nonhomologous acrocentric chromosomes with breakpoints close to centromeres could evolve into a B-like chromosome. With the recent development of the comparative chromosome painting technique to reconstruct karyotype evolution, it is becoming clear that chromosome number reductions are often accompanied by pericentromeric inversions and translocations between acrocentric chromosomes (64).

Although the minichromosomes formed from Robertsonian translocation events are mainly composed of centromeric sequences, they are frequently lost because of the lack of essential genes and their failure to pair and segregate properly during meiosis. Centromeric regions are also highly dynamic in sequence composition and display a low recombination frequency (65), and recent findings by Hall et al. (60) point to (peri)centromeres as genomic regions that may experience selective pressures distinct from those acting on euchromatin. They can tolerate rapid changes in structure and sequence content, such as large insertions of B chromosome-typical sequences, e.g. mobile elements, rDNA arrays, and satellite arrays. When a nonessential centromeric fragment survives, rapid sequence alteration may prevent meiotic pairing with the As, and the gain of a drive (by an unknown mechanism) may put it on the evolutionary pathway to a proto-B.

In addition, tertiary trisomics, which appear in the progenies of translocation heterozygotes, have been hypothesized, under certain circumstances (e.g., suppressed crossing-over, rapid loss of genetic activity to overcome genetic imbalance, and positive selection for plants with an extra chromosome), to be suited for B chromosome formation (e.g. in the garden pea (66)).

The most widely used approaches to study the evolution and DNA composition of B chromosomes are based on the isolation and characterization of only a single or a small group of mainly repetitive sequences. These approaches have been valuable in tracing the fate of various repeats in a wide range of species. However, they do not allow for the global comparative analysis of sequence profiles required for elucidating evolutionary trends on the whole genome and B chromosome level. To overcome this obstacle, in the future, new low-cost sequencing technologies such as 454 pyrosequencing (67) should be used for B chromosome sequence analysis. Most recently, 454-sequencing has been used to shotgun-sequence flow-sorted plant chromosomes. As few as 10,000 copies of barley chromosome 1H were flow-sorted and used as a template to assess gene content and genomic composition of this chromosome (K. Mayer, T. Wicker, J. Dolezel, N. Stein et al., personal communication). Most Bs are smaller than As and can easily be sorted by flow cytometry (68). For instance, one rye B encodes around 800 Mbp, thus a single 454-run could identify ~50% of sequence information of flow-sorted Bs. Thus, the application of sorted Bs for 454-sequencing is the most efficient approach for the sequence analysis of Bs. Comparative sequence analysis will then significantly improve our knowledge of the origin of Bs and hence that of the evolution of genomes.

3. Non-Mendelian Segregation Behavior of B Chromosomes

Bs fail to pair with any members of the A chromosome set during meiosis, although they may pair and form chiasmata among themselves. Because Bs appear to be devoid of essential genes and have no known adaptive advantage, their persistence in natural populations depends on their mechanisms of drive that they have evolved to ensure their survival. In many cases, maintenance is engendered by their transmission at higher than Mendelian frequencies, which allows their successful spread and accumulation in populations.

The drive mechanisms of maize and rye Bs are well-studied examples that allow B chromosome accumulation. Bs of rye undergo a directed nondisjunction, into the generative nucleus, at the first mitosis of the pollen. The generative nucleus then produces two sperm nuclei, each with an unreduced number of Bs. Essentially, the B chromosome fails to separate its chromatids at this first division of the pollen, placing both chromatids of each B in the generative nucleus and thereby in the next generation.

Notably, the B chromosome centromeres appear to divide normally, but on either side of the centromere are "sticking sites", which prevent normal anaphase separation of the chromatids, leading to nondisjunction at an average frequency of about 86% in rye (69). Nondisjunction works equally well when the rye B is introduced as an addition chromosome into hexaploid wheat (70–72) or *Secale vavilovii* Grossh. (73). Therefore, the segregation behavior of the B is mainly autonomous and independent of the background genotype. The B itself controls the process of nondisjunction and B-transmission frequency (74). The accumulation mechanism of the rye B requires a factor located on the end of its long arm, which may also act *in trans*. Bs that lack this terminal region, where two repeat families E3900 and D1100 reside, undergo normal disjunction (72, 75), but if a standard B (76), or the terminal region of the long arm of the B (72) is also present in the same cell, the standard B mediates nondisjunction of both itself and the deficient B.

For analysis of the function of the terminal B-region in more detail, different rye B-wheat (*Triticum aestivum* L.) chromosome translocations and B deletions were generated using the wheat gametocidal system ((72) also see contribution of T. Endo). No whole-arm translocations were found between rye B and wheat A chromosomes, so the B centromere might have a unique structure that prevents centromeric fusion with the wheat centromere. Bs with deficiencies in the short arm retained the capacity for nondisjunction, albeit at lower frequencies than the standard B, so the size of the pericentromeric B-region might regulate the action of nondisjunction. Analysis of Bs with deficiencies in the long arm indicated that a critical nondisjunction element might be located within the region between the E3900- and D1100-positive chromosome regions. Alternatively, the number of the repetitive sequences themselves could be the critical factor for nondisjunction. If the B-terminal region is translocated to a wheat chromosome, a balanced number of B centromeres and terminal regions seems to be required for the regulation of nondisjunction (72).

The nondisjunction process in maize differs from that in rye. At least three properties allow the maize B to increase in numbers: nondisjunction at the second pollen mitosis, preferential fertilization of the egg by the sperm containing the B (77–81), and suppression of meiotic loss when the Bs are unpaired (82).

As in rye, the B-accumulation mechanism in maize requires a factor located on the end of the long arm of the B that may act in *trans* (83–85). Furthermore, nondisjunction of Bs takes place in the endosperm and tapetum, and in binucleated tapetal cells, Bs mediate A chromosome instability (86, 87). One A-located "gene" seems to codetermine maize B accumulation by preferential fertilization, and another "gene(s)" determines the meiotic loss of Bs (88). Sperm nuclei containing deletion derivatives of B-9 (translocations lines involving the B and chromosome 9), which lack the centric heterochromatin and possibly some adjacent euchromatin of the B chromosome, no longer have the capacity for preferential fertilization (89).

So far no gene sequence, or nondisjunction element, on any B has been characterized that plays a role in B accumulation, and the speculation is therefore tempting that noncoding RNA acts on the process of B chromosome nondisjunction. In fission yeast, for example, the repeats flanking the kinetochore are essential for sister chromatid cohesion and are maintained in a proper heterochromatic state by the RNAi machinery (90). Similarly, pericentromeric heterochromatin is required for proper chromosome cohesion and disjunction in flies and also in other organisms (91, 92). Notably, forced accumulation of human centromeric noncoding satellite transcripts leads to defects in separation of sister chromatids (93). Also, for plants, RNA molecules have been shown to play a role in establishing centromeric heterochromatin domains (94, 95). In this context, the transcriptional activity of repeats located in the B chromosome-nondisjunction-controlling region of rye (50) is striking. The unique chromatin conformation and transcriptional activity of the B-terminal region could be involved in the trans-acting mechanism of directed nondisjunction characteristic of B transmission.

4. Centromere Organization of B Chromosomes

An understanding of the structure and regulation of A and B centromeres is a prerequisite for a better understanding of the unique segregation behavior of Bs. The B-specific repeat ZmBs has been used to describe extensively the centromere of maize Bs (24, 96–98), which are among the best-characterized plant centromeres. The centromeres of the maize Bs contains several megabases of ZmBs, a 156-bp satellite repeat (CentC), and centromere-specific retrotransposons (CRM elements). Only a small fraction of the ZmBs repeats interacts with kinetochore protein CENH3 (also called CENPA), the histone H3 variant specific to functional centromeres. CentC, which marks the CENH3-associated chromatin in maize A centromeres, is restricted to a similar 700-kb domain within the larger context of the ZmBs repeats (99). Centromere specification must have an epigenetic component, as dicentric A–B translocation chromosomes are characterized by stable inheritance of an inactive state of one of the centromeres over several generations (100).

A comparison of maize A and B chromosomes seems to show that Bs are enriched with DNA elements that are normally found at or near A centromeres (101). A similar tendency has been described for the rye B, which is characterized by a higher copy number of the rye retrotransposon-like centromeric repeat pAWRC.1 (102). In contrast to maize Bs (101), the rye kinetochore protein CENH3 is present in equal amounts on both As and Bs (Houben, unpublished data).

The centromeric region of *B. dichromosomatica* standard Bs (of cytodeme A1, A2, and A4) is enriched with a B-specific tandem repeat (Bd49) that is not microscopically detectable on A chromosomes (103, 104). Initially, the predominantly centromeric location of the Bd49 repeat suggested a possible role for this sequence in the drive process, but a noncentromeric Bd49 signal in *B. dichromosomatica* cytodeme A3 and differences in signal sizes among all the Bs of different cytodemes do not support this assumption (21).

5. Effects and Transcripts Associated with B Chromosomes

Although Bs are not essential, some phenotypic effects have been reported, and these effects are usually cumulative, depending upon the number and not the presence or absence of Bs. In low number, Bs have little if any influence on the phenotype, but at high numbers, they often have a negative influence on fitness and fertility of the organism (reviewed in refs. 2, 4, 9, 81).

Evidence exists that Bs directly or indirectly influence the behavior of A chromosomes. One of the most striking of such effects is the potential impact of Bs on diploidization in allopolyploid hybrids, e.g., *Lolium temulentum* × *L. perenne* + B (105), where Bs prevent or suppress the homologous pairing of As. In wheat × *Aegilops* hybrids, Bs contributed by the *Aegilops* parent seem to be able to substitute for the *ph* locus of the hexaploid wheat (further examples are reviewed by refs. 15, 106, 108).

Indirect evidence for weak transcriptional activity of Bs results from comparative analysis of esterase isozyme activity in plants with and without Bs in *Scilla autumnalis* L. (107) and rye (108). In B-positive plants, additional bands were detected by protein electrophoresis, but in both cases, whether the additional bands were caused by a B-located gene or whether Bs influenced the transcription behavior of an A-located gene remains unclear. For grasshoppers, Bs have been demonstrated to alter the expression of A chromosome genes (109).

Except for the B-located 45S rRNA gene of *C. capillaris*, in which one of two B-specific members of the rRNA gene family was weakly transcribed (44), there was no direct molecular evidence for transcription of B chromosome genes in plants until the transcriptional activity of B-specific repetitive sequences has been demonstrated. In maize and rye, retrotransposon-derived high-copy elements (27, 28), and in rye two repeat families, E3900 and D1100, clustered at the B chromosome long arm are transcriptionally active in a tissue-specific manner (50). The function of these B-transcripts and the mechanism of transcription of B-repeats are unknown at present. It has been hypothesized that these transcripts could have a structural function in the organization and regulation of Bs (27, 50).

Recently, the general transcription activity of rye Bs has been analyzed by comparative cDNA-AFLP (28). In addition to weak B chromosome transcription, evidence has been found that Bs are able to downregulate A chromosome-localized sequences in a genotype-dependent manner. It is likely that Bs may result in a variety of epigenetic effects, including the differential regulation of A-localized transposable elements through mechanisms such as homology-dependent RNA interference pathways (110). Since the rye B most likely originated from the A chromosome complement, it seems to be reasonable that the transcription alterations of A-located sequences are caused by homology-dependent mechanisms (111), as has been proposed for the remodelling of gene-activities in newly formed hybrids and allopolyploids (112). Another hypothesis for explaining how the Bs exert control over the rest of the genome postulates their effects on the spatial organization of the genome itself. Recent work suggests that spatial positioning of genes and chromosomes can influence gene expression (113). Indeed, Delgado et al. (114) observed that in interphase nuclei, A chromosomal ribosomal DNA displays a more compact distribution in cells with Bs compared to cells without Bs. A more compact distribution of rDNA sites suggests a lower level of rRNA gene activity. A similar effect of an almost gene deficient chromosome has been demonstrated for *Drosophila*. Lemos et al. (115) demonstrated that the Y chromosome of *D. melanogaster* regulates the activity of hundreds of genes located on other chromosomes.

6. Potential Applications of B Chromosomes

B chromosomes have been employed in mapping the A genome, modulating recombination, and exploring the structure of the centromere and the process of nondisjunction, as discussed by (3, 5, 116). In the future, B chromosomes could become even important for the generation of chromosome-based vectors for gene transfer. Telomere-mediated chromosome truncation has recently been adapted for A and B chromosomes of maize (117–120). With respect to the possible use of Bs as a vector for transgenes, recall that Bs have little or no effect on an individual's phenotype, and this issue is only of concern where a high number of Bs can reduce vigor (19). Constitutive transgene expression from B-derived minichromosomes suggests that inactivation of transgenes on B chromosomes (118), if it occurs, is at least not a rapid process. Because of the intrinsic postmeiotic drive of intact Bs, a B-chromosome-derived vector might potentially reveal an increase of transmission frequency above Mendelian expectation, and this feature would have to be silenced (121).

Acknowledgments

AH has been supported by the DFG (1779/10-1, Germany) and the JSPS (S-09092, Japan). SN has been supported by the JSPS and the DFG under Japan–Germany Research Cooperative Program. We would like to thank Neil Jones for discussions and helpful comments on the manuscript.

References

1. Liehr T, Mrasek K, Kosyakova N, Ogilvie CM, Vermeesch J, Trifonov V, et al. (2008) Small supernumerary marker chromosomes (sSMC) in humans; are there B chromosomes hidden among them. *Mol Cytogenet* **1**,12.
2. Bougourd SM, Jones RN. B chromosomes: a physiological enigma. (1997) *New Phytol* **137**, 43–54.
3. Jones RN, Viegas W, Houben A. (2008) A century of B chromosomes in plants: So what? *Ann Bot (Lond)* **101**, 767–775.
4. Jones RN, Rees H. (1982) B chromosomes. 1st Edition. Academic Press, London & New York.
5. Jones RN, Gonzalez-Sanchez M, Gonzalez-Garcia M, Vega JM, Puertas MJ. (2008) Chromosomes with a life of their own. *Cytogenet Genome Res* **120**, 265–280.
6. Jones RN. (1995) B chromosomes in plants. *New Phytol* **131**, 411–434.
7. Jones RN. (1991) B-chromosome drive. *Am Nat* **137**, 430–442.
8. Jones RN. (2003) B chromosomes. In: B. Thomas (ed), *Encyclopaedia of Plant & Crop Science*, Marcel Dekker Inc., New York. pp. 71–74.
9. Jones RN. (1995) Tansley review No. 85 – B chromosomes in plants. *New Phytol* **131**, 411–434.
10. Jones N, Houben A. (2003) B chromosomes in plants: escapees from the A chromosome genome? *Trends Plant Sci* **8**, 417–423.
11. Camacho JPM, Sharbel TF, Beukeboom LW. (2000) B-chromosome evolution. *Philos Trans R Soc Lond B Biol Sci* **355**, 163–178.
12. Camacho JPM. (2005) B chromosomes. In: TR. Gregory (ed), *The Evolution of the Genome*, Elsevier, San Diego. pp 223–285.
13. Camacho JPM. (2004) B chromosomes in the eukaryote genome. *Cytogenet Cell Genet* **106**, 147–410.
14. Jenkins G, Jones RN. (2004) B chromosomes in hybrids of temperate cereals and grasses. *Cytogenet Genome Res* **106**, 314–319.
15. Palestis BG, Trivers R, Burt A, Jones RN. (2004) The distribution of B chromosomes across species. *Cytogenet Genome Res* **106**, 151–158.
16. Levin DA, Palestris BG, Jones RN, Trivers R. (2005) Phyletic hot spots for B chromosomes in angiosperms. *Evolution* **59**, 962–969.
17. Leach CR, Houben A, Timmis JN. (2004) The B chromosomes in Brachycome. *Cytogenet Genome Res* **106**,199–209.
18. Trivers R, Burt A, Palestis BG. (2004) B chromosomes and genome size in flowering plants. *Genome* **47**, 1–8.
19. Puertas MJ. (2002) Nature and evolution of B chromosomes in plants: A non-coding but information-rich part of plant genomes. *Cytogenet Genome Res* **96**, 198–205.
20. Jones RN, Puertas MJ. (1993) The B-chromosomes of rye (*Secale cereale* L.). In: KK. Dhir, TS. Sareen (eds), *Frontiers in Plant*

Science Research, Bhagwati Enterprises, Delhi, India. pp. 81–112.

21. Houben A, Thompson N, Ahne R, Leach CR, Verlin D, Timmis JN. (1999) A monophyletic origin of the B chromosomes of *Brachycome dichromosomatica* (Asteraceae). *Plant Syst Evol* **219**, 127–135.
22. Houben A, Verlin D, Leach CR, Timmis JN. (2001) The genomic complexity of micro B chromosomes of *Brachycome dichromosomatica*. *Chromosoma* **110**, 451–459.
23. Peng SF, Lin YP, Lin BY. (2005) Characterization of AFLP sequences from regions of maize B chromosome defined by 12 B-10L translocations. *Genetics* **169**, 375–388.
24. Alfenito MR, Birchler JA. (1993) Molecular characterization of a maize B chromosome centric sequence. *Genetics* **135**, 589–97.
25. Cheng YM, Lin BY. (2003) Cloning and characterization of maize B chromosome sequences derived from microdissection. *Genetics* **164**, 299–310.
26. Cheng YM, Lin BY. (2004) Molecular organization of large fragments in the maize B chromosome: Indication of a novel repeat. *Genetics* **166**, 1947–1961.
27. Lamb JC, Riddle NC, Cheng YM, Theuri J, Birchler JA. (2007) Localization and transcription of a retrotransposon-derived element on the maize B chromosome. *Chromosome Res* **15**, 383–398.
28. Carchilan M, Kumke K, Mikolajewski S, Houben A. (2009) Rye B chromosomes are weakly transcribed and might alter the transcriptional activity of A chromosome sequences. *Chromosoma* **118**, 607–616.
29. SanMiguel P, Tikhonov A, Jin YK, Motchoulskaia N, Zakharov D, Melake-Berhan A, et al. (1996) Nested retrotransposons in the intergenic regions of the maize genome. *Science* **274**, 765–768.
30. Lo KL, Peng SF, Chen LJ, Lin BY. (2009) Tandem organization of StarkB element (22.8 kb) in the maize B chromosome. *Mol Genet Genomics* **282**, 131–139.
31. Kynast RG, Galatowitsch MW, Huettl PA, Phillips RL, Rines HW. (2007) Adding B-chromosomes of *Zea mays* L. to the genome of *Avena sativa* L. *Maize Genet Coop Newsl* **81**, 17–19.
32. McAllister BF. (1995) Isolation and characterization of a retroelement from B chromosome (PSR) in the parasitic wasp *Nasonia vitripennis*. *Insect Mol Biol* **4**, 253–262.
33. Ziegler CG, Lamatsch DK, Steinlein C, Engel W, Schartl M, Schmid M. (2003) The giant B chromosome of the cyprinid fish *Alburnus alburnus* harbours a retrotransposon-derived repetitive DNA sequence. *Chromosome Res* **11**, 23–35.
34. Cohen S, Segal D. (2009) Extrachromosomal circular DNA in eukaryotes: possible involvement in the plasticity of tandem repeats. *Cytogenet Genome Res* **124**, 327–338.
35. Cohen S, Houben A, Segal D. (2008) Extrachromosomal circular DNA derived from tandemly repeated genomic sequences in plants. *Plant J* **53**, 1027–1034.
36. Green DM. (1990) Muller's rachet and the evolution of supernumerary chromosomes. *Genome* **33**, 818–824.
37. Dhar MK, Friebe B, Koul AK, Gill BS. (2002) Origin of an apparent B chromosome by mutation, chromosome fragmentation and specific DNA sequence amplification. *Chromosoma* **111**, 332–340.
38. Csonka E, Cserpan I, Fodor K, Hollo G, Katona R, Kereso J, et al. (2000) Novel generation of human satellite DNA-based artificial chromosomes in mammalian cells. *J Cell Sci* **113**, 3207–3216.
39. Borisjuk N, Borisjuk L, Komarnytsky S, Timeva S, Hemleben V, Gleba Y, et al. (2000) Tobacco ribosomal DNA spacer element stimulates amplification and expression of heterologous genes. *Nat Biotechnol* **18**, 1303–1306.
40. Schubert I, Wobus U. (1985) In situ hybridisation confirms jumping nucleolus organizing regions in Allium. *Chromosoma* **92**, 143–148.
41. Abirached-Darmency M, Prado-Vivant E, Chelysheva L, Pouthier T. (2005) Variation in rDNA locus number and position among legume species and detection of 2 linked rDNA loci in the model *Medicago truncatula* by FISH. *Genome* **48**, 556–561.
42. Dubcovsky J, Dvorak J. (1995) Ribosomal RNA multigene loci: nomads of the Triticeae genomes. *Genetics* **140**, 1367–1377.
43. Datson PM, Murray BG. (2006) Ribosomal DNA locus evolution in Nemesia: transposition rather than structural rearrangement as the key mechanism? *Chromosome Res* **14**, 845–857.
44. Leach CR, Houben A, Field B, Pistrick K, Demidov D, Timmis JN. (2005) Molecular evidence for transcription of genes on a B chromosome in *Crepis capillaris*. *Genetics* **171**, 269–278.
45. Donald TM, Houben A, Leach CR, Timmis JN. (1997) Ribosomal RNA genes specific to the B chromosomes in *Brachycome dichromosomatica* are not transcribed in leaf tissue. *Genome* **40**, 674–681.

46. Marschner S, Meister A, Blattner FR, Houben A. (2007) Evolution and function of B chromosome 45S rDNA sequences in *Brachycome dichromosomatica*. *Genome* **50**, 638–644.
47. Kumke K, Jones RN, Houben A. (2008) B chromosomes of *Puschkinia libanotica* are characterized by a reduced level of euchromatic histone H3 methylation marks. *Cytogenet Genome Res* **121**, 266–270.
48. Marschner S, Kumke K, Houben A. (2007) B chromosomes of *B. dichromosomatica* show a reduced level of euchromatic histone H3 methylation marks. *Chromosome Res* **15**, 215–222.
49. Jin WW, Lamb JC, Zhang WL, Kolano B, Birchler JA, Jiang JM. (2008) Histone modifications associated with both A and B chromosomes of maize. *Chromosome Res* **16**, 1203–1214.
50. Carchilan M, Delgado M, Ribeiro T, Costa-Nunes P, Caperta A, Morais-Cecilio L, et al. (2007) Transcriptionally active heterochromatin in rye B chromosomes. *Plant Cell* **19**, 1738–1749.
51. Dover G. (1994) Concerted evolution, molecular drive and natural selection. *Curr Biol* **4**, 1165–1166.
52. Lim KY, Kovarik A, Matyasek R, Bezdek M, Lichtenstein CP, Leitch AR. (2000) Gene conversion of ribosomal DNA in *Nicotiana tabacum* is associated with undermethylated, decondensed and probably active gene units. *Chromosoma* **109**, 161–172.
53. Dadejova M, Lim KY, Souckova-Skalicka K, Matyasek R, Grandbastien MA, Leitch A, et al. (2007) Transcription activity of rRNA genes correlates with a tendency towards intergenomic homogenization in Nicotiana allotetraploids. *New Phytol* **174**, 658–668.
54. van Vugt J, de Nooijer S, Stouthamer R, de Jong H. (2005) NOR activity and repeat sequences of the paternal sex ratio chromosome of the parasitoid wasp Trichogramma kaykai. *Chromosoma* **114**, 410–419.
55. Sapre B, Deshpande S. (1987) Origin of B chromosomes in Coix L. through spontaneous interspecific hybridisation. *J Hered* **78**, 191–196.
56. Riddle NC, Birchler JA. (2003) Effects of reunited diverged regulatory hierarchies in allopolyploids and species hybrids. *Trends Genet* **19**, 597–600.
57. Gernand D, Rutten T, Varshney A, Rubtsova M, Prodanovic S, Bruss C, et al. (2005) Uniparental chromosome elimination at mitosis and interphase in wheat and pearl millet crosses involves micronucleus formation, progressive heterochromatinization, and DNA fragmentation. *Plant Cell* **17**, 2431–2438.
58. Perfectti F, Werren JH. (2001) The interspecific origin of B chromosomes: Experimental evidence. *Evolution* **55**, 1069–1073.
59. Zhang P, Li WL, Friebe B, Gill BS. (2008) The origin of a "zebra" chromosome in wheat suggests nonhomologous recombination as a novel mechanism for new chromosome evolution and step changes in chromosome number. *Genetics* **179**, 1169–1177.
60. Hall AE, Kettler GC, Preuss D. (2006) Dynamic evolution at pericentromeres. *Genome Res* **16**, 355–364.
61. Lysak MA, Berr A, Pecinka A, Schmidt R, McBreen K, Schubert I. (2006) Mechanisms of chromosome number reduction in *Arabidopsis thaliana* and related Brassicaceae species. *Proc Natl Acad Sci USA* **103**, 5224–5229.
62. Schubert I. (2007) Chromosome evolution. *Curr Opin Plant Biol* **10**, 109–115.
63. Patton JL. (1977) B chromosome systems in pocket mouse, *Perognathus baileyi* – Meiosis and C-band studies. *Chromosoma* **60**, 1–14.
64. Mandakova T, Lysak MA. (2008) Chromosomal phylogeny and karyotype evolution in x = 7 crucifer species (Brassicaceae). *Plant Cell* **20**, 2559–2570.
65. Gaut BS, Wright SI, Rizzon C, Dvorak J, Anderson LK. (2007) Recombination: an underappreciated factor in the evolution of plant genomes. *Nat Rev Genet* **8**, 77–84.
66. Berdnikov VA, Gorel FL, Kosterin OE, Bogdanova VS. (2003) Tertiary trisomics in the garden pea as a model of B chromosome evolution in plants. *Heredity* **91**, 577–583.
67. Margulies M, Egholm M, Altman WE, Attiya S, Bader JS, Bemben LA, et al. (2005) Genome sequencing in microfabricated high-density picolitre reactors. *Nature* **437**, 376–380.
68. Kubalakova M, Valarik M, Barto J, Vrana J, Cihalikova J, Molnar-Lang M, et al. (2003) Analysis and sorting of rye (*Secale cereale* L.) chromosomes using flow cytometry. *Genome* **46**, 893–905.
69. Matthews RB, Jones RN. (1983) Dynamics of the B chromosome polymorphism in rye. II. Estimates of parameters. *Heredity* **50**, 119–137.
70. Müntzing A. (1970) Chromosomal variation in the Lindström strain of wheat carrying accessory chromosomes in rye. *Hereditas* **66**, 279–286.
71. Niwa K, Horiuchi G, Hirai Y. (1997) Production and characterization of common wheat with B chromosomes of rye from Korea. *Hereditas* **126**, 139–146.

72. Endo TR, Nasuda S, Jones N, Dou Q, Akahori A, Wakimoto M, et al. (2008) Dissection of rye B chromosomes, and nondisjunction properties of the dissected segments in a common wheat background. *Genes Genet Syst* **83**, 23–30.
73. Puertas MJ, Romera F, Delapena A. (1985) Comparison of B-chromosome effects on *Secale cereale* and *Secale vavilovii. Heredity* **55**, 229–234.
74. Romera F, Jimenez MM, Puertas MJ. (1991) Genetic control of the rate of transmission of rye B chromosomes .1. Effects in 2Bx0B crosses. *Heredity* **66**, 61–65.
75. Müntzing A. (1948) Cytological studies of extra fragment chromosomes in rye. V. A new fragment type arisen by deletion. *Hereditas* **34**, 435–442.
76. Lima-de-Faria A. (1962) Genetic interaction in rye expressed at the chromosome phenotype. *Genetics* **47**, 1455–1462.
77. Roman H. (1948) Selective fertilization in maize. *Genetics* **33**, 122.
78. Roman H. (1948) Directed fertilization in maize. *Proc Natl Acad Sci USA* **34**, 36–42.
79. Carlson WR. (1969) Factors affecting preferential fertilization in maize. *Genetics* **62**, 543–554.
80. Rusche ML, Mogensen HL, Shi L, Keim P, Rougier M, Chaboud A, et al. (1997) B chromosome behavior in maize pollen as determined by a molecular probe. *Genetics* **147**, 1915–1921.
81. Carlson W. (2009) The B chromosome of maize. In: J.L. Bennetzen, S. Hake (eds), *Maize Handbook – Volume II: Genetics and Genomics*, Springer, New York. pp. 459–480.
82. Carlson WR, Roseman RR. (1992) A new property of the maize B-chromosome. *Genetics* **131**, 211–223.
83. Roman H. (1947) Mitotic nondisjunction in the case of interchanges involving the B-type chromosome in maize. *Genetics* **32**, 391–409.
84. Carlson WR. (1978) B-chromosome of corn. *Ann Rev Genet* **12**, 5–23.
85. Lamb JC, Han F, Auger DL, Birchler J. (2006) A trans-acting factor required for nondisjunction of the B chromosome is located distal to the TB-4Lb breakpoint on the B chromosome. *Maize Genet Coop Newsl* **80**, 51–54.
86. Chiavarino AM, Rosato M, Manzanero S, Jimenez G, Gonzalez-Sanchez M, Puertas MJ. (2000) Chromosome nondisjunction and instabilities in tapetal cells are affected by B chromosomes in maize. *Genetics* **155**, 889–897.
87. Gonzalez-Sanchez M, Rosato M, Chiavarino M, Puertas MJ. (2004) Chromosome instabilities and programmed cell death in tapetal cells of maize with B chromosomes and effects on pollen viability. *Genetics* **166**, 999–1009.
88. Gonzalez-Sanchez M, Gonzalez-Gonzalez E, Molina F, Chiavarino AM, Rosato M, Puertas MJ. (2003) One gene determines maize B chromosome accumulation by preferential fertilisation; another gene(s) determines their meiotic loss. *Heredity* **90**, 122–129.
89. Carlson WR. (2007) Locating a site on the maize B chromosome that controls preferential fertilization. *Genome* **50**, 578–587.
90. Volpe T, et al. (2003) RNA interference is required for normal centromere function in fission yeast. *Chromosome Res* **11**, 137–146.
91. Pidoux AL, Allshire R. (2005) The role of heterochromatin in centromere function. *Philos Trans R Soc Lond B Biol Sci* **360**, 569–579.
92. Vos LJ, Famulski JK, Chan GK. (2006) How to build a centromere: from centromeric and pericentromeric chromatin to kinetochore assembly. *Biochem Cell Biol* **84**, 619–639.
93. Bouzinba-Segard H, Guais A, Francastel C. (2006) Accumulation of small murine minor satellite transcripts leads to impaired centromeric architecture and function. *Proc Natl Acad Sci USA* **103**, 8709–8714.
94. Topp CN, Zhong CX, Dawe RK. (2004) Centromere-encoded RNAs are integral components of the maize kinetochore. *Proc Natl Acad Sci USA* **101**, 15986–15991.
95. May BP, Lippman ZB, Fang Y, Spector DL, Martienssen RA. (2005) Differential regulation of strand-specific transcripts from Arabidopsis centromeric satellite repeats. *PLoS Genet* **1**, e79.
96. Kaszas E, Birchler JA. (1996) Misdivision analysis of centromere structure in maize. *EMBO J* **15**, 5246–5255.
97. Kaszas E, Birchler JA. (1998) Meiotic transmission rates correlate with physical features of rearranged centromeres in maize. *Genetics* **150**, 1683–1692.
98. Kaszas E, Kato A, Birchler JA. (2002) Cytological and molecular analysis of centromere misdivision in maize. *Genome* **45**, 759–768.
99. Jin WW, Lamb JC, Vega JM, Dawe RK, Birchler JA, Jiang J. (2005) Molecular and functional dissection of the maize B chromosome centromere. *Plant Cell* **17**, 1412–1423.

100. Han F, Lamb JC, Birchler JA. (2006) High frequency of centromere inactivation resulting in stable dicentric chromosomes of maize. *Proc Natl Acad Sci USA* **103**, 3238–3243.
101. Lamb JC, Kato A, Birchler JA. (2005) Sequences associated with A chromosome centromeres are present throughout the maize B chromosome. *Chromosoma* **113**, 337–349.
102. Wilkes TM, Francki MG, Langridge P, Karp A, Jones RN, Forster JW. (1995) Analysis of rye B-chromosome structure using fluorescence in situ hybridization (FISH). *Chromosome Res* **3**, 466–472.
103. Leach CR, Donald TM, Franks TK, Spiniello SS, Hanrahan CF, Timmis JN. (1995) Organization and origin of a B chromosome centromeric sequence from *Brachycome dichromosomatica*. *Chromosoma* **103**, 708–714.
104. Franks TK, Houben A, Leach CR, Timmis JN. (1996) The molecular organisation of a B chromosome tandem repeat sequence from *Brachycome dichromosomatica*. *Chromosoma* **105**, 223–230.
105. Evans GM, Davies EW. (1985) The genetics of meiotic chromosome pairing in *Lolium temulentum* × *Lolium perenne* tetraploids. *Theor Appl Genet* **71**, 185–192.
106. Tanaka M, Kawahara T. (1982) Cytogenetical effects of B chromosomes in plants – a review. *Rep Plant Germplasm Inst Kyoto Univ* **5**, 1–18.
107. Ruiz-Rejon M, Posse F, Oliver JL. (1980) The B-chromosome system of *Scilla autumnalis* (Liliaceae) – Effects at the isoenzyme level. *Chromosoma* **79**, 341–348.
108. Bang JW, Choi HW. (1990) Genetic analysis of esterase izozymes in rye (*Secale cereale* L.). *Korean J Genet* **12**, 87–94.
109. Teruel M, Cabrero J, Perfectti F, Camacho JP. (2007) Nucleolus size variation during meiosis and NOR activity of a B chromosome in the grasshopper *Eyprepocnemis plorans*. *Chromosome Res* **15**, 755–765.
110. Slotkin RK, Martienssen R. (2007) Transposable elements and the epigenetic regulation of the genome. *Nat Rev Genet* **8**, 272–285.
111. Matzke MA, Aufsatz W, Kanno T, Mette MF, Matzke AJ. (2002) Homology-dependent gene silencing and host defense in plants. *Adv Genet* **46**, 235–275.
112. Comai L. (2005) The advantages and disadvantages of being polyploid. *Nat Rev Genet* **6**, 836–846.
113. Misteli T. (2007) Beyond the sequence: cellular organization of genome function. *Cell* **128**, 787–800.
114. Delgado M, Caperta A, Ribeiro T, Viegas W, Jones RN, Morais-Cecilio L. (2004) Different numbers of rye B chromosomes induce identical compaction changes in distinct A chromosome domains. *Cytogenet Genome Res* **106**, 320–324.
115. Lemos B, Araripe LO, Hartl DL. (2008) Polymorphic Y chromosomes harbor cryptic variation with manifold functional consequences. *Science* **319**, 91–93.
116. Birchler JA, Gao Z, Han F. (2009) A tale of two centromeres – diversity of structure but conservation of function in plants and animals. *Funct Integr Genomics* **9**, 7–13.
117. Yu W, Han F, Birchler JA. (2007) Engineered minichromosomes in plants. *Curr Opin Biotechnol* **18**, 425–431.
118. Yu W, Han F, Gao Z, Vega JM, Birchler JA. (2007) Construction and behavior of engineered minichromosomes in maize. *Proc Natl Acad Sci USA* **104**, 8924–8929.
119. Yu WC, Lamb JC, Han FP, Birchler JA. (2006) Telomere-mediated chromosomal truncation in maize. *Proc Natl Acad Sci USA* **103**, 17331–17336.
120. Birchler JA, Yu W, Han F. (2008) Plant engineered minichromosomes and artificial chromosome platforms. *Cytogenet Genome Res* **120**, 228–232.
121. Houben A, Schubert I. (2007) Engineered plant minichromosomes: a resurrection of B chromosomes? *Plant Cell* **19**, 2323–2327.

Chapter 6

Telomere Truncation in Plants

Chunhui Xu and Weichang Yu

Abstract

Telomeres are highly repetitive sequences at the ends of chromosomes that act as protection structure for chromosome stability. The integration of telomere sequences into the genome by genetic transformation can create chromosome instability because the integrated telomere sequences tend to form de novo telomeres at the site of integration. Thus, telomere repeats can be used to generate minichromosomes by telomere-mediated chromosome truncation in both plants and animals for chromosome studies as well as the applications in genetic engineering as engineered minichromosomes or artificial chromosomes. This protocol describes the procedure for telomere truncation of maize chromosomes by genetic transformation of telomere-containing constructs by both *Agrobacterium*- and biolistic-mediated transformations.

Key words: Maize, Telomere, Telomere truncation, Genetic transformation, Chromosome, Minichromosome

1. Introduction

Telomere-mediated chromosomal truncation has been successfully used in the creation of minichromosomes as potential vectors for gene therapy in medicine and multiple gene expression platforms for genetic engineering in agriculture (1, 2). Telomeres are the ends of chromosomes that ensure the stability of linear structures. Telomeric DNA consists of highly conserved tandem repeats. It has been found in mammals that cloned telomere sequence could be reintroduced into the cells (3), and DNA replication would terminate at the location where the telomere sequence integrated and produce functional new telomeres (4). Thus, by introducing telomere-containing sequences, chromosome truncation can be achieved. This technology has been used in the creation of human minichromosomes (5–9). Similarly, a conserved telomere repeat

James A. Birchler (ed.), *Plant Chromosome Engineering: Methods and Protocols*, Methods in Molecular Biology, vol. 701, DOI 10.1007/978-1-61737-957-4_6, © Springer Science+Business Media, LLC 2011

in plants (TTTAGGG) has been manipulated to create chromosome truncations in maize (10). This technology has been successfully used in the construction of maize minichromosomes and can be applied to the generation of minichromosomes for other plants because the telomere sequence in plants is conserved. The maize minichromosomes can be inherited stably in the offspring (11, 12). In this protocol, we describe the procedure for maize chromosome truncation with telomere-containing constructs (11–13) by *Agrobacteria* and bombardment transformations.

2. Materials

2.1. Maize Embryo Isolation and Callus Induction

1. Ears of maize Hi-II hybrid, gathered from greenhouse at 10–12 days after pollination.
2. Sterile water.
3. Bleach, commercial bleach containing 5.25% sodium hypochlorite.
4. Tween-20.
5. Callus initiation medium (PHI-C) (14): Add 4.0 g of N6 salts (Sigma) (15), 30 g of sucrose, 2.76 g of l-proline, 0.1 g of casein hydrolysate, 0.1 g of myo-inositol, 2.0 ml of 1 mg/ml 2,4-D into 0.8 l H_2O, adjust pH to 5.8 with KOH, add H_2O to 1 l, and add 2.5 g of gelrite (Sigma). Autoclave for 20 min, cool to 55°C, and add 1 ml of 1,000× N6 vitamins and 0.1 ml of 8.5 mg/ml silver nitrate (both are filter sterilized). Pour plates (see Note 1).

2.2. Maize Transformation

2.2.1. Plasmids and Agrobacteria

1. Telomere-containing constructs: pWY76, pWY86, and a control construct pWY96 without telomere sequence (Fig. 1) as reported (11).
2. *Agrobacterium tumefaciens* strain EHA101 (16) carrying the respective plasmids.

2.2.2. Agrobacterium-Mediated Maize Transformation

1. YEP medium: Dissolve 10 g of trypton, 10 g of yeast extract, and 5 g of sodium chloride in H_2O, and make up the volume to 1 l. Autoclave for 20 min. For the solid medium, add 15 g of agar before autoclaving.
2. PHI-A (14): Add 2.0 g of N6 salts, 68.5 g of sucrose, 36.0 g of glucose, 0.7 g of l-proline, 0.5 g of MES [2-(*N*-Morpholino) ethanesulfonic acid sodium salt] (Sigma), 0.1 g of myo-inositol, 1.5 ml of 1 mg/ml 2,4-D, 1 ml of 1,000× N6 vitamins, and 0.1 ml of 8.5 mg/ml silver nitrate into 0.8 l of H_2O, adjust pH to 5.2 with KOH, add H_2O to 1 l, and filter-sterilize. Add 1 ml of 100 mM acetosyringone (Sigma) before use.

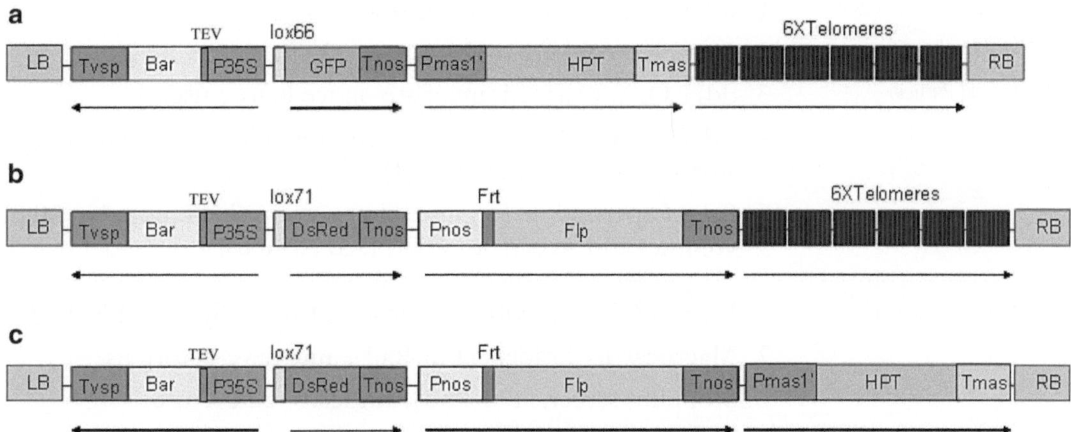

Fig. 1. Constructs for telomere truncation. Telomere truncation constructs pWY76 (**a**), pWY86 (**b**), and the control construct pWY96 (**c**). LB, T-DNA left border; RB, T-DNA right border; Tvsp, terminator from soybean vegetative storage protein gene; Bar, bialaphos resistance gene as a selection marker gene; TEV, tobacco etch virus 5′ untranslated region; P35S, 2× 35 S promoter from cauliflower mosaic virus; Tnos, Nos terminator from *Agrobacterium*; Tmas, Mas terminator from *Agrobacterium*; Pnos, Nos promoter from *Agrobacterium*; Pmas1′, Mas promoter from *Agrobacterium*; lox and FRT, site-specific recombination sites; HPT, hygromycin B resistance gene; GFP, green fluorescent protein gene; DsRed, red fluorescent protein gene; FLP, recombinase gene; Telomeres, telomere units of pAtT4 isolated from *Arabidopsis thaliana*. Modified from ref. (11), Copyright 2006 National Academy of Sciences, USA.

3. PHI-B (14): Add 2.0 g of N6 salts, 30 g of sucrose, 0.7 g of l-proline, 0.5 g of MES, 0.1 g of myo-inositol, 1.5 ml of 1 mg/ml 2,4-D into 0.8 l of H_2O, adjust pH to 5.8 with KOH, add H_2O to 1 l, and add 8 g of agar. Autoclave for 20 min, cool to 55°C, and add 1 ml of 1,000× N6 vitamins, 0.1 ml of 8.5 mg/ml silver nitrate, 0.4 g of l-cysteine, 1 ml of 100 mM acetosyringone. Pour plates.
4. PHI-C: The same as used for callus induction.
5. Selection medium I: PHI-C medium plus 250 mg/l cefotaxime and 1.5 mg/l bialaphos (Gold Biotechnology, St. Louis, MO).
6. Selection medium II: the same as selection medium I, except the bialaphos concentration is 3 mg/l.
7. Regeneration medium: Add 4.3 g of MS salts (Sigma) (17), 60 g of sucrose, and 0.1 g of myo-inositol into 0.8 l of H_2O, adjust pH to 5.8 with KOH, add H_2O to 1 l, and add 3 g of gelrite. Autoclave for 20 min, cool to 55°C, and add 1 ml of 1,000× MS vitamins, 2.5 ml of 100 mg/ml cefotaxime, and 0.5 ml of 3 mg/ml bialaphos. Pour plates.
8. Rooting medium: Add 2.9 g of MS salts, 30 g of sucrose, 0.1 g of myo-inositol into 0.8 l of H_2O, adjust pH to 5.8 with KOH, add H_2O to 1 l, and add 3 g of gelrite. Autoclave for 20 min, cool to 55°C, and add 1 ml of 1,000× MS vitamins. Pour into culture vessels.

2.2.3. Biolistic Transformation

2.2.3.1. Gold Particle Preparation

1. 0.6 μm Gold particles (Bio-Rad).
2. 100% Ice-cold ethanol, store at –20°C (see Note 2).
3. ddH_2O, autoclaved, incubate on ice before use.
4. 2.5 M $CaCl_2$, filter-sterilized, aliquoted in 50-μl volume, and stored at –20°C (see Note 3).
5. 0.1 M Spermidine (Sigma), filter-sterilized, aliquoted in 20 μl volume, and stored at –20°C (see Note 3).

2.2.3.2. Bombardment

1. Biolistic® PDS-1,000/He Particle Delivery System (Bio-Rad).
2. Macrocarrier holders (Bio-Rad), autoclave before use.
3. Stopping screen (Bio-Rad), autoclave before use.
4. 70% Ethanol.
5. 70% Isopropanol.
6. Rupture disk (Bio-Rad, 650 psi), sterilized with 70% isopropanol in an autoclaved beaker for at least 15 min before use.
7. Macrocarriers (Bio-Rad), sterilized with 70% ethanol in an autoclaved beaker for at least 15 min before use.
8. Waterman filter papers, autoclaved.

2.2.3.3. Culture Media

1. Osmotic medium: Add 4.0 g of N6 salts, 30 g of sucrose, 36.4 g of mannitol, 36.4 g of sorbitol, 0.7 g of l-proline, 0.1 g of myo-inositol, 2.0 ml of 1 mg/ml 2,4-D into 0.8 l of H_2O, adjust pH to 5.8 with KOH, add H_2O to 1 l, and add 2.5 g of gelrite. Autoclave for 20 min, cool to 55°C, and add 1 ml of 1,000× N6 vitamins and 0.5 ml of 8.5 mg/ml silver nitrate. Pour plates.
2. All other media used are the same as those in *Agrobacterium*-mediated transformation, except that the cefotaxime is eliminated.

2.3. FISH

2.3.1. Chromosome Preparation

1. Nitrous oxide gas chamber (18) (Fig. 2).
2. 90% (Store at 4°C) and 100% Acetic acid.
3. 70% Ethanol.
4. 5× Citric buffer: Add 14.7 g of sodium citrate (50 mM) and 14.6 g of EDTA (50 mM) into 0.8 l of H_2O, adjust to pH 5.5 with citric acid, add H_2O to 1 l.
5. Enzyme solution: Make on ice bath. Weight 0.1 g of pectolyase Y-23 (Yakult Pharmaceutical Ind. Co. Ltd, Tokyo, Japan), 0.2 g of cellulose Onozuka R-10 (Yakult), and 9.7 g of 5× citric buffer. Mix by stirring on ice to completely dissolve the enzymes, and make up the volume to 10 ml with H_2O. Quickly aliquot into 0.5-ml tubes, 20 μl in each tube. Store at –20°C.

Fig. 2. Gas chamber for nitrous oxide treatment of root tips. Designed and manufactured by the University of Missouri Instrument Shop (Columbia, MO). Scale bar = 1 cm.

6. 2% Aceto-orcein: Dissolve 2.0 g of orcein in 55 ml boiling acetic acid. Cool the solution, add 45 ml of distilled water, and filter (see Note 4).

2.3.2. Probe Labeling

1. 10× Nick translation buffer: Dissolve 6.05 g of Tris base (500 mM) and 476.0 mg of $MgCl_2$ (50 mM) in 60 ml of water, adjust the pH to 7.8 with HCl, and make up the volume to 100 ml. Add 701.0 μl of 2-mercaptoethanol (100 mM) (Sigma) and 100 μg/ml of bovine serum albumin fraction V (Sigma).
2. Labeled dNTPs: ChromaTide Alexa Fluor 594-5-dUTP or Alexa Fluor 488-5-dUTP (Invitrogen).
3. DNA polymerase I (10 U/μl) (Invitrogen).
4. DNase I (100 mU/μl): First, prepare 2× DNase buffer by 1:5 diluting from 10× Nick translation buffer. Then, dissolve 2,000 U DNase I (Roche, grade I) in 0.5 ml of 2× DNase buffer on ice, add 0.5 ml of glycerol and mix gently, store this 2 U/μl solution at −20°C. Dilute 10 μl of 2 U/μl DNase I solution in 190 μl 1× DNase buffer (with 50% glycerol) to make 100 mU/μl working stock.
5. 10× TE: Dissolve 12.1 g of Tris base and 2.9 g of EDTA in H_2O, adjust to pH 7.8 with HCl, and make up the volume to 1 l.
6. Salmon sperm DNA (Sigma): Dissolve salmon sperm DNA in 1× TE at the concentration of 10 mg/ml and shake on a shaker overnight. Autoclave for 30 min. Run sample on agarose gel to check the length of the DNA fragments. The proper size should be 100–300 bp. Sonication or passing

several times through a 17-gauge needle may be needed to shear the DNA to desirable size.

7. 50× TAE: Dissolve 242.0 g of Tris base with H_2O, add 57.1 ml of glacial acetic acid and 100 ml 0.5 M EDTA (pH 8.0), and make up the volume to 1 l.
8. 3 M Sodium acetate: Dissolve 49.2 g of sodium acetate in H_2O, adjust pH to 5.2 with glacial acetic acid, and make up the volume to 200 ml.
9. 20× SSC: Dissolve 175.3 g of NaCl and 88.2 g of sodium citrate in H_2O, adjust the pH to 7.0 with HCl, and add H_2O to make 1 l in volume.

2.3.3. Signal Detection

1. Coplin jar.
2. Vectashield containing 4′,6-diamidino-2-phenylindole (DAPI) (Vector Laboratories, Burlingame, CA).

2.4. Southern Blot Analysis

2.4.1. Genomic DNA Preparation

1. Plant DNAzol® Reagent (Invitrogen).
2. 100% Ethanol.
3. 1× TE buffer.
4. Chloroform.
5. 1× and 50× TAE (see Subheading 2.3.2.7).
6. UltraPure Agarose (Invitrogen).
7. Restriction enzymes: *Hin*dIII, *Eco*RV, and *Sma*I (New England BioLabs).
8. Denaturing buffer: Dissolve 20 g of NaOH (0.5 N) and 87.6 g of NaCl (1.5 M) in H_2O, and make up to the volume of 1 l.
9. Neutralizing buffer: Dissolve 63.50 g of Tris base (0.5 M) and 87.60 g of NaCl (1.5 M) in H_2O, adjust pH to 7.5 with HCl, and make up to the volume of 1 l.
10. Transfer buffer: 10× SSC. Dilute from 20× SSC.
11. Qiabiane nylon membrane (QIAGEN, Valencia, CA).

2.4.2. Prehybridization, Probe Labeling, and Hybridization

1. 20× SSC (Subheading 2.3.2.9).
2. 20% SDS.
3. Prime-It II Random Primer Labeling Kit (Stratagene).
4. ^{32}P-dCTP (Perkin-Elmer).
5. 1 M Tris–HCl (pH 7.5 and 8.0): Dissolve 121.1 g of Tris base in 800 ml of H_2O, adjust pH to 7.5 or 8.0 with HCl. Make up the volume to 1 l and autoclave.
6. 0.5 M EDTA (pH 8.0): Dissolve 186.1 g of EDTA in 800 ml of H_2O. Adjust the pH to 8.0 with NaOH pellets (about 20 g), make up the volume to 1 l, and autoclave.

7. Prehybridization buffer: Mix 70 ml of H_2O, 5.0 ml of 1 M Tris–HCl (pH 8.0), 2.0 ml of 0.5 M EDTA (pH 8.0), 25 ml of 20× SSC, 1 ml of 100× Denhardt's solution, and 1.0 ml of 20% SDS. Add 1 ml of denatured 10 mg/ml salmon sperm DNA (Subheading 2.3.2.6) before use.
8. Hybridization buffer: Mix 50 ml of H_2O, 20 ml of 50% dextran sulfate, 5 ml of 1 M Tris–HCl (pH 8.0), 2 ml of 0.5 M EDTA (pH 8.0), 25 ml of 20× SSC, 1 ml of 100× Denhardt's solution, and 1 ml of 20% SDS. Add 1 ml of denatured 10 mg/ml salmon sperm DNA before use.

3. Methods

3.1. Maize Embryo Isolation and Callus Induction

1. Surface-sterilize dehusked maize Hi-II ears by submerging in 50% commercial bleach supplemented with 0.1% Tween-20 in a big glass jar for 20 min. Wash the ears with sterile water for three times.
2. In a large sterile petri dish (150× 15 mm), insert the tip of a straight forceps into one end of the cob to secure the ear, cut off the top-half of the kernels, and pick out the 1.2–1.5 mm immature embryos with a sterile spatula.
3. The embryos can be used directly for transformation by *Agrobacteria* or bombardment.
4. The embryos can be cultured on PHI-C medium for callus initiation. After culturing on PHI-C medium in the dark at 28°C for 2–4 weeks, embryonic calli will emerge. Subculture the calli with new PHI-C medium every 2–3 weeks. The calli can be used as explants for transformation by either *Agrobacteria* or bombardment.

3.2. Maize Transformation

3.2.1. Agrobacterium-Mediated Transformation

1. Streak *Agrobacteria* carrying the telomere truncating plasmids stored in glycerol stock at −80°C on YEP solid medium containing antibiotics. Culture at 28°C for 2 days to let single colonies develop.
2. Pick a single colony and streak it on YEP medium containing antibiotics. Culture at 22°C for 2–3 days.
3. Take two full loops of bacteria from the plate and suspend in 5 ml of PHI-A in a 14-ml Falcon tube. Shake vigorously to suspend the bacteria well.
4. Adjust the OD_{550} of the bacteria suspension to 0.35.
5. Culture at room temperature in a shaker at 150 rpm for 4 h.
6. Aliquot 1-ml suspension into 1.5-ml microcentrifuge tubes.
7. Put about 100 immature embryos or calli into one tube. Invert the tube 20 times to mix the explants and bacteria.

8. Stand the tube for 10 min. Pour into a plate containing PHI-B medium that was overlaid with sterile filter paper. Draw off the bacteria, lay the embryos facing down on the filter.
9. Culture in dark at 22°C for 2 days.
10. Transfer the explants to PHI-C medium supplemented with antibiotics to inhibit the *Agrobacteria* growth and culture in dark at 28°C for 7 days. The infected materials can be transferred onto selection medium for selection.

3.2.2. Biolistic-Mediated Maize Transformation

3.2.2.1. Washing of Gold Particles

1. Weigh 15 mg of 0.6 μm gold particles in a 1.5-ml siliconized sterile microcentrifuge tube. This is 10× quantities.
2. Add 500 μl of cold absolute ethanol to the tube and sonicate in water bath for 15 s with an Ultrasonic Jewelry Cleaner.
3. Close the tube and gather the droplets to the bottom of the tube by tapping it on the bench. Set the tube in an ice bath to let the particles settle (about 10–20 min).
4. Centrifuge at 3,000 rpm (850 g) for 60 s and remove the supernatant with sediments facing down to avoid being disturbed. Add 1 ml of ice-cold sterile water, drop the water down along the sides of the tube.
5. Finger-vortex well and let the particles settle out in an ice bath and then centrifuge at 3,000 rpm (850 g) for 60 s. Repeat the rinse steps for another two times and centrifuge at 5,000 rpm (2,400 g) for 15 s at the last time.
6. Remove the supernatant, add 500 μl of cold sterile H_2O, finger-vortex, sonicate for 15 s, and then place the tube on a vortexer. Keep the vortexer shaking during the next step.
7. Prepare ten 1.5-ml tubes. Aliquot 25 μl of gold suspension into every tube from the first one to the last one, then another 25 μl from the last one to the first one. This will ensure that the quantities of the gold particles in each tube are equal.
8. Store the washed gold particles at –20°C.

3.2.2.2. Osmotic Treatment of the Explants

1. Place the osmotic plate on the center of the target plate shelf of the gene gun and draw a circle with a marker on the bottom of the plate along the center hole of the target shelf.
2. Transfer the immature embryos or calli that have been subcultured for 7–10 days onto the osmotic medium in the area of the circle. We usually prepare ten plates one time. Treat for 4 h before bombardment.

3.2.2.3. Coating of Gold Particles

1. At 1 h before bombardment, hand-thaw washed gold particles, $CaCl_2$, and spermidine, each one tube, for about ten plates to be bombarded, and put them in an ice bath.

2. Ultrasonicate the gold particles for 15 s in a water bath and then add 1 μg of the plasmid containing telomere repeat. Finger-vortex to mix.
3. Tap to gather the droplets at the bottom of the tube. Add 50 μl of 2.5 M $CaCl_2$ and suck the suspension up and down once with a pipette. Keep the tube shaking on a vortexer and add 20 μl of 0.1 M spermidine. Close the tube and finger-vortex well. Then, return to the vortexer and shake for 10 min.
4. Put the tube in an ice bath and let the gold particles settle out for 3–5 min. Centrifuge at 5,000 rpm (2,400 g) for 10 s. Remove the supernatant. Add 250 μl of ice-cold absolute ethanol, and finger-vortex well (see Note 5) until the suspension looks silty.
5. Centrifuge at 5,000 rpm (2,400 g) for 10 s. Remove the supernatant and add 140 μl of cold absolute ethanol. Finger-vortex thoroughly and place it in ice bath before bombardment.

3.2.2.4. Bombardment

1. Clean the bombardment chamber wall, rupture disk retaining cap, gold particle launch assembly, and target plate shelf with 70% ethanol.
2. In the hood, surface-sterilize rupture disk and macrocarrier in 70% isopropanol and 70% ethanol, respectively, for at least 15 min, and then spread them on sterilized filter papers and let them dry (15–30 min).
3. Carefully fix the macrocarriers into macrocarrier holders in a sterile plate. Vortex the coated gold particles briefly; quickly pipette 10 μl and add onto the macrocarrier by moving the tip spirally from the center outward until it arrives at the edge of the hole of the holder. This will ensure even distribution of the gold particles.
4. Place the rupture disk into the recess of retaining cap and screw it firmly onto the end of the gas acceleration tube at the top of the bombardment chamber.
5. Place a stopping screen onto the stopping screen support. After the macrocarrier has been dried, fix it into the launch assembly with the gold particle toward the stopping screen. Cover the macrocarrier cover lid and place the launch assembly into the top slot of the bombardment chamber with the gold particles facing down.
6. Place the target plate onto the target shelf, fit the circle on the bottom to the center hole of the shelf, and remove the cover of the plate. Place them into the 6-cm slot of the chamber. Close the door of bombardment chamber.
7. Open the valve of helium tank. Turn the adjustment handle of the helium regulator until the pressure is 200 psi over the burst pressure of the rupture disk.

8. Power on the vacuum pump and the gene gun, turn the vacuum switch to VAC position, and evacuate the chamber to 28 mm of mercury. Then, quickly press the vacuum switch to the bottom HOLD position, and as a result, the pressure of the bombardment chamber will be held.
9. Press and hold the FIRE switch. The pressure in the gas acceleration tube will increase until the rupture disk bursts. Release the FIRE switch. Turn vacuum switch to VEN position. Wait until the pressure in the bombardment chamber decrease to nearly zero.
10. Open the door of bombardment chamber, pull out the target plate, and cover the plate. Replace the rupture disk, stopping screen, and the macrocarrier and bombard the next plate.
11. After all the bombardment have been finished, clean the bombardment chamber. Close the valve of the helium tank and evacuate the bombardment chamber again. Press the vacuum switch to HOLD and press the FIRE switch repeatedly until the pressure in the gas acceleration tube decreases to zero. Turn off the system.
12. Keep the bombarded explants on the osmotic medium overnight in the dark before transferring them onto selection medium.

3.2.2.5. Selection and Regeneration

1. Subculture the transformed embryos or calli onto selection medium I with 1.5 mg/l bialaphos. Place the flat face of the embryos facing down. Culture in the dark at 28°C for 2 weeks.
2. Subculture in the selection medium II containing 3.0 mg/l bialaphos. Culture in dark at 28°C every 2 weeks until bialaphos-resistant calli grow large enough.
3. The positive calli then could be transferred to regeneration medium. Culture in dark at 25°C for 2–3 weeks.
4. Transfer the calli to rooting medium. Culture at 25°C under 16 h light and 8 h dark for 2 weeks. Transfer the plantlets to new rooting medium in culture tube and culture in the same condition until the plants are big enough.
5. Take the plantlets out of the agar medium, wash the roots with tap water to get rid of the culture media (19) and then transfer the plants into soil; place the plants in greenhouse and be careful to keep moisture for the plants for the first few days before lifting off the cover.
6. Transgenic plants are kept in the greenhouse until maturity and screened by FISH and Southern analysis.
7. A first sign of chromosome truncation may be provided by a pollen abortion test (20), in which freshly shed pollen is

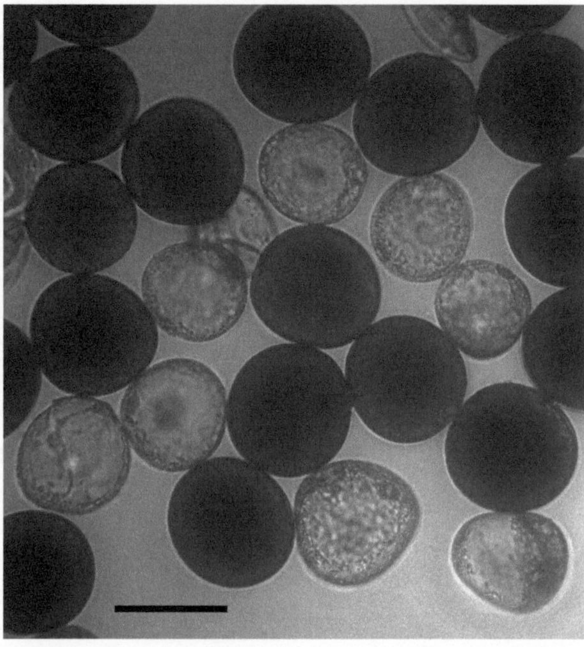

Fig. 3. Pollen abortion. Normal pollen is black. Aborted pollen is doughnut-shaped and transparent. Scale bar = 100 μm.

checked with a pocket microscope. Aborted pollens are doughnut-shaped and transparent, while normal ones look black (Fig. 3).

3.3. FISH

3.3.1. Chromosome Preparation

1. Root tips from transgenic plants are collected from vigorously growing newly emerged roots. Pick the root tips of 1–2 cm long roots, place into 0.5 ml tubes, and spray water into tubes to keep it moist. Punch a hole in the top of each tube and place them into the nitrous oxide gas chamber. Close the lid of the chamber tightly.
2. Connect the gas chamber with the nitrous oxide tank and adjust the pressure in the chamber higher than 800 kPa.
3. Wait for 1 min and close the valve of the chamber tightly. Turn off the gas tank and release the pressure in the vessel between the chamber and the tank.
4. Treat the root tips with high-pressure nitrous oxide for 2–3 h.
5. Add ice-cold 90% acetic acid into each tube. Incubate in ice bath for 10 min to fix the cells.
6. Remove 90% acetic acid. The root tips could now be digested with enzyme. Or, add 70% ethanol and store them at –20°C.

7. Thaw and place the enzyme solution in ice bath. Wash the root tips with H_2O. Roll the root tip on filter paper to remove the sticky transparent stuff. Cut and discard the root cap.
8. Cut the roots at 2 mm to the tip. Put the 2-mm fragment into enzyme solution, one to two root tips in 20 μl of enzyme solution. This fragment contains the growing point.
9. Digest in 37°C water bath for 60 min. Return to ice bath.
10. Vortex the root tips for about 15 s. If the root tips were sufficiently digested, the cells could be totally separated after vortex, and only a transparent ring, which was the residue of the root tip, will be left in the tube. Pick the residues out and discard.
11. Briefly centrifuge at 5,000–6,000 rpm (2,400–3,400 g) for 15 s. Remove supernatant and add 200 μl of 70% ethanol. Vortex for 15 s to resuspend the protoplasts.
12. Centrifuge at 5,000–6,000 rpm (2,400–3,400 g) for 15 s. Remove supernatant and add 30 μl of 100% acetic acid. Vortex for 15 s to resuspend the protoplasts.
13. Prepare a humidity box with layers of paper tissue on the bottom. Wet the paper and the box by spraying water into the box.
14. Place some slides on the paper tissue and make sure that the slides are horizontal. Drop 5 μl of protoplast suspension onto each slide. Cover the box until the acetic acid stops expanding. Open the box and wait until all of the acetic acid vaporizes. The protoplast will be broken at this step and chromosomes will be released from the nucleus.
15. Check the situation of the dividing phase by staining with 2% aceto-orcein. If satisfied, go to the next step with unstained slides.
16. Place the slides directly under UV light in a UV cross-linker. Expose with an optimized energy (120 mJ/cm^2) to cross-link the chromosome with the slides.

3.3.2. Probe Labeling and Purification

1. Prepare the pWY96 plasmid with a QIAGEN miniprep kit. Label it by nick translation with Alexa Fluor 594-dUTP as a FISH probe for detection of transgenes because pWY96 has very high homology with pWY76 and pWY86 constructs except lacking the telomere sequence. Avoid light in the following steps by using dark tubes or wrapping the tubes with aluminum foil.
2. On ice, mix 2 μg of purified plasmid, 2.0 μl of 10× nick translation buffer, 0.5 μl of 1 mM a labeled dNTP, and 2.0 μl of 2 mM each nonlabeled dNTPs. Then, add 8.0 μl of 10 U/μl DNA polymerase I and 0.4 μl of 100 mU/μl DNase I and mix by gently pipetting; do not vortex.
3. Incubate at 15°C for 2.5 h.

4. Add 175 μl of 5× TAE (pH 5.2) + 140 ng/μl autoclaved salmon sperm DNA. Vortex and transfer to a 1.5-ml tube and incubate on ice.
5. Add 1/10 volume of 3 M sodium acetate (pH 5.2) and two volumes of 100% ethanol. Mix by vortexing. Store at −20°C for at least 2 h or overnight to precipitate DNA.
6. Centrifuge at 15,000 rpm (22,000 g) for 30 min. Remove the supernatant.
7. Wash the DNA with 70% ethanol. Remove 70% ethanol and wash with 100% ethanol. Remove the ethanol, but beware of not to lose the labeled DNA pellet.
8. Air-dry for 30 min in total darkness. Dissolve the labeled DNA in 100 μl of 2× SSC. The concentration is 20 ng/μl. Use 5 μl (100 ng) per slide in the following hybridization. The transgene detection probe can be mixed with karyotyping probes (21) in the hybridization.

3.3.3. Signal Detection

1. Preheat a food storage container with a few layers of Kimwipes wet with 2× SSC at 55°C. Prepare a boiling water vapor bath. Do not let the chromosomes dry in the following steps.
2. Add 5–10 μl of probes to each slide. Cover with cover glass.
3. Denature the slides with boiling water vapor for 5 min.
4. Place the slides into the preheated humid container immediately. Seal the container quickly and return to 55°C.
5. Hybridize overnight.
6. Preheat a Coplin jar containing 2× SSC to 55°C.
7. Put the slides into the jar and pull up and down to drop the cover glasses. Incubate for 10 min at 55°C.
8. Drain off and wipe the excess liquid at the edges of slides. Add 10 μl of Vectashield Mounting Medium with DAPI. Cover with cover glass. Detect the signals under a fluorescence microscope. For the truncated chromosomes, the signals will locate at the end of the chromosomes. Compare the chromosome with its homologue or normal karyotypes to determine if there is any truncation (Fig. 4). Suspected chromosome truncations can be confirmed by a Southern analysis described in the next section.

3.4. Southern Blot Analysis

3.4.1. Genomic DNA Extraction, Digestion, Electrophoresis, and Blotting

1. Grind 3–5 g of leaves of transgenic plants in liquid nitrogen.
2. Transfer the frozen powder to centrifuge tubes containing Plant DNAzol. Volume of DNAzol should be three times the weight of the leaves. Mix thoroughly by gently inverting the tubes. Incubate at 25°C for 5 min.
3. Add equal volume of chloroform and shake vigorously. Then, incubate at 25°C and shake for 5 min.

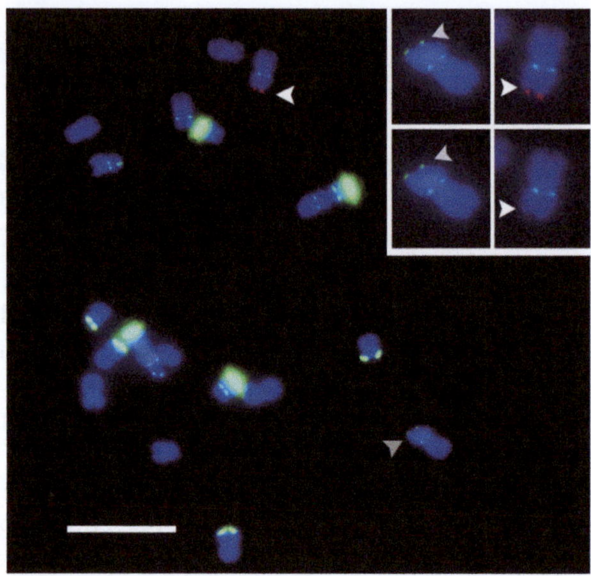

Fig. 4. Chromosome truncation. Metaphase chromosomes were hybridized with pWY96 probe (*red*) and mixtures of CentC (*green*) and knob (*green*) repetitive sequence probes. Arrows denote the transgene truncation sites (*white arrows*) and the corresponding sites on the homologues (*gray arrows*). A chromosome 1 short arm terminal knob truncation is shown. Insert shows chromosome 1 homologues with (*upper*) and without (*lower*) the transgene (*red*). Scale bar = 10 μm. Modified from (11), Copyright 2006 National Academy of Sciences, USA.

4. Centrifuge at 12,000 rpm (16,000 g) for 10 min. Transfer the supernatant to new tubes.
5. Add two volumes of 100% ethanol and mix thoroughly.
6. Store at room temperature for 5 min. Centrifuge at 5,000 rpm (3,000 g) for 4 min. Remove supernatant.
7. Prepare DNAzol–ethanol washing solution by mixing one volume of DNAzol with 0.75 volume of 100% ethanol. Add washing solution with the volume equal to that of DNAzol used for DNA extraction. Vortex and stand the samples for 5 min. Centrifuge at 5,000 rpm (3,000 g) for 4 min.
8. Remove supernatant. Wash with 70% ethanol and centrifuge at 5,000 rpm (3,000 g) for 4 min.
9. Remove supernatant thoroughly. Air-dry and dissolve DNA with TE buffer. If the DNA is difficult to dissolve, use 8 mM NaOH instead of TE.
10. Determine the concentration of the DNA using a UV spectrophotometer. For each sample, digest 20 μg of DNA with 100 U of restriction enzyme in 30 μl volume overnight.
11. Electrophorese at 25 V overnight with 0.7% agarose in 1× TAE.

12. Denature the gel for 45 min with denaturation buffer.
13. Neutralize the gel for 30 min with neutralization buffer.
14. Transfer the DNA onto a Qiabiane nylon membrane with 10× SSC overnight.
15. Wash the nylon membrane in 2× SSC for 15 min, gently shake to remove the gel particles.
16. Cross-link the DNA to the membrane with a UV cross-linker. Place the membrane between filter papers and dry in hood. Store the membrane at 4°C or proceed to hybridization immediately.

3.4.2. Prehybridization, Probe Labeling, Hybridization, and Signal Detection

1. Before prehybridization, denature 10 mg/ml salmon sperm DNA at 95°C for 20 min and incubate in ice bath until use. Preheat the prehybridization and hybridization buffer to 65°C.
2. Incubate the blot with 15 ml of prehybridization buffer for 3 h at 65°C by rotating in a hybridization tube. Drive the air bubbles between the blot and the hybridization tube.
3. During prehybridization, label the probe with ^{32}P-dCTP using a Random Primer Labeling kit, purify the product with a column, and elute the probe with 100 μl of TE. Shield and protective devices should be used for handling radioactive agents.
4. Denature the probe by heating at 95°C for 10 min. Incubate in ice bath for 5 min and briefly spin. Put on ice before use.
5. Replace the prehybridization buffer with hybridization buffer. Add the probe into the hybridization buffer.
6. Hybridize at 65°C overnight by rotating in a hybridization tube.
7. Wash the blot twice at room temperature with 2× SSC containing 0.5% SDS, 5 and 10 min, respectively.
8. Wash the blot twice with 0.1× SSC containing 0.1% SDS. For the first time, wash at room temperature for 20 min. For the second time, wash at 65°C for 30 min.
9. Drain off all the liquids from the blot. Wrap it in plastic wrap.
10. Fix the blot and X-ray film into an X-ray box in the dark room. Expose in –70°C freezer for the proper time.
11. Develop the film in the dark room. Samples containing truncated chromosomes will display smeared bands (Fig. 5, lanes 1–3 and 7–9) because telomerase adds different numbers of telomere repeats at the newly seeded chromosomal termini in each cell, and the seeded telomeres are heterogeneous in size.

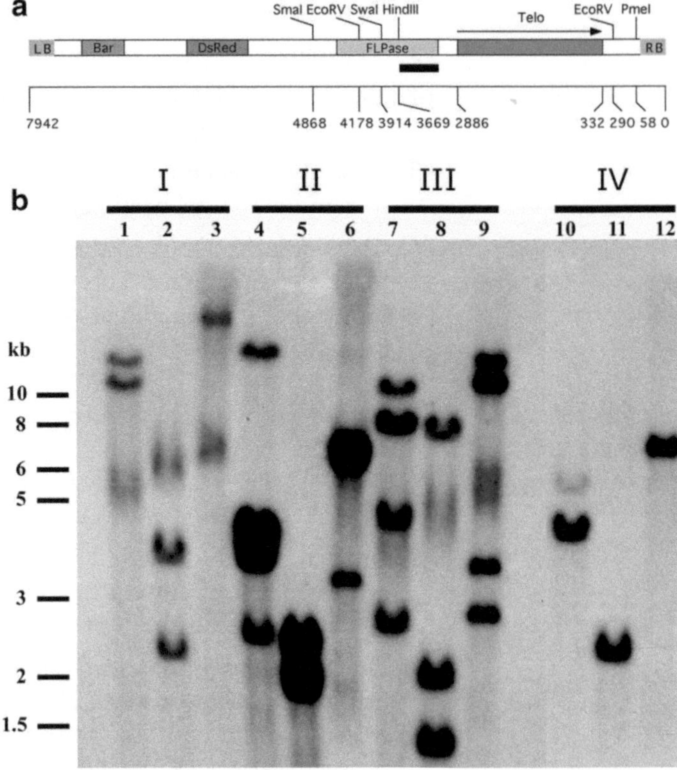

Fig. 5. Southern analysis of chromosome truncations. (**a**) Restriction map of the T-DNA region of the pWY86 construct with restriction sites (*top line*) and their positions relative to the right border (RB) (*lower line*). Bar shows the position of a FLP probe used for hybridization. (**b**) Restriction mapping of the positions of four transgenes by a Southern blot. Genomic DNAs from four events, I (lanes 1–3), II (lanes 4–6), III (lanes 7–9), and IV (lanes 10–12) are digested with *Hin*dIII (lanes 1, 4, 7, 10), *Eco*RV (lanes 2, 5, 8, 11), and *Sma*I (lanes 3, 6, 9, 12). DNA fragment sizes are indicated at the left side. Telomere smears (events I and III) are shown in lanes 1–3 and 7–9. Modified from ref. (11), Copyright 2006 National Academy of Sciences, USA.

4. Notes

1. For medium solidified with gelrite, it is better to pour plate before solidification because it is difficult to be redissolved by heating if it has solidified.
2. All the cold absolute ethanol should be taken straight from the freezer in this protocol.
3. $CaCl_2$ and spermidine should be newly prepared or stored at −20°C. Do not use those stored at 4°C because they will make the gold particles float on the liquid and stick on the tube wall above the liquid.

4. Centrifuge 1 ml in a 1.5-ml tube at 12,000 rpm (14,000 g) for 10 min before use. Transfer the supernatant into new tubes and then use for staining. This step is to eliminate the sediments in the solution, which badly influence the results.
5. Gold particles that stick on the wall of the tube could be pushed down with sterile tips.

Acknowledgments

This protocol was developed in James A. Birchler's lab at the University of Missouri. The authors would like to thank Professor Birchler for his insightful advice during the protocol development.

References

1. Basu, J., and Willard, H. F. (2005) Artificial and engineered chromosomes: non-integrating vectors for gene therapy. *Trends Mol. Med.* **11**, 251–8.
2. Yu, W., Han, F., and Birchler, J. A. (2007) Engineered minichromosomes in plants. *Curr. Opin. Biotechnol.* **18**, 425–31.
3. Farr, C., Fantes, J., Goodfellow, P., and Cooke, H. (1991) Functional reintroduction of human telomeres into mammalian cells. *Proc. Natl Acad. Sci. USA* **88**, 7006–10.
4. Barnett, M. A., Buckle, V. J., Evans, E. P., Porter, A. C., Rout, D., Smith, A. G., and Brown, W. R. (1993) Telomere directed fragmentation of mammalian chromosomes. *Nucleic Acids Res.* **21**, 27–36.
5. Farr, C. J., Bayne, R. A., Kipling, D., Mills, W., Critcher, R., and Cooke, H. J. (1995) Generation of a human X-derived minichromosome using telomere-associated chromosome fragmentation. *EMBO J.* **14**, 5444–54.
6. Heller, R., Brown, K. E., Burgtorf, C., and Brown, W. R. (1996) Mini-chromosomes derived from the human Y chromosome by telomere directed chromosome breakage. *Proc. Natl Acad. Sci. USA* **93**, 7125–30.
7. Mills, W., Critcher, R., Lee, C., and Farr, C. J. (1999) Generation of an approximately 2.4 Mb human X centromere-based minichromosome by targeted telomere-associated chromosome fragmentation in DT40. *Hum. Mol. Genet.* **8**, 751–61.
8. Yang, J. W., Pendon, C., Yang, J., Haywood, N., Chand, A., and Brown, W. R. (2000) Human mini-chromosomes with minimal centromeres. *Hum. Mol. Genet.* **9**, 1891–902.
9. Saffery, R., Wong, L. H., Irvine, D. V., Bateman, M. A., Griffiths, B., Cutts, S. M., Cancilla, M. R., Cendron, A. C., Stafford, A. J., and Choo, K. H. (2001) Construction of neocentromere-based human minichromosomes by telomere-associated chromosomal truncation. *Proc. Natl Acad. Sci. USA* **98**, 5705–10.
10. Richards, E. J., and Ausubel, F. M. (1988) Isolation of a higher eukaryotic telomere from Arabidopsis thaliana. *Cell* **53**,127–36.
11. Yu, W., Lamb, J. C., Han, F., and Birchler, J. A. (2006) Telomere-mediated chromosomal truncation in maize. *Proc. Natl Acad. Sci. USA* **103**, 17331–6.
12. Yu, W., Han, F., Gao, Z., Vega, J. M., and Birchler, J. A. (2007) Construction and behavior of engineered minichromosomes in maize. *Proc. Natl Acad. Sci. USA* **104**, 8924–9.
13. Frame, B. R., Shou, H., Chikwamba, R. K., Zhang, Z., Xiang, C., Fonger, T. M., Pegg, S. E., Li, B., Nettleton, D. S., Pei, D., and Wang, K. (2002) *Agrobacterium tumefaciens*-mediated transformation of maize embryos using a standard binary vector system. *Plant Physiol.* **129**, 13–22.
14. Zhao, Z. Y., and Ranch, J. (2006) Transformation of maize via *Agrobacterium tumefaciens* using a binary co-integrate vector system. In: Loyola-Vargas, V. M. and Vázquez-Flota, F., Eds. 2nd edition, *Methods in Molecular Biology, Plant Cell Culture Protocols.* Vol 318 Humana Press, New Jersey, pp. 315–23.
15. Chu, C.C., Wang, C. C., Sun, C. S., Hsu, K. C., Yin, K. C., Chu, C. Y., and Bi, F. Y. (1975) Establishment of an efficient medium for anther culture of rice, through comparative

experiments on the nitrogen sources. *Sci. Sin.* **18**, 659–68.

16. Hood, E. E., Helmer, G. L., Fraley, R. T., and Chilton, M. D. (1986) The hypervirulence of *Agrobacterium tumefaciens* A281 is encoded in a region of pTiBo542 outside of T-DNA. *J. Bacteriol.* **168**, 1291–301.
17. Murashige, T., and Skoog, F. (1962) A revised medium for rapid growth and bioassays with tobacco tissue cultures. *Physiol. Plant* **15**, 473–97.
18. Kato, A., Albert, P. S., Vega, J. M., and Birchler, J. A. (2006) Sensitive FISH signal detection in maize using directly labeled probes produced by high concentration DNA polymerase nick translation. *Biotech. Histochem.* **81**, 71–8.
19. Frame, B. R., Zhang, H. Z., Cocciolone S. M., Sidorenko, L. V., Dietrich, C. R., Pegg, S. E., Zhen, S., Schnable, P. S., Wang, K. (2000) Production of transgenic maize from bombarded type II callus: effect of gold particle and callus morphology on transformation efficiency. *In Vitro Cell. Dev. Biol.-Plant* **36**, 21–9.
20. Phillips, R. R. (1994) Classification of pollen abortion in the field. *The Maize Handbook*, eds Freeling M, Walbot V (Springer, New York), pp297–8.
21. Kato, A., Lamb, J. C., and Birchler, J.A. (2004) Chromosome painting using repetitive DNA sequences as probes for somatic chromosome identification in maize. *Proc. Natl Acad. Sci. USA.* **101,** 13554–9.

Chapter 7

Engineered Plant Minichromosomes

Robert T. Gaeta and Lakshminarasimhan Krishnaswamy

Abstract

The advent of transgenic technologies has met many challenges, both technical and political; however, these technologies are now widely applied, particularly for crop improvement. Bioengineering has resulted in plants carrying resistance to herbicides, insects, and viruses, as well as entire biosynthetic pathways. Some of the technical challenges in generating transgenic plant or animal materials include: an inability to control the location and nature of the integration of transgenic DNA into the host genome, and linkage of transformed genes to selectable antibiotic resistance genes used in the production of the transgene cassette. Furthermore, successive transformation of multiple genes may require the use of several selection genes. The coordinated expression of multiple stacked genes would be required for complex biosynthetic pathways or combined traits. Engineered nonintegrating minichromosomes can overcome many of these problems and hold much promise as key players in the next generation of transgenic technologies for improved crop plants. In this review, we discuss the history of artificial chromosome technology with an emphasis on engineered plant minichromosomes.

Key words: Engineered minichromosome, Artificial chromosome, Telomere-mediated truncation, Site-specific recombination

1. Introduction

Minichromosomes describe a broad range of small chromosomes that contain some or all of the elements essential for their replication and autonomous existence within a cell. In *E. coli*, minichromosomes can exist as small circular plasmids with little more than an origin of replication (*ori*C) (1). In eukaryotes such as plants and animals, they are often linear and require functional centromeres and telomeres in addition to origins of replication and exist separately from the normal karyotype of a cell. Extrachromosomal genetic entities such as minichromosomes (natural or engineered) have been reported in bacteria, yeast, mammals, birds, amphibians,

James A. Birchler (ed.), *Plant Chromosome Engineering: Methods and Protocols*, Methods in Molecular Biology, vol. 701, DOI 10.1007/978-1-61737-957-4_7, © Springer Science+Business Media, LLC 2011

protozoans, and plants (2–9). B chromosomes, which are common in nearly 10% of plants species, represent the largest class of such entities (10). Studies of minichromosomes have contributed significantly to our understanding of chromatin, chromosome structure, and replication. Most recently, the term minichromosome has been used to define small, engineered (artificial) chromosomes in human cell lines (7, 11) and in the crop plant corn (*Zea mays*) (8, 9, 12). In this review, we use the term "engineered minichromosome" to refer to artificial plant chromosomes derived from the truncation of endogenous chromosomes.

2. Engineered (Artificial) Chromosomes

Artificial chromosomes are nonintegrating vectors designed to contain specific genetic components and have the capability for stably harboring large amounts of DNA (13). As with minichromosomes, they exist and segregate independent of the normal chromosome complement (karyotype) of the host cell. In lower organisms like bacteria and yeast, artificial chromosomes have been engineered for large-insert molecular cloning and utilized for a multitude of genomic and genetic analyses. In yeast, *E. coli*, and mammalian systems, artificial chromosomes have been available for decades; however, engineered plant chromosomes are recent developments. There are two approaches for engineering a chromosome: the bottom-up approach and the top-down approach (reviewed in ref. 14). The bottom-up approach involves assembling the essential elements of a chromosome piece by piece (using telomere and or centromere sequences) and relies on de novo assembly. This method has been proven successful in yeast and has been utilized in mammalian systems (15–17). In the top-down approach, artificial chromosomes are derived from the breakage and truncation of preexisting endogenous chromosomes. This method has been proven successful in both animal and plant systems (7–9, 11, 16, 18, 19).

Linear yeast artificial chromosomes (YACs) were constructed over 20 years ago in *Saccharomyces cerevisiae* using telomere, centromere, and origin of replication sequences, and are capable of maintaining large DNA inserts of up to 2 Mb in length (15, 20). Bacterial artificial chromosomes (BACs) were created in *E. coli*, and P1 artificial chromosomes (PACs) were engineered in the P1 bacteriophage, both of which are capable of maintaining up to 300 kb of DNA (21–25). The utility of these early engineered chromosomes (YACs, BACs, and PACs) as vectors for large DNA inserts cannot be understated, and they have played an essential role in the sequencing, mapping, and characterization of many large genomes. In addition, the early studies conducted by Murray and Szostak (15) were instrumental in determining the minimal requirements for chromosome maintenance and stability (i.e., centromeres, telomeres, and replication origins) in budding yeast. Unfortunately, these early artificial chromosomes did not function autonomously in mammalian and plant cells, which have unique requirements for proper replication, segregation, and transmission. However, one study successfully utilized YACs for vectoring mammalian genes into mice (26).

Circular human artificial episomal chromosomes (HAECs) were first reported by Sun et al. (27) and were used for cloning and analyzing large genomic inserts up to 330 kb. While this system allowed for the random cloning of human DNA in human cells, it relied on viral sequences for replication and was not an efficient platform for the development of engineered chromosomes for use in gene therapy. Harrington et al. (16) constructed human artificial chromosomes (HACs) by introducing human centromere, telomere, and genomic DNA into human cells, which resulted in both de novo artificial chromosomes (bottom-up method) and minichromosomes. These were the first linear, mitotically stable chromosomes developed for mammalian systems. YACs containing human centromere and telomere sequences were later utilized for assembling mammalian artificial chromosomes (MACs) (17). In other studies, telomere-mediated chromosome truncation was used in human cell lines to generate minichromosomes (7, 11, 18, 19, 28). In these studies, human telomere arrays (TTAGGG) were introduced into human cell lines and truncated X and Y-chromosomes were selected. The research by Heller et al. (11) resulted in Y-chromosome derived minichromosomes that contained little more than a centromere. In these studies, truncating constructs were detectable at the termini of truncated chromosomes by fluorescent in situ hybridization (FISH) and/or RFLP analysis, suggesting that telomere sequences resulted in healing of the chromosome ends. In mammals, minichromosomes may prove to be effective vectors for gene therapies (reviewed in ref. 29).

Most recently, telomere truncation was applied successfully in plants to generate engineered minichromosomes in maize (8, 9). In these experiments, minichromosomes were derived from A-chromosomes and supernumerary B chromosomes. The minichromosomes transmit both mitotically and meiotically. The truncating constructs contained visual marker (e.g., DsRed) and selection transgenes, as well as functional site-specific recombination sites (i.e., LoxP and FRT sites). This marked the first demonstration that transgenes could be attached to engineered plant minichromosomes along with specific sequences for facilitating targeted recombination. The limitations of plant engineered chromosomes are yet to be fully tested, and they promise to further our understanding of basic chromosome biology, while simultaneously providing a platform for the next generation of transgenic technologies in agriculture. These qualities will be useful for facilitating the agricultural needs of the growing human population and for increasing the plasticity of crop germplasm in the face of an uncertain global climate.

3. Methods for Engineering Minichromosomes in Plants

Eukaryotic chromosomes have three essential regions: telomere, centromere, and origins of replication (30). Telomeres consist of tandem repeats of simple, short, G-rich sequence (e.g., TTTAGGG in plants), and associated proteins. The primary structure is conserved over a wide range of organisms. The telomere cap protects the ends of chromosomes from sticking to each other, and also from degradation by the action of exonucleolytic enzymes. Centromeres are also characterized by tandem repetitive sequences and associated proteins. In contrast to the telomere, the centromeres consist of several types of repetitive DNA sequences, which are not widely conserved among diverse organisms. In addition to the primary structure of DNA, epigenetic factors also have a significant role in determining centromere function. The centromere region is also where the kinetochore assembles. Kinetochores are composed of a dynamic complex of proteins that provide the sites of attachment to spindle fibers. The centromere, therefore, is essential for segregation of the chromosomes to the spindle poles during mitosis and meiosis. Origins of replication are the locations where DNA replication is initiated, and eukaryotic chromosomes have numerous origins along their length. The sequences of replication origins have been characterized in yeast (31, 32); however, initiation of replication in higher eukaryotes does not seem to correlate with specific DNA sequences, but rather it appears to be determined epigenetically (30).

In order for engineered chromosomes to replicate, be stably maintained, and perpetuate over successive generations of cell division, it is imperative that they have functional centromeres, origins of replication, and telomeres. As stated above, minichromosomes can be generated by two different approaches (Fig. 2; reviewed in ref. 33): the "top-down approach," where minichromosomes are generated by inducing truncation of endogenous chromosome followed by de novo telomere synthesis, and the "build up" or "bottom-up approach," where minichromosomes are generated through de novo assembly of centromere repeat sequence, telomere repeat sequence, and filler DNA that are introduced into the cell (Figs. 1 and 2). Both techniques have been applied in plant and animal systems. One advantage of generating minichromosomes by the "top-down approach" is

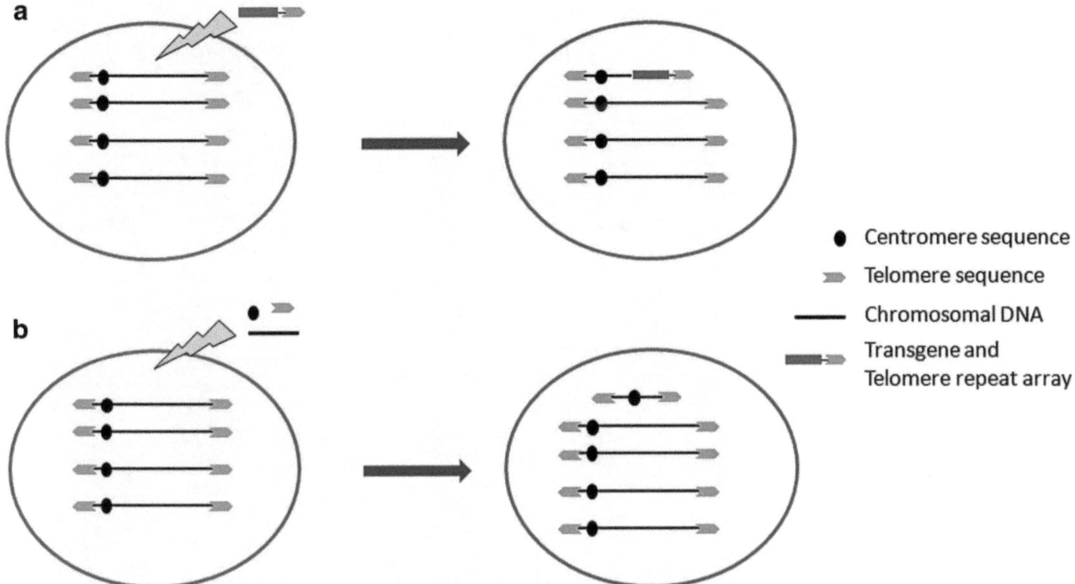

Fig. 1. Generation of minichromosome by top-down (**a**) and bottom-up (**b**) strategies. (**a**) In the top-down approach, minichromosomes are formed by inducing truncation of endogenous chromosomes. In telomere-mediated truncation, a transgene construct carrying telomere repeat arrays is transformed into the cell. Integration of the transgene could truncate the chromosome, and the introduced telomere sequence serves as seed for de novo telomere synthesis at the site of truncation. (**b**) In the bottom-up strategy, minichromosomes are assembled piece by piece from individual chromosome elements. When DNA sequences of centromeric repeats, telomeric repeats, and filler DNA are transformed into the cell, these may assemble into a functional minichromosome.

that endogenous chromosome-derived minichromosomes have a defined structure and organization and are therefore more stable and amenable to further modifications. In contrast, the structure of de novo assembled minichromosomes is variable and poorly organized because of random incorporation of genomic DNA sequences from the native chromosomes.

3.1. Top-Down Approaches to Minichromosome Assembly in Plants

There are several ways to generate minichromosomes from endogenous chromosomes. Brock and Pryor generated minichromosomes from maize chromosome 10 following gamma irradiation of pollen (34). The broken chromosomes generated by this method were thought to undergo "healing" during DNA repair and result in relatively unstable minichromosomes. Zheng and colleagues described a range of stable minichromosomes in a study of the chromosome type breakage–fusion–bridge cycle (BFB cycle) (35–37). When the two centromeres of a dicentric chromosome are pulled towards opposite poles during meiotic or mitotic anaphase, a bridge is formed across the metaphase plate. The chromosome bridge breaks at random, and de novo repair of the broken chromosome ends can result in the formation of minichromosomes. Kato and colleagues adapted this genetic tool as a top-down approach to generate several B chromosome-derived minichromosomes,

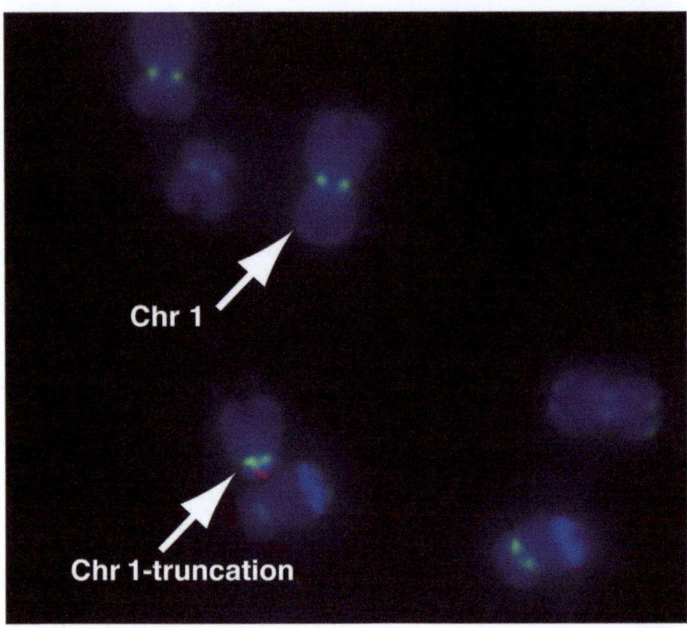

Fig. 2. Telomere-mediated truncation of chromosome 1 in maize. A partial metaphase spread is shown in which a normal chromosome 1 (Chr 1) and truncated chromosome 1 (Chr 1-trunc) are shown (see *arrows*). Fluorescent in situ hybridization (FISH) was performed using centromere (CentC) and subtelomeric repeat (4-12-1) probes (labeled *green*), as well as a probe homologous to the truncating construct (labeled *red*). The truncated chromosome was derived from telomere-mediated truncation using the construct WY86 as described by Yu et al. (8). In this example, the short arm of chromosome 1 has been lost, and the truncating construct labeled in *red* can be seen.

which were analyzed structurally and cytologically (38). In each of these examples, minichromosomes were generated entirely from endogenous chromosomes without introducing any foreign DNA sequences. However, these minichromosomes require subsequent targeting with transgenes if they are to be used as vectors or platforms for biosynthetic pathways.

In contrast, telomere-mediated truncation using an engineered construct offers the distinct advantage that minichromosome formation is coincident with transgene delivery. In a strategy similar to what has seen success earlier in mammalian cell lines, Yu et al. (8, 9) transformed maize embryos with DNA constructs containing telomere repeat arrays. Both biolistic and *Agrobacterium*-mediated transformation methods were used. In these experiments, several transformation events were recovered in which the integrating transgene resulted in truncation of the chromosome (Fig. 2). Chromosome truncation and minichromosome formation was not observed when a similar construct lacking telomere sequence was used for transformation. While the exact mechanism of telomere-mediated truncation is not known, it is presumed that during transgene integration into the genome, the telomere repeat array serves as a seed for de novo telomere synthesis, leading to truncation and capping of the chromosome.

In these studies, minichromosomes were frequently recovered from the truncation of B chromosomes (9). In maize and many other plants, B chromosomes are naturally occurring supernumerary chromosomes whose presence has little effect on phenotype when present in low numbers (39–41). These minichromosomes were meiotically transmissible through male gametes at a frequency of 12–39%. Because B chromosomes are not known to carry any genes, truncation was unlikely to result in deficiencies. Bar and GUS genes showed cytological colocalization with the truncating construct on the B-derived minichromosomes. Expression of Bar was evidenced by transgene-induced resistance to bialaphos selection, and GUS protein expression was detectable in several tissues including leaves, roots, shoots, mature kernel embryos, and endosperm. These data suggest that B chromosomes will provide useful platforms for assembling minichromosomes in plants.

Minichromosomes were also generated by telomere-mediated truncation of normal chromosomes (A chromosomes) in maize diploids (8, 9). When A chromosome truncations were detected in regenerated plants, large truncations were rare, possibly because the loss of large chromosomal segments was selected against during tissue culture and plant regeneration. Furthermore, A-derived minichromosomes were not transmitted meiotically in diploids, presumably because of deficiencies caused by loss of genes during truncation; however, in a spontaneous autotetraploid plant, a minichromosome derived from truncation of the long arm of chromosome 7, (named R2), was successfully recovered in progenies (9). This was done by backcrossing the minichromosome-containing line into diploid plants for three generations. This minichromosome could be stably inherited through both mitosis and meiosis.

Minichromosome platforms will have greater utility if they can be modified in vivo. For example, several genes for desirable traits could be introduced in successive generations and targeted to the transgene cassettes on the minichromosome. To test this possibility Yu et al. (8, 9) utilized a site-specific recombination system (Cre–lox (see ref. 42)). The telomere truncation constructs used included a lox71 sequence upstream of a promoterless marker gene, DsRed. When crossed to a plant carrying 35S–lox66–Cre expression cassette at the tip of chromosome arm 3L, expression of the site-specific recombinase facilitated recombination between the lox71 on the truncated chromosome and lox66 on chromosome 3. This placed the 35S promoter upstream of DsRed gene located on the minichromosomes and activated expression of the gene. This demonstrated the feasibility of using site-specific recombination systems to introduce new sequence to minichromosomes.

3.2. The Bottom-Up Approach to Minichromosome Assembly in Plants

There have been two reports of minichromosomes claiming to be engineered by the "bottom-up" approach in maize. Carlson et al. introduced a circular DNA with maize centromeric repeat sequences, a selection marker gene, a phenotypic marker gene, and filler DNA (12). It was claimed that the introduced DNA functioned in vivo to form autonomous circular chromosomes referred as "Maize minichromosome" (MMCs). However, the FISH, genetic, and molecular data did not clearly establish the autonomous nature of the transgene, but rather the high frequency of meiotic transmission (~50% as a hemizygote and ~93% as a homozygote) is inconsistent with the circular nature of such a claimed tiny chromosome and suggests the transgenes may actually have integrated into the maize genome (reviewed in ref. 43). Furthermore, previous studies have suggested that unpaired monosomic chromosomes and other small chromosomes tend to be lost at a high frequency (44, 45). In addition, previous studies have indicated that circular chromosomes suffer loss and instability with initiation of the BFB cycle (36, 37).

Ananiev et al. transformed maize with native centromeric segments, origins of replication, selectable marker genes, and telomeric repeats (46). Although the data clearly confirm the presence of minichromosomes in this case, the true cause of their formation, whether de novo assembly or telomere-mediated truncation resulting from the introduced telomere sequences, was not resolved. Indeed, the introduction of circular constructs without telomere repeats only resulted in stable integrations into the chromosomes without centromere function. Further studies are needed to confirm the feasibility of this approach in plant systems.

4. Methods for Transforming Minichromosome-Generating DNA Constructs

The two widely adapted techniques for transformation in plants include biolistic (particle bombardment) and *Agrobacterium*-mediated transformation (reviewed in ref. 47). Both techniques are effective for generating minichromosomes by telomere-mediated truncation in maize (8, 9). However, both techniques pose certain challenges. Biolistics is a valuable transformation tool especially for plants that are recalcitrant to *Agrobacterium*-mediated transformation and involves bombarding plant cells with gold particles coated with DNA to be transformed. The frequency of transformation and truncation is more efficient by the biolistics method (9); however, an inherent disadvantage with this technique is that transgenes often insert as tandem repeats and at multiple locations in the genome. Such an arrangement could lead to silencing of genes on the transgene cassette. It is possible that such tandem arrays can be resolved. For example,

Srivastava et al. demonstrated that when transgenes contain a *lox* site, concatameric transgene insertions could be resolved to single copy during site-specific recombination between the outermost sites in the array (48). A similar strategy could be applied to resolve tandem copies of transgene cassettes used for telomere-mediated truncation. The issue of transgene insertion at multiple locations in the genome can be resolved easily by backcrossing the transformant to a parental genotype followed by self-pollination in successive generations.

Agrobacterium-mediated transformation is fairly efficient in a wide range of dicotyledonous plants, and several monocotyledonous plants. The advantages of *Agrobacterium*-mediated transformation include lower transgene copy number and stable gene expression (49–51). However, the transformation efficiency is much lower in comparison to particle bombardment. In addition, no B-chromosome truncations were recovered in experiments using *Agrobacterium*-mediated transformation of maize (8). Nevertheless, it is promising that the list of plants transformable by *Agrobacterium* is steadily growing. This includes several agronomical and commercially important plants such as rice, wheat, maize, sorghum, and barley (52).

5. Detection and Confirmation of Minichromosomes

When a truncating DNA construct is introduced into a plant cell, there are several possible outcomes: the transgene could get inserted into the chromosome, it could truncate a chromosome while inserting, or the introduced DNA could get enzymatically degraded. Yu et al. (8) reported that when truncation constructs were transformed into maize, 231 independent transgenic plants were recovered, of which 118 insertions were at a distal position on the chromosome. Fifty-six distal insertions were further tested, and nine resulted in truncation. With a variety of possible outcomes, it is imperative to test and confirm putative truncation events.

The process of localizing transgenes has been simplified with the recent advance in fluorescent in situ hybridization (FISH) in maize (53, 54). DNA probes targetting transgenes and sequences specific to each of the maize chromosomes can be labeled with fluorescent molecules and hybridized onto metaphase chromosome spreads from root-tip preparations, allowing the location of a transgene in the genome to be determined. However, the current limits of resolution of the FISH technique cannot distinguish between a terminal location of the transgene (as in case of true truncation) and a nontruncating distal insertion very close to the chromosome end. For these reasons, other methods of verification are helpful. Southern blot hybridization can be applied to determine whether a distal insertion is a truncation event. Genomic DNA from the plants are digested with a restriction enzyme that cuts the truncating construct closer to the telomere array, and hybridization is performed using a probe specific to the DNA distal to the enzyme site. Truncation events in which the transgene cassette lies adjacent to the newly formed telomere will appear as a smear on the autoradiograph, reflecting the variable size of telomeres in the minichromosome (8). On the other hand, a nontruncating transgene insertion will likely appear as a distinct band.

The terminal position of the truncating transgene could also be tested by a PCR-based technique called "Primer extension telomere repeat amplification" (PETRA) (55). In this two-step technique, first an adapter oligo with the 3′ end complementary to the G-rich single strand overhang at the end of the telomere is used to generate a copy of the complementary strand. This product is PCR-amplified in the second step using an oligo with the adapter sequence and a primer complementary to the truncating construct. The presence of the truncating construct in the terminal position alone will amplify the 3′ end of the construct and telomere.

Plants carrying minichromosomes will be of commercial and/ or agronomic interest only if the minichromosome is stable through successive mitotic cycles and if it can pass through the gametophyte. Genetic and cytological analysis at different stages of development of the plant and at successive generations can help determine the utility of any given minichromosome. Biotechnological utility crucially depends on the transcriptionally active or inactive state of the transgenes on the minichromosome, stable transmission, and amenability (see below).

6. Next-Generation Engineered Minichromosomes in Plants and Their Applications

Genetic engineering has tremendous potential for revolutionizing agriculture. Plant biotechnology has resulted in the development of plants with resistance to herbicides, insects, and viruses, as well as plants carrying entire biosynthetic pathways (56). Transgenic expression of multiple genes is required for complex biosynthetic pathways or combined traits, and both genes and regulatory elements need to be stacked for these purposes. For instance, ideal plants might carry resistance for multiple insect and fungal pests, as well as tolerance to temperature extremes, water stress, or salt concentrations. However, traditional transformation techniques have drawbacks. They tend to generate random genomic integrations that may suffer from position effect variegation or insertions that may disrupt endogenous genes. Consequently, transgenic events require intensive screening for stable integration, meiotic transmission, and function, as well as extensive backcrossing for purification. Movement of transgenes from one genetic background to another is limited by linkage to undesirable genes or QTL. Currently, several alternatives to traditional transformation exist, including: episomal viral vectors (57), organellar transformation (58, 59), and engineered minichromosomes (8, 9, 12). In the following sections, we discuss the distinct advantages of engineered chromosomes.

Engineered minichromosome platforms derived by top-down or bottom-up approaches will be useful in facilitating the next generation of transgenic crop plants and will offer new solutions to some of the challenges associated with classical transformation. Minichromosomes could be used as biofactories for metabolite production, as well as for basic studies of chromosome biology. They can be developed in one line or cultivar, and introgressed into related cultivars without the problems associated with cis linkage drag or position effects. As an example, a naturally occurring (nontransgenic) minichromosome discovered in *Arabidopsis* was easily transferred from one ecotype to another by simple backcrossing (60).

Using top-down approaches they may be synthesized from chromosomes of the normal karotype (A chromosomes) or supernumery B chromosomes in maize (8, 9); however, A-chromosome truncations are difficult to recover meiotically from diploids, and successful recoveries have occurred only in spontaneous tetraploids (see above). It is possible that trisomics and addition lines could serve as targets for deriving specific A-chromosome derived minichromosomes. In such lines, the diploid chromosome balance would be restored upon truncation of the extra chromosome, possibly favoring their recovery. B chromosomes offer a unique and efficient platform for assembling minichromosomes by the top-down method, in that they exist autonomously from the Karyotype (A genome), are essentially phenotypically inert, and exist in hundreds of plant species (61, 62). Minichromosomes engineered from B chromosomes in maize have the added benefit of nondisjunction during the second pollen mitosis, and preferential fertilization of the egg cell during pollination (reviewed ref. 63). These properties would facilitate rapid accumulation or segregation of minichromosomes, allowing for copy number control and rapid removal when they are no longer needed in a particular genetic background. Control over dosage would allow for maximal production of some biomolecules that may have industrial, pharmaceutical, or nutritional value. It is possible that high copy numbers of transgenes on minichromosomes would experience cosuppression, but currently, there are no data on these limitations. Minichromosomes could also be outfitted with pollen-lethal genes, which would prevent them from spreading unwittingly by pollen dispersal.

Minichromosomes also have the potential to be modified in a site-specific manner. The next generation of maize minichromosomes will need to be amendable if they are to be widely applied among crop plants and for multiple purposes (reviewed in ref. 64). Outfitting future minichromosomes with multiple specific genes, regulatory elements, and site-specific recombination sequences (e.g., LoxP, FRT, aatt, etc.) will allow for versatility across a spectrum of applications (65). Lines carrying the specific recombinases in these cases would also have to be developed for use as "modifier" lines when performing such manipulations. Using a complex array of transformation constructs containing different site-specific recombination systems, selection genes could be removed and recycled, and genes may be added or subtracted during successive transformations. It should be possible to remove all unnecessary DNA utilized during the cloning and engineering phases of chromosome construction.

In general, homologous recombination is very efficient in mammalian systems, but poor in plants (66, 67); however, recent studies have demonstrated that inducing double-stranded DNA breaks through the use of I-*Sce*I in *Arabidopsis* (68), and synthetic zinc-finger nucleases in maize can lead to targeted recombination (69–71). For modifying transgene cassettes in plants, the Cre/lox system has already demonstrated its utility. It has been proven effective for the removal of transgenes (72, 73), as well as for the integration of new genes into preexisting *lox* sites in both *Arabidopsis* (74) and rice (75). The Cre/*lox* system has also demonstrated its ability for resolving complex tandem insertions down to single copy in both wheat and maize (48, 76). This benefit would be realized when minichromosomes are generated by particle bombardment, a procedure that often results in tandem transgene arrays. Yu et al. demonstrated somatic recombination with *lox* sites on minichromosomes in the presence of Cre recombinase (9). These studies suggest the feasibility of amending minichromosomes using site-specific recombination. However, meiotic transmission of recombinants between minichromosomes and endogenous chromosomes or introduced plasmids has not yet been demonstrated in plants. It is also possible that entire chromosome arms or single genes could be transferred to or from minichromosomes, permitting functional genomic analysis and studies of genome dosage.

Several studies have now demonstrated evidence that minichromosomes can be produced in maize by both top-down and bottom-up approaches (see above). However, these technologies are still in their infancy, and extensive improvement is needed before they can be created, modified, and applied with ease to other crops. To date, there are no published reports of success in generating minichromosomes in other plant species, although it is likely to be only a matter of time before the technology becomes as widely applied to crop improvement as traditional transgenic methods. Telomere-mediated truncation of endogenous plant chromosomes will probably be quite an amenable method for minichromosome production, since the telomere-repeat sequence is highly conserved. Using the top-down approach, initial minichromosomes could potentially be "trimmed" further through successive truncation to contain little more than centromeres, transgene cassettes, and telomeres (reviewed in ref. 63). In a similar way, the bottom-up approach would have the benefit of minimization and control over the overall genetic content of the entire synthesized chromosome. However, because centromere sequences are often species-specific, a single platform would be difficult to move from one species to another.

7. Conclusions and Future Questions

In conclusion, engineered (artificial) chromosomes have come a long way in the last quarter century. From yeast and bacteria to mammals and plants, chromosome vectors have played an important role in our understanding of chromosome biology and have proven invaluable in genetic and genomic analyses. Future engineered minichromosomes hold great promise for furthering these understandings and promise to expand upon classical transgenic research; however, there remain many unanswered questions regarding the basic biology of chromosomes, both "normal" and "engineered." Some of the questions future studies can address include: (1) what conditions contribute to optimal formation, stability, and transmissibility of engineered plant minichromosomes; (2) what are the benefits and drawbacks to minichromosomes derived via the top-down or bottom-up approach, and how can these methods realize their greatest potential; (3) what are the size restrictions for engineered chromosomes; (4) what is the maximum copy number of minichromosomes that can be maintained by a cell, and how does copy number affect the expression of transgene cassettes and transmissibility; (5) what is the effect of genetic background, i.e., do different host genotypes contain genetic variation contributing to the formation and functioning of minichromosomes; (6) what methods will prove most fruitful for amending minichromosomes through the in vivo removal and replacement of selection genes, genes of interest, and promoters and other regulatory elements; (7) can minichromosomes be built up to contain extensive and elaborate biosynthetic pathways, in which regulator and response genes faithfully integrate to recapitulate the production of complex metabolites; and (8) will it be possible to transfer genes and chromosomal fragments between minichromosome and host genome in a site-specific manner?

Acknowledgments

Research supported by National Science Foundation grant DBI 0701297 to James Birchler.

References

1. Dasgupta, S., and Lobner-Olesen, A. (2004) Host controlled plasmid replication: *Escherichia coli* minichromosomes. *Plasmid* **52**, 151–68.
2. Rush, M.G., and Misra, R. (1985) Extrachromosomal DNA in eukaryotes. *Plasmid* **14**, 177–91.
3. Ryoji, M., Tominna, E., and Yasui, W. (1989) Minichromosome assembly accompanying repair-type DNA synthesis in *Xenopus* oocytes. *Nucleic Acids Res* **24**, 10243–58.
4. Bitgood, J.J., and Shoffner, R.N. (1990) Cytology and cytogenetics. In: Poultry breeding and genetics (Crawford, R.D., ed). Amsterdam: Elsevier, 401–27.
5. Gibson, S.I., Surosky, R.T., and Tye, B.-K. (1990) The phenotype of the minichromosome maintenance mutant mcm3 is characteristic of mutants defective in DNA replication. *Mol Cell Biol* **10**, 5707–20.
6. Weiden, M., Oshheim, Y.N., Beyer, A.L., and Van der Ploeg, L.H.T. (1991) Chromosome structure: DNA nucleotide sequence elements of a subset of the minichromosomes of the protozoan *Trypanosoma brucei. Mol Cell Biol* **11**, 3823–34.
7. Farr, C.J., Bayne, R.A.L., Kipling, D., Mills, W., Critcher, R., and Cooke, H.J. (1995) Generation of a human X-derived minichromosome using telomere-associated chromosome fragmentation. *EMBO J* **14**, 5444–54.
8. Yu, W., Lamb, J.C., Han, F., and Birchler, J.A. (2006) Telomere-mediated chromosomal truncation in maize. *Proc Natl Acad Sci USA* **103**, 17331–36.
9. Yu, W., Han, F., Gao, Z., Vega, J.M., and Birchler, J.A. (2007) Construction and behavior of engineered minichromosomes in maize. *Proc Natl Acad Sci USA* **104**, 8924–29.
10. Jones, N., and Rees, H. (1982) Chromosomes. Academic Press, London.
11. Heller, R., Brown, K.E., Burgtorf, C., and Brown, W.R.A. (1996) Mini-chromosomes derived from the human Y chromosome by telomere directed chromosome breakage. *Proc Natl Acad Sci USA* **93**, 7125–30.
12. Carlson, S.R., Rudgers, G.W., Zieler, H., Mach, J.M., Luo, S., Grunden, E., Krol, C., Copenhaver, G.P., and Preuss, D. (2007) Meiotic transmission of an in vitro-assembled autonomous maize minichromosome. *PLOS Genet* **3**, e179. Doi:10.1371/journal.pgen.0030179.
13. Duncan, A., and Gyula, H. (2007) Chromosomal engineering. *Curr Opin Biotechnol* **18**, 420–24.
14. Houben, A., and Schubert, I. (2007) Engineered plant minichromosomes: a resurrection of B chromosomes. *Plant Cell* **19**, 2323–27.
15. Murray, W.M., and Szostack, J.W. (1983) Construction of artificial chromosomes in yeast. *Nature* **305**, 189–93.
16. Harrington, J.J., Van Bokkelen, G., Mays, R.W., Gustashaw, K., and Willard, H.F. (1997) Formation of de novo centromeres and construction of first-generation human artificial microchromosomes. *Nat Genet* **15**, 345–55.
17. Ikeno, M., Grimes, B., Okazaki, T., Nakano, M., Saitoh, K., Hoshino, H., McGill, N.I., Cooke, H., and Masumoto, H. (1998) Construction of YAC-based mammalian artificial chromosomes. *Nat Biotechnol* **16**, 431–39.
18. Farr, C., Fantes, J., Goodfellow, P., and Cooke, H. (1991) Functional reintroduction of human telomeres into mammalian cells. *Proc Natl Acad Sci USA* **88**, 7006–10.
19. Farr, C.J., Stevanovic, M., Thomson, E.J., Goodfellow, P.N., and Cooke, H.J. (1992) Telomere-associated chromosome fragmentation: applications in genome manipulation and analysis. *Nat Genet* **2**, 275–82.
20. Burke, D.T., Carle, G.F., and Olson, M.V. (1987) Cloning of large segments of exogenous DNA into yeast by means of artificial chromosome vectors. *Science* **236**, 806–12.
21. O'Connor, M., Peifer, M., and Bender, W. (1989) Construction of large DNA segments in *Escherichia coli. Science* **16**, 1307–12.
22. Hosoda, F., Nishimura, S., Uchida, H., and Ohki, M. (1990) An F factor based cloning system for large DNA fragments. *Nucleic Acids Res* **18**, 3863–69.
23. Shizuya, H., Birren, B., Kim, U.J., Mancino, V., Slepak, T., Tachiiri, Y., and Simon, M. (1992) Cloning and stable maintenance of 300-kilobase-pair fragments of human DNA *Escherichia coli*

using an F-factor-based vector. *Proc Natl Acad Sci USA* **89**, 8794–97.

24. Sternberg, N. (1990). Bacteriophage P1 cloning system for the isolation, amplification, and recovery of DNA fragments as large as 100 kilobase pairs. *Proc Natl Acad Sci USA* **87**, 103–7.
25. Ioannou, P.A., Amemiya, C.T., Garnes, J., Kroisel, P.M., Shizuya, H., Chen, C., Batzer, M.A., and de Jong, P.J. (1994). A new bacteriophage P1-derived vector for the propagation of large human DNA fragments. *Nat Genet* **6**, 84–9.
26. Jakobovits, A., Moore, A.L., Green, L.L., Vergara, G.J., Maynard-Currie, C.E., Austin, H.A., and Klapholz, S. (1993) Germ-line transmission and expression of a human-derived yeast artificial chromosome. *Nature* **362(6417)**, 255–8.
27. Sun, T.-Q., Fenstermacher, D.A., and Vos, J.-M. (1994) Human artificial episomal chromosomes for cloning large DNA fragments in human cells. *Nature* **8**, 33–41.
28. Itzhaki, J.E., Barnett, M.A., MacCarthy, A.B., Buckle, V.J., Brown, W.R.A., and Porter, A.C.G. (1992) Targetted breakage of a human chromosome mediated by cloned human telomeric DNA. *Nat Genet* **2**, 283–7.
29. Basu, J., and Willard, H.F. (2005) Artificial and engineered chromosomes: non-integrating vectors for gene therapy. *Trends Mol Med* **11**, 251–8.
30. Baird, D.M., and Farr, C.J. (2006) The organization and function of chromosomes. *EMBO Rep* **7**, 372–6.
31. Stinchcomb, D.T., Struhl, K., and Davis, R.W. (1979) Isolation and characterisation of a yeast chromosomal replicator. *Nature* **282**, 39–43.
32. Marahrens, Y., and Stillman, B. (1992) A yeast chromosomal origin of DNA replication defined by multiple functional elements. *Science* **255**, 817–23.
33. Grimes, B., and Cooke, H. (1998) Engineering mammalian chromosomes. *Hum Mol Genet* **7**, 1635–40.
34. Brock, R.D., and Pryor, A.J. (1996) An unstable minichromosome generates variegated oil yellow maize seedlings. *Chromosoma* **104**, 575–84.
35. Zheng, Y.Z., Roseman, R.R., and Carlson, W.R. (1999) Time course study of the chromosome-type breakage-fusion-bridge cycle in maize. *Genetics* **153**, 1435–44.
36. McClintock, B. (1939) The behavior in successive nuclear divisions of a chromosome broken at meiosis. *Proc Natl Acad Sci USA* **25**, 405–16.
37. McClintock, B. (1941) The stability of broken ends of chromosomes in *Zea mays*. *Genetics* **26**, 234–82.
38. Kato, A., Zheng, Y.Z., Auger, D.L., Phelps-Durr, T., Bauer, M.J., Lamb, J.C., and Birchler, J.A. (2005) Minichromosomes derived from the B chromosome of maize. *Cytogenet Genome Res* **109**, 156–65.
39. Carlson, W.R. (1978) The B chromosome of corn. *Annu Rev Genet* **12**, 5–23.
40. Jones, N.R., and Rees, H. (1982) B Chromosomes. Academic, London, 266 pp.
41. Jones, N., and Houben, A. (2003) B chromosomes in plants: escapees from the A chromosome genome? *Trends Plant Sci* **8**, 417–23.
42. Albert, H., Dale, E.C., Lee, E., and Ow, D.W. (1995) Site-specific integration of DNA into wild-type and mutant lox sites placed in the plant genome. *Plant J* **7**, 649–59.
43. Houben, A., Dawe, R.K., Jiang, J., and Schubert, I. (2008) Engineered plant minichromosomes: a bottom-up success? *Plant Cell* **20**, 8–10.
44. Dawe, R.K. (1998) Meiotic chromosome organization and segregation in plants. *Annu Rev Plant Physiol Plant Mol Biol* **49**, 371–95.
45. Phelps-Durr, T.L., and Birchler, J.A. (2004) An asymptotic determination of minimum centromere size for the maize B chromosome. *Cytogenet Genome Res* **106**, 309–13.
46. Ananiev, E.V., Wu, C., Chamberlin, M.A., Svitashev, S., Schwartz, C., Gordon-Kamm, W., and Tingey, S. (2009) Artificial chromosome formation in maize (*Zea mays* L.). *Chromosoma* **118**, 157–77.
47. Newell, C.A. (2000) Plant transformation technology: developments and applications. *Mol Biotechnol* **16**, 53–65.
48. Srivastava, V., Anderson, O.D., and Ow, D.W. (1999) Single-copy transgenic wheat generated through the resolution of complex integration patterns. *Proc Natl Acad Sci USA* **96**, 11117–21.
49. Zhao, Z.Y., Gu, W., Cai, T., Tagliani, L.A., Hondred, D.A., Bond, D., Krell, S., Rudert, M.L., Bruce, W.B., and Pierce, D.A. (1998) Molecular analysis of T0 plants transformed by *Agrobacterium* and comparison of *Agrobacterium*-mediated transformation with bombardment transformation in maize. *Maize Genet Coop Newsl* **72**, 34–37.
50. Dai, S., Zheng, P., Marmey, P., Zhang, S., Tian, W.Z., Chen, S.Y., Beachy, R.N., and Fau, C. (2001) Comparative analysis of transgenic rice plants obtained by *Agrobacterium*-mediated transformation and particle bombardment. *Mol Breed* **7**, 25–33.
51. Shou, H., Frame, B.R., Whitham, S.A., and Wang, K. (2004) Assessment of transgenic maize events produced by particle bombardment or *Agrobacterium*-mediated transformation. *Mol Breed* **13**, 201–8.

52. Cheng, M., Lowe, B.A., Spencer, M., Ye, X., and Armstrong, C.L. (2004) Factors influencing *Agrobacterium*-mediated transformation of monocotyledonous species *In Vitro Cell Dev Biol Plant* **40**, 31–45.
53. Kato, A., Albert, P.S., Vega, J.M., and Birchler, J.A. (2006) Sensitive fluorescence in situ hybridization signal detection in maize using directly labeled probes produced by high concentration DNA polymerase nick translation. *Biotech Histochem* **81**, 71–8.
54. Kato, A., Vega, J.M., Han, F., Lamb, J.C., and Birchler, J.A. (2005) Advances in plant chromosome identification and cytogenetic techniques. *Curr Opin Plant Biol* **8**, 148–54.
55. Heacock, M., Spangler, E., Riha, K., Puizina, J., and Shippen, D.E. (2004) Molecular analysis of telomere fusions in Arabidopsis: multiple pathways for chromosome end-joining. *EMBO J* **23**, 2304–13.
56. Halpin, C. (2005) Gene stacking in transgenic plants-the challenge for 21st century plant biotechnology. *Plant Biotechnol J* **3**, 141–55.
57. Gleba, Y., Klimyuk, V., and Marillonnet, S. (2007) Viral vectors for the expression of proteins in plants. *Curr Opin Biotechnol* **18**, 134–41.
58. Maliga, P. (2004) Plastid transformation in higher plants. *Annu Rev Plant Biol* **55**, 289–313.
59. Bock, R. (2007) Plastid biotechnology: prospects for herbicide and insect resistance, metabolic engineering and molecular farming. *Curr Opin Biotechnol* **18**, 100–6.
60. Murata, M., Shibata, F., and Yokota, E. (2006) The origin, meiotic behavior, and transmission of a novel minichromosome in *Arabidopsis thaliana*. *Chromosoma* **115**, 311–19.
61. Carlson, W.R. (1970) Nondisjunction and isochromosome formation in the B chromosome of maize. *Chromosoma* **30**, 356–65.
62. Ricci, G.L., Silva, N., Pagliarini, M.S., and Scapim, C.A. (2007) B chromosomes in popcorn (*Zea mays* L.). *Genet Mol Res* **6**, 137–43.
63. Birchler, J.A., Yu, W., and Han, F. (2008). Plant engineered minichromosomes and artificial chromosome platforms. *Cytogenet Genome Res* **120**, 228–32.
64. Yu, W., Han, F., and Birchler, J.A. (2007) Engineered minichromosomes in plants. *Curr Opin Biotechnol* **18**, 425–31.
65. Ow, D.W. (2007) GM maize from site-specific recombination technology, what next? *Curr Opin Biotechnol* **18**, 115–20.
66. Puchta, H. (2002) Gene replacement by homologous recombination in plants. *Plant Mol Biol* **48**, 173–82.
67. Reiss, B. (2003) Homologous recombination and gene targeting in plant cells. *Int Rev Cytol* **228**, 85–139.
68. Pacher, M., Schmidt-Puchta, W., and Puchta, H. (2007) Two unlinked double-strand breaks can induce reciprocal exchanges in plant genomes via homologous recombination and nonhomologous end joining. *Genetics* **175**, 21–9.
69. Wright, D.A., Townsend, J.A., Winfrey, R.J., Irwin, P.A., Rajagopal, J., Lonosky, P.M., Hall, B.D., Jondle, M.D., and Voytas, D.F. (2005) High-frequency homologous recombination in plants mediated by zinc-finger nucleases. *Plant J* **44**, 693–705.
70. Townsend, J.A., Wright, D.A., Winfrey, R.J., Fu, F., Maeder, M.L., Joung, J.K., and Voytas, D.F. (2009) High-frequency modification of plant genes using engineered zinc-finger nucleases. *Nature* **459**, 442–5.
71. Shukla, V.K., Doyon, Y., Miller, J.C., DeKelver, R.C., Moehle, E.A., Worden, S.E., Mitchell, J.C., Arnold, N.L., Gopalan, S., Meng, X., Choi, V.M., Rock, J.M., Wu, Y.Y., Katibah, G.E., Zhifang, G., McCaskill, D., Simpson, M.A., Blakeslee, B., Greenwalt, S.A., Butler, H.J., Hinkley, S.J., Zhang, L., Rebar, E.J., Gregory, P.D., and Urnov, F.D. (2009) Precise genome modification in the crop species *Zea mays* using zinc-finger nucleases. *Nature* **459**, 437–41.
72. Hoa, T.T.C., Bong, B.B., Huq, E., and Hodges, T.K. (2002) Cre/lox site-specific recombination controls the excision of a transgene from the rice genome. *Theor Appl Genet* **104**, 518–25.
73. Li, Z., Xing, A., Moon, B.P., Burgoyne, S.A., Guida, A.D., Liang, H., Lee, C., Caster, C.S., Barton, J.E., Klein, T.M., and Falco, S.C. (2007) A Cre/loxP-mediated self-activating gene excision system to produce marker gene free transgenic soybean plants. *Plant Mol Biol* **65**, 329–41.
74. Vergunst, A.C., Jansen, L.E., Fransz, P.F., de Jong, J.H., and Hooykaas, P.J. (2000) Cre/lox-mediated recombination in Arabidopsis: evidence for transmission of a translocation and a deletion event. *Chromosoma* **4**, 287–97.
75. Srivastava, V., Ariza-Nieto, M., and Wilson, A.J. (2004) Cre-mediated site-specific gene integration for consistent transgene expression in rice. *Plant Biotechnol J* **2**, 169–79.
76. Srivastava, V., and Ow, D. (2001) Single-copy primary transformants of maize obtained through co-introduction of recombinase-expressing construct. *Plant Mol Biol* **46**, 561–6.

Chapter 8

Method for Bxb1-Mediated Site-Specific Integration *In Planta*

Yuan-Yeu Yau, Yueju Wang, James G. Thomson, and David W. Ow

Abstract

Gene targeting in plants through homologous recombination has been sparsely reported, although notable breakthroughs have been achieved in recent years (1). In particular, the use of zinc finger nucleases to promote homologous end joining has revived the promise that homologous gene targeting could someday become practical for plant genetic engineering (2, 3). An alternative and complementary approach that has progressed steadily over the years has been recombinase-mediated site-specific integration (4). In this approach, a first recombination site is introduced into the genome to serve as a target for inserting subsequent DNA. Here, we describe the method for generating the chromosomal target and the subsequent insertion of new DNA into the chromosomal target by Bxb1-mediated site-specific integration. This method would permit the comparison of different molecular constructs at the same genomic locations.

Key words: GMO, Gene transfer, Gene expression, Transgene, Recombination, Cre–*lox*

1. Introduction

The first description of recombinase-based site-specific integration *in planta* was achieved using the Cre–*lox* recombination system, where Cre is a 38-kDa recombinase and *lox* is its 34-bp recognition site (5). The recombination reaction between *lox* sites is reversible; hence, a circular DNA molecule inserting into another molecule can readily excise out. To enrich for insertion events, mutant recombination sites less prone to reversion have often been used, as well as displacement schemes that downregulate recombinase expression upon site-specific integration (6, 7). These experiments have shown that transgenes site-specifically placed into a chromosome location can be expressed with fidelity over the generations (8–10). Another method to obtain stable

James A. Birchler (ed.), *Plant Chromosome Engineering: Methods and Protocols*, Methods in Molecular Biology, vol. 701, DOI 10.1007/978-1-61737-957-4_8, © Springer Science+Business Media, LLC 2011

site-specific integration is through the use of a cassette exchange strategy that does not result in flanking the insert with active recombination sites of the same orientation. Success has been reported for the Cre–*lox* system (11) as well as two other similar systems: R–*RS* (12) and FLP–*FRT* (13), with R and FLP being the recombinases and *RS* and *FRT* being their respective recognition sites. Like Cre–*lox*, their recombination reactions are also reversible.

Despite the practicality of site-specific integration, the use of reversible recombination systems has a limitation. Specifically, it would be difficult to append DNA to the same chromosome location, as repeated rounds of insertions would introduce additional sites that are recombinogenic (14). Hence, from the late 1990s, this laboratory has sought to develop site-specific integration based on an irreversible type of recombination. The first system found to operate in eukaryotes was the phiC31 (ϕC31) system, where the integrase (recombinase) recombines with the phage and bacterial attachment sites known as *attP* and *attB* ((15); in this paper abbreviated as PP′ and BB′, respectively). The product sites, *attL* and *attR*, are not substrates for phiC31 integrase, unless an accompanying excisionase is available. Hence, in practice, the reaction can be considered unidirectional. Insertion into chloroplast DNA through use of this system has been reported (16), as have intermolecular recombination events *in planta* (17). More recently, this laboratory has described additional systems with properties analogous to phiC31 (18). In this chapter, we describe using the Bxb1 system to generate transgenic tobacco plants that harbor site-specifically integrated DNA.

2. Materials

2.1. Recombinant DNA

1. LB agar medium: LB Agar (Lennox L Agar, Invitrogen, Carlsbad, CA, Cat. No. 22700-041). Add 32 g per 1 L dH_2O and autoclave.
2. LB liquid medium: Luria Broth (Miller's LB Broth Base, Invitrogen, Cat. No. 12795-084). Add 25 g per 1 L dH_2O and autoclave.
3. T4 polynucleotide kinase: From New England Biolabs Inc., Ipswich, MA (Cat. No. M0201). Store at –20°C.
4. T4 DNA ligase: From New England Biolabs Inc. (Cat. No. M0202S). Store at –20°C.
5. Restriction enzymes: From New England Biolabs Inc. Store at –20°C.

6. DNA Polymerase I, Large (Klenow) Fragment: From New England Biolabs Inc. (Cat. No. M0210S). Store at –20°C.
7. High-Fidelity PCR kit: PfuTurbo® DNA Polymerase (Stratagene, La Jolla, CA, Cat. No. 600250) or Phusion™ High-Fidelity DNA Polymerase (New England Biolabs Inc., Cat. No. F-553).
8. pNMT-TOPO (Invitrogen, Cat. No. K18101).
9. Ampicillin: Dissolve ampicillin sodium salt (Sigma, St. Louis, MO, Cat. No. A9518) in dH_2O, filter-sterilize (0.22 μm Millex® GP Syringe Driven Filter Unit, Millipore, Billerica, MA, Cat. No. SLGP033RS).
10. Kanamycin: Dissolve kanamycin sulfate (from *Streptomyces kanamyceticus*) (Sigma, Cat. No. K4000) in dH_2O and filter-sterilize (0.22-μm Millex® GP Syringe Driven Filter Unit).
11. Hygromycin B: In phosphate-buffered saline (PBS), 50 mg/mL, filter-sterilized (Roche, Indianapolis, IN, Cat. No. 10 843 555 001). Store at ~4°C.
12. *Escherichia coli* DH5α competent cells: Genotype: F^-ϕ80*lac*ZΔM15 Δ(*lac*ZYA-*arg*F)U169 *rec*A1 *end*A1 *hsd*R17(r_k^-, m_k^+) *pho*A *sup*E44 *thi*-1 *gyr*A96 *rel*A1 λ^-. Subcloning Efficiency™ DH5α™ Competent Cells (Invitrogen, Cat. No. 18265-017), or electroporation-grade DH5α competent cells prepared as described by Miller and Nickoloff (19).
13. QIAprep® Spin Miniprep Kit (Qiagen, Valencia, CA, Cat. No. 27106).
14. QIAGEN® Plasmid Midi Kit (Qiagen, Cat. No. 12143).

2.2. PCR Analysis

1. Extraction buffer: 200 mM Tris–HCl, pH 5.7, 250 mM NaCl, 25 mM EDTA, and 0.5% SDS.
2. Isopropanol: 2-Propanol, 99.5%, HPLC grade (Sigma, Cat. No. 270490).
3. TE (Tris–EDTA) buffer: 10 mM Tris and 1 mM EDTA, pH 8.0.
4. DNA polymerase set: GoTaq® Flexi DNA Polymerase with 5× Colorless GoTaq® Flexi buffer and 25 mM $MgCl_2$ solution (Promega, Madison, WI, Cat. No. M829B).
5. 2.5 mM deoxynucleotide triphosphate (dNTP) mix: Dilute from 10 mM dNTP mix (10 mM each of dATP, dCTP, dGTP, dTTP, pH 7.5; Promega, Cat. No. U151B).
6. Primers: Oligomers (Bioneer, Inc., Alameda, CA). Dilute the oligomer pellet to a working concentration of 10 μM with autoclaved dH_2O; store at –20°C.
7. DMSO: Dimethyl sulfoxide (Sigma, Cat. No. D-5879).

8. 50× TAE (Tris–Acetate–EDTA): 242 g Tris base, 57.1 mL glacial acetic acid, 100 mL 0.5 M EDTA (pH 8.0), adjusted to 1 L with dH_2O.
9. Agarose: UltraPure™ agarose (Invitrogen, Cat. No. 15510-027).
10. DNA size markers: 1 kb DNA Ladder (Promega, Cat. No. G5711).
11. QIAquick® Gel Extraction Kit (Qiagen, Cat. No. 28704).

2.3. Tobacco Tissue Culture

1. MS salts: Murashige and Skoog Basal Salt Mixture (MS) (Sigma, Cat. No. M5524).
2. Vitamins: Gamborg's vitamin solution, 1,000×, plant cell culture tested (Sigma, Cat. No. G1019).
3. α-Naphthaleneacetic acid (NAA), 1 mg/mL, plant cell culture tested (Sigma, Cat. No. N1641).
4. Benzyladenine (BA or BAP), also known as 6-benzylaminopurine hydrochloride (Sigma, Cat. No. B5920). A few drops of concentrated HCl dissolves the powder.
5. 2,4-Dichlorophenoxyacetic acid (2,4-D), plant cell culture tested (Sigma, Cat. No. D7299), dissolved in ethanol.
6. Agar: Select agar® (Invitrogen, Cat. No. 30391-049).
7. MSNT Agar medium: 1× MS salts, 87.6 mM sucrose, 1× Gamborg's vitamin solution. For 1 L, add 4.3 g MS salts, 30.0 g sucrose, and 1.0 mL 1,000× Gamborg's vitamin solution into dH_2O, adjust pH to 5.8 with 1 N KOH, 8 g of Select agar®, adjust volume, and autoclave.
8. MSNTS Agar medium: MSNT medium supplemented with 537 nM NAA and 4.44 nM BA. For 1 L, add 4.3 g MS salts, 30.0 g sucrose, 1.0 mL 1,000× Gamborg's vitamin solution, 100 µL 1 mg/mL NAA, 50 µL 20 mg/mL BA, adjust pH to 5.8 with 1 N KOH, 8 g of Select agar®, adjust volume, and autoclave.
9. *Nicotiana tabacum* cv. Wisconsin 38: Plants harboring the target construct are grown in MSNT Agar medium in Phytatray™ II boxes (Sigma, Cat. No. P5929) at 26°C, 16 h light in a growth chamber. Plant stem apices are cut and transferred to fresh MSNT Agar medium every 4–6 weeks for vegetative propagation.

2.4. Particle Bombardment

1. Gold particles: Biolistic® 1.0 Micron Gold particles (Bio-Rad, Hercules, CA, Cat. No. 1652263), in 60 µg/µL aliquots in dH_2O, prepared according to the manufacturer's instructions. Stored at −20°C.
2. Rupture disks: Biolistic® 1,100 psi rupture disk (Bio-Rad, Cat. No. 1652329).

3. Stopping screens: Biolistic® stopping screens (Bio-Rad, Cat. No. 1652336).
4. 100% Ethanol: 200 Proof, HPLC/spectrophotometric grade (Sigma, Cat. No. 459828).
5. Spermidine: Free base (Sigma, Cat. No. S0266).
6. Macrocarriers (Bio-Rad, Cat. No. 1652335).
7. Macrocarrier holders (Bio-Rad, Cat. No. 1652322).

2.5. Agrobacterium Transformation

1. ElectroMax™ *A. tumefaciens* strain LBA4404 competent cells (Invitrogen, Cat. No. 18313-015). Store at –80°C.
2. Gene Pulser® Cuvette (Bio-Rad, Cat. No. 165-2089).
3. Cefotaxime solution: Dissolve cefotaxime sodium salt (Agri-Bio, Inc., Bay Harbour, FL, Cat. No. 2000) in dH_2O and filter-sterilize with a 0.22-μm Millex® GP Syringe Driven Filter Unit.
4. Acetosyringone: 3′,5′-Dimethoxy-4′-hydroxyacetophenone (Aldrich, Milwaukee, WI, Cat. No. D13440-6).

2.6. Protoplast Transformation

1. Cellulase: Cellulase "Onozuka RS," derived from *Trichoderma viride* (*Phyto*Technology Laboratories™, Shawnee Mission, KS, Cat. No. C214).
2. Macerozyme: Macerozyme R-10, derived from *Rhizopus* sp. (Serva Electrophoresis GmbH, Heidelberg, Germany, Cat. No. 28302).
3. D-mannitol (Sigma, Cat. No. M1902).
4. D-xylose (Fisher Scientific, Fair Lawn, NJ, Cat. No. X9-25).
5. MES: 2-(*N*-morpholino)ethanesulfonic acid (Sigma, Cat. No. M8250).
6. Polyethylene glycol (PEG), MW = 3350 (Sigma, Cat. No. P3640).
7. Carrier DNA: 10 mg/mL Solution of calf thymus DNA (Sigma, Cat. No. D1501). Sonicated to average size of 0.5–2 kb, extracted with phenol, then chloroform, and precipitated with ethanol. Alternatively, Sigma Cat. No. D8661 may be used as supplied.
8. Protoplasting enzyme solution: Dissolve 1% cellulase and 0.2% macerozyme in K_3AS solution (see below), adjust the pH to 5.8, filter-sterilize the solution with a Millipore 0.22-μm vacuum-driven disposable filtration Stericup® and Steritop™ setup (Millipore, Cat. No. SCGPU01RE). Prepare fresh for each use.
9. K_3A medium: 1× MS salts, 3.13 mM $CaCl_2{\cdot}2H_2O$, 1.67 mM xylose, 300 mM sucrose, 400 mM D-mannitol, 1× Gamborg's

vitamin solution, 5.38 μM NAA, 888 nM BA, and 452 nM 2,4-D. For 1 L, add 4.3 g MS salts, 0.46 g $CaCl_2 \cdot 2H_2O$, 0.25 g xylose, 102.96 g sucrose, 72.8 g D-mannitol, 1.0 mL 1,000× Gamborg's vitamin solution, 1 mL 1 mg/mL NAA, 10.0 μL 20 mg/mL BA, 0.5 mL 200 μg/mL 2,4-D into dH_2O, adjust pH to 5.8 with 1 N KOH, filter-sterilize with a Millipore 0.22-μm vacuum-driven disposable filtration Stericup® and Steritop™ setup. Solution stable at room temperature for at least a month.

10. K_3AS solution: 1× MS salts, 3.13 mM $CaCl_2 \cdot 2H_2O$, 1.67 mM xylose, 400 mM sucrose, 1× Gamborg's vitamin solution, 10.8 μM NAA, and 2.22 μM BA. For 1 L, add 4.3 g MS salts, 0.46 g $CaCl_2 \cdot 2H_2O$, 0.25 g xylose, 137.0 g sucrose, 1.0 mL 1,000× Gamborg's vitamin solution, 2.0 mL 1 mg/mL NAA, 25.0 μL 20 mg/mL BA, dH_2O, adjust pH to 5.8 with 1 N KOH, filter-sterilize with a Millipore 0.22-μm vacuum-driven disposable filtration Stericup® and Steritop™ setup. Solution stable at room temperature for at least a month.
11. W5 solution: 154 mM NaCl, 125 mM $CaCl_2 \cdot 2H_2O$, 5 mM KCl, and 5 mM glucose. For 1 L, add 9 g NaCl, 18.37 g $CaCl_2 \cdot 2H_2O$, 0.37 g KCl, and 0.90 g glucose to dH_2O, adjust pH to 5.8 with 1 N KOH, filter-sterilize with a Millipore 0.22-μm vacuum-driven disposable filtration Stericup® and Steritop™ setup. Solution stable at room temperature for at least a month.
12. 2× MaMg solution: 0.8 M D-mannitol, 30 mM $MgCl_2$, 0.2% (w/v) MES. For 100 mL, add 14.6 g D-mannitol, 0.61 g $MgCl_2$, and 0.20 g MES to dH_2O, adjust pH to 5.6 with 1 N KOH, filter-sterilize with a Millipore 0.22-μm vacuum-driven disposable filtration Stericup® and Steritop™ setup. 2× MaMg is used to prepare the PEG solution. 1× MaMg is diluted from 2× MaMg.
13. 40% (w/v) PEG solution: Dissolve PEG in 2× MaMg solution, adjust volume to 1× MaMg, check pH 5.6–7.0; adjust if needed with KOH or HCl, filter-sterilize with a Millipore 0.22-μm vacuum-driven disposable filtration Stericup® and Steritop™ setup. May be frozen and stored in 1.5-mL aliquots at –20°C for future use.
14. Wash solution: 0.2 M $CaCl_2 \cdot 2H_2O$, 0.5% (w/v) MES. For 500 mL, add 14.7 g $CaCl_2 \cdot 2H_2O$ and 2.5 g MES to dH_2O, adjust pH to 5.8 with 1 N KOH, filter-sterilize with a Millipore 0.22-μm vacuum-driven disposable filtration Stericup® and Steritop™ setup. Solution stable at room temperature for at least a month.

3. Methods

To conduct recombinase-mediated site-specific integration, a target line must first be established harboring a target construct with a target recombination site. Typically, transgenic lines are screened for the random integration of a nonrearranged single copy of a target construct, such as pYWP72 or pYWB73 (Fig. 1a, e). Upon establishing a target line, an integrating construct such as pYWJTSB2 or pYWSP3 (Fig. 1b, f) is introduced into the target line. Site-specific recombination between the episomal and genomic recombination sites is promoted by a site-specific recombinase, which can be provided by the transient expression of a cointroduced DNA construct (Fig. 1c) that expresses the Bxb1 recombinase gene.

Prior to conducting this integration *in planta*, an initial test can be conducted in bacteria to ascertain that two participating DNA constructs can indeed recombine to form a cointegrate molecule and that a PCR assay can detect the recombinant junctions formed by the recombination between *attP* and *attB*. Upon successful detection of recombination, plant target lines derived from *Agrobacterium* transformation can also be tested through a transient assay to ascertain that the introduced DNA recombines with the chromosomally situated target. In this chapter, we describe (a) the DNA constructs, (b) the generation of target lines, (c) the transient assays for detecting site-specific recombination in bacteria and plant cells, and (d) the subsequent stable selection for transformed cells leading to transgenic site-specific integrants.

3.1. Recombinant DNA Constructs

Standard recombinant DNA methods as described in Sambrook et al. (20) were used throughout. Bacterial transformations were carried out using *E. coli* DH5α. Plasmid DNA was isolated and purified using QIAprep® Spin Miniprep and QIAGEN® Plasmid Midi Kits.

Target constructs pYWP72, pYWB73. As targets for site-specific integration, an array of *attP* or *attB* sites from ϕC31, TP901, U153, Bxb1 recombination systems was assembled by the stepwise addition of each synthetic recombination site. The ϕC31 *attP* attachment site was assembled from four overlapping oligonucleotides (for all oligonucleotides, see Note 1). Oligonucleotides were phosphorylated by T4 polynucleotide kinase (20 U), ATP (5 μM), 37°C, 1 h. After inactivating the kinase at 80°C for 15 min, 1 μL (10 pmol) of each of four phosphorylated oligonucleotides was annealed in 6 μL of water, at 95°C for 10 min and cooled to room temperature. The annealed product with *Apa*I and *Xma*I overhangs was inserted between the *Apa*I and *Xma*I sites of pRB140-ApaI [from Robert Blanvillain, comprises

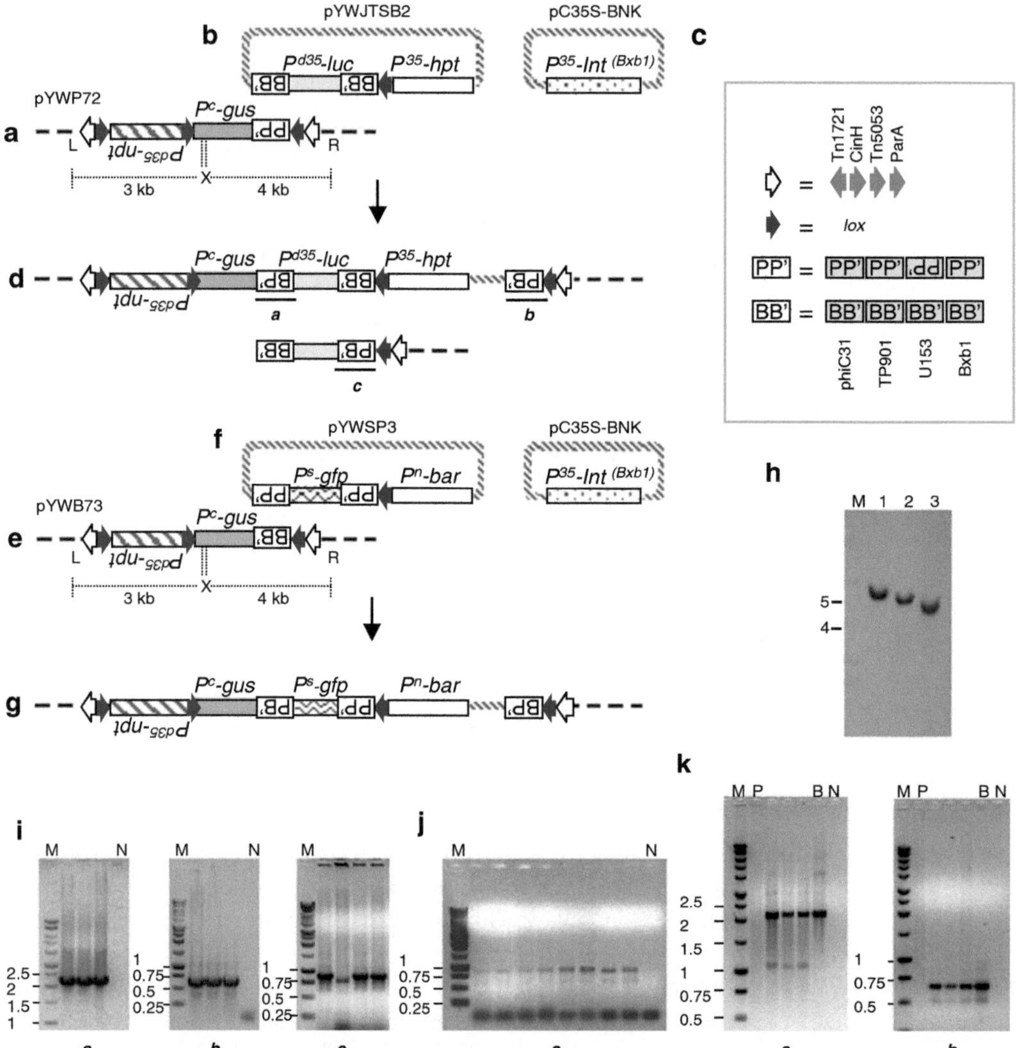

Fig. 1. Site-specific integration by *attP* × *attB* recombination. Schematic not to scale representation of target constructs (**a**) pYWP72 and (**e**) pYWB73; integration constructs (**b**) pYWJTSB2 and (**f**) pYWSP3, (**c**) Bxb1 recombinase-expression construct pC35S-BNK or p35S-BNK, and expected integration structures (**d, g**). Symbols for recombination sites indicated. P^c CoYMV promoter, P^n *nos* promoter, P^s ScBV promoter, P^{35} CaMV 35S promoter, P^{d35} CaMV 35S promoter with duplicated enhancer, *npt* neomycin phosphotransferase II gene, *gus* β-d-glucuronidase gene, *luc* firefly luciferase gene, *hpt* hygromycin phosphotransferase II gene, *gfp* green fluorescent protein gene, *bar* bialaphos resistance gene, *int* integrase gene, from mycobacteriophage Bxb1. X, *Xba*I site; L, R, T-DNA left and right borders, respectively. Genes transcribed left to right except for *npt*, gene terminators not shown. (**h**) Representative Southern blot for single copy pYWP72 target construct in the genome showing a single band >4 kb (distance from *Xba*I to *R* = 4 kb) hybridizing to a *gus* DNA probe for three independent plant lines. Representative PCR analysis for site-specific recombination junctions in bacteria (**i**), bombarded plant leaf tissue (**j**), and transgenic leaf tissue (**k**). Recombination with *luc* upstream and downstream *attB* sites leads to predicted PCR junction products a (2.2 kb) and c (0.85 kb), respectively, as indicated in (**d**) and labeled below the gels. PCR junction product b (0.7 kb) confirms recombination with *luc* upstream *attB*. Individual lanes represent samples tested, except for *lane M*, size markers (in kb); *lane N*, no template negative control; P, target plant DNA; B, bacterial assay product used as positive control (N, P, B not shown in all *panels*).

(*left border -loxP-P*^{d35}*-npt-loxP-P*^{C}*-gus-ϕC31 attB-loxP (inverted)-right border*)] to yield pYWP1 (*see* Fig. 1 for abbreviations). The TP901 *attP* recombination site, assembled from four overlapping oligonucleotides as described for the ϕC31 site, was inserted into pYWP1 through *Bam*HI and *Apa*I sites to yield pYWP2. Likewise, the U153 assembled *attP* recombination site was inserted into pYWP2 through the *Bam*HI sites to generate pYWP3. The *Bam*HI site between U153 and TP901 *attP* sites was abolished by generating a 5′-AGATCC-3′ sequence. The assembled Bxbl *attP* site was inserted into the *Bam*HI site of pYWP3 to yield pYWP4, with *Bam*HI site between Bxbl and U153 *attP* sites was abolished by a 5′-AGATCC-3′ formation.

For the *attB* array, a Bxbl *attB* from overlapping oligonucleotides with GATC overhangs was inserted into the *Bam*HI sites of pRB140 to regenerate a *Bam*HI site only next to the ϕC31 *attB* and which also included an adjacent *Apa*I site (see Note 1). The resulting plasmid, pYWB1 was used for insertion of an assembled U153 *attB* site with *Bam*HI and *Apa*I overhangs to yield pYWB2. Finally, an assembled TP901 *attB* with GATC overhangs was inserted into the pYWB2 *Bam*HI site to regenerate a single *Bam*HI site next to ϕC31 *attB* to form pYWB3.

An array of recognition sites from recombination systems CinH, ParA, Tn1721, and Tn5053 for site-specific deletion of the transgenic locus was assembled in pP50C17. Primers with unique restriction sites at the ends were used to amplify each of the four recognition sites using PfuTurbo DNA Polymerase (see Note 1). The PCR amplified CinH site was inserted into pNMT-TOPO to yield pCinHres, which was cleaved with *Pac*I and *Not*I, for insertion of the ParA site to form pCP, followed by *Pac*I and *Bam*HI cleavage for insertion of the Tn1721 site to generate pC17P, and finally followed by cleavage of pC17P *Bam*HI and *Pst*I sites for insertion of the Tn5053 site. However, *Pst*I cleavage also removed the CinH-Tn1721 sites, which had to be replaced by a *Pst*I fragment from pC17P to generate pP50C17. The fragment containing the ParA, Tn5053, CinH, and Tn1721 recombination sites was retrieved from pP50C17 from *Pml*I cleavage and blunt end ligated into the *Pme*I site, near the RB of pYWP4 to yield pYWP5, and to pYWB3 to yield pYWB4. The same fragment was then ligated into the *Pml*I site, near the LB of pYWP5 or to pYWB4 to yield the target construct pYWP72 or pYWB73 (Fig. 1b).

Integrating constructs pYWJTSB2, pYWSP3. A *Bst*XI to *Ase*I fragment comprising of the *35S* promoter-*hptII-35S* terminator was isolated from pCambia1300 (http://www.openinnovation.org/daisy/cambia/585.html) and made blunt by treatment with DNA polymerase I, Large (Klenow) Fragment, and dNTPs, and inserted into the *PmaC* I of pRB160 (from Robert Blanvillain) to yield pYWS0.

A fragment containing the array of ϕC31-TP901-U153-Bxb1 *attB* sites was PCR-amplified with PfuTurbo® DNA Polymerase from pYWB73 using primers (5′-ATAGCGGCCGC ATAGAAA ACAAAATATAGCGCGCAAACA-3′ and 5′-ATATTAATTAA TTA TGACGGTCTCGAAGCCGCGGT-3′) that included restriction sites *Not*I and *Pac*I (underlined) to insert into pYWS0 to produce pYWSB1. The same fragment with four *attB* recombination sites is also PCR-amplified from pYWB73 using primers (5′-ATACTGCAGATAGAAAACAAAATATAGCGCG CAAAC-3′and5′-ATAGTTAACTTATGACGGTCTCGAAGCCGCGGT-3′) that included *Pst*I and *Hpa*I sites (underlined) to insert into pYWSB1 to form pYWSB2. However, a mutation was found in *hpt*II, necessitating a replacement of a *Hpa*I to *Spe*I fragment with a corresponding (blunt end)-*Kpn*I-*loxP* (inverted)-*35S* promoter-*hptII-35S* terminator-*Spe*I fragment amplified from pCambia1300 by Phusion™ High-Fidelity DNA Polymerase to yield pYWJTSB2 (Fig. 1b). Primers used were: 5′-AGTCGGTACC*ATAACTTC GTATAGCATACATTATACGAAGTTAT*TTGGCTA GAGCAGCTTGC-3′ [*Kpn*I site underlined; *loxP* (inverted) in italic]; 5′-AGTCACTAGTCTGTATCGAGTGGTGATTTTG-3′ (*Spe*I site underlined).

Construct pYWSP3 is analogous to pYWJTSB2 except that it contains *attP* instead of *attB* arrays, selectable marker *bar* instead of *hpt*, and reporter gene *gfp* instead of *luc*. Its assembly consists of similar stepwise additions or replacement of existing DNA in progenitor plasmids of pYWSB2. In sum, its relevant DNA from left to right in Fig. 1f comprises: a *Sac*I to *Kpn*I fragment of the *attP* array (PCR amplified from pYWP72), a *Kpn*I to *Xho*I fragment of the sugarcane bacilliform badnavirus ScBV promoter (21), a *Xho*I to *Not*I fragment encoding GFP, a *Not*I to *Pst*I fragment comprising an octopine synthase gene (*ocs*) terminator, a second *Pst*I to *Pst*I *attP* array, and a *Pst*I to *Eco*RI (blunted) fragment comprising an inverted *loxP* site and a *nos* promoter-*bar*-*nos* terminator.

Bxb1 recombinase-expressing constructs pC35S-BNK and p35S-BNK. pC35S-BNK was derived from the *cre*-expression construct pMM23 (*22*). An *Asc*I site replaced the *Xba*I site between *35S* promoter and *cre*, and a *Spe*I site was added to the *Sph*I site between *cre* and *nos3′* to yield pMM23AS. A *Hin*dIII-P^{35}-*cre*-*nos3′*-*Sac*I fragment was retrieved from pMM23AS and inserted into pCambia-2300 (http://www.openinnovation.org/daisy/cambia/585.html), an *Agrobacterium* binary vector with kanamycin genes for selection (*nptII* for plant and *nptIII* for bacteria), leading to pC35S-Cre. The *Bxb1* ORF was PCR-amplified from pNMTBxb1 (*18*) with Phusion™ High-Fidelity DNA Polymerase using a forward primer (5′-AGTCGGCGCGCC*GCCACC*ATG AGAGCCCTGGTAGTCATCC-3′) that included a *Asc*I site (underlined) and a Kozak sequence (italic), and a reverse primer

(5′-ACTACTAGTCTAAGCGGCCGC*CGGGTCCTCGACCTT TCGCTTCTTCTTCGGGGCCG*CCGACATCCCGGTGTGTA GCC-3′) with a *Spe*I site (underlined) and a nuclear localization signal from SV40 (NLS; italic), and inserted between the *Asc*I and *Spe*I sites of pC35S-Cre, replacing the *cre* ORF with the *Bxb1* ORF to yield pC35S-BNK. The same Bxb1 ORF was later transferred from pC35S-BNK into the pMM23AS backbone to yield p35S-BNK.

3.2. Generation of Target Lines

1. pYWP72 or pYWB73 was introduced into *A. tumefaciens* LBA4404 by electroporation in a Gene Pulser® Cuvette, with a 0.1-cm electrode gap. The *Bio-Rad* Gene Pulser™ device was set to 25 μFD (capacitance), 1.8 kV (field strength), and the Gene Pulser™ Controller was set to 200 Ω (resistance). Transformants were selected on LB agar plates with 50 μg/mL kanamycin at 28°C. Single colonies were grown in LB liquid medium for PCR analysis to confirm the plasmid presence.
2. LBA4404 harboring pYWP72 or pYWB73 was used for leaf disc transformation of tobacco (23). Surface-sterilized tobacco seeds were germinated on MSNT agar medium. Aseptically excised leaf explants from 7 to 8 cm tall plants were dipped into *A. tumefaciens* (OD_{600} = 0.8) culture for 45 min, blotted dry with sterile filter paper, and then placed on MSNT agar medium supplemented with 1 μg/mL BA and 100 μg/mL acetosyringone. After 3 days, infected explants were transferred to MSNT agar medium supplemented with 1 μg/mL BA, 0.1 μg/mL NAA, 500 μg/mL cefotaxime, and 100 μg/mL kanamycin.
3. Putative transgenic shoots were observed after 6 weeks, which were excised and planted to root on hormone-free MSNT agar medium containing 250 μg/mL cefotaxime.
4. Standard Southern hybridization was used to determine target lines with a single precise copy insertion of the introduced target construct (Fig. 1h).

3.3. Functional Recombination in Bacteria

1. Plasmid DNA (~0.25 μg) representing the target construct (marked by a bacterial kanamycin resistance gene), the integration construct (marked by a bacterial hygromycin resistance gene), and the recombinase-expressing construct (marked by a bacterial ampicillin resistance gene) were cotransformed into *E. coli* DH5α by electroporation (voltage at 1.5 kV). Transformed bacteria were selected on LB plates supplemented with ampicillin (100 μg/mL), kanamycin (50 μg/mL), and hygromycin B (50 μg/mL).
2. Individual 2-day-old colonies were picked with a sterile toothpick and directly added to PCR tubes. PCR was carried out in 20-μL reactions containing 1× GoTaq® Flexi buffer,

2.5 mM $MgCl_2$, 0.25 mM each of dGTP, dATP, dTTP, and dCTP, 0.5 μM of each primer, 1 U of GoTaq® Flexi DNA polymerase, and 0.25 μL of dimethyl sulfoxide (DMSO) (see Note 2), in a Techne™ Thermal cycler (Techne Inc., Burlington, NJ), with conditions: One cycle at 94°C for 4 min; 38 cycles at 94°C (30 s), 57°C (1 min), and 72°C (1 min); and then finally one cycle at 72°C for 2 min.

3. PCR products were resolved on a 1% 1× TAE agarose gel supplemented with 0.125 μg/mL ethidium bromide (EtBr), and photographed under a UV light (Fig. 1i).
4. Gel bands were excised and purified with a QIAquick® Gel Extraction Kit for sequencing, following the instructions provided by the manufacturer. The purified bands were also used as positive controls for subsequent PCR detection of site-specific integration *in planta*.
5. The same primers used to amplify the recombination junctions were used in sequencing the PCR product according to the manufacturer's BigDye® Terminator v. 3.1 Cycle Sequencing Kit Protocol with an ABI Prism 3100 Genetic Analyzer (Applied Biosystems, Foster, CA). Sequencing data were analyzed with the Sequencher™ software (Gene Codes Corporation, Ann Arbor, MI) and the program Basic Local Alignment Search Tool (BLAST) from the National Center for Biotechnology Information (NCBI, http://www.ncbi.nlm.nih.gov).

3.4. Functional Recombination In Planta (Biolistic Transient Assay)

1. Leaves of approximately 5-cm long from a tobacco target line were excised from tissue-cultured plants with a sterile surgical blade (USA Scientific, Woodland, CA) and arranged in the center of a Petri dish with MSNTS agar medium.
2. DNA comprising the integration construct and the recombinase-expressing construct were added to 50 μL gold particle suspension, under continuous vortexing in the order of 5 μL of DNA (1 μg/μL, consisting of 4 μg of the recombinase-expressing construct and 1 μg of the integration construct), 50 μL 2.5 M $CaCl_2$, and 20 μL 0.1 M spermidine. After continuous vortexing for 3 more minutes, gold particles were centrifuged at 9,500×*g* for 10 s and as much of the supernatant as possible was removed. The gold particles were then washed with 250 μL of 100% ethanol. Washed gold particles were resuspended in 60 μL of 100% ethanol, and typically, 10 μL of the DNA-coated particles were placed onto the center of the macrocarrier installed in the macrocarrier holder, and allowed to air-dry. Each leaf sample was bombarded once with the Biolistic™ PDS-1000/He Particle Delivery System (Bio-Rad, Hercules, CA) with bombardment conditions as following: helium pressure 1,100 psi, stopping plate to target

tissue distance 90 mm, vacuum pressure 27 in. Hg. Bombarded areas on the leaf were marked, and the Petri dish was sealed with Parafilm and incubated at 30°C in the dark for 24 h prior to the analysis of site-specific recombination.

3. Particle-bombarded leaf tissue (~1 cm diameter, or approximately 10 mg) was placed in a 1.5-mL microcentrifuge tube with 400 μL of extraction buffer and ground with a Kontes pellet pestle® (VWR, Batavia, IL), manually or by a pellet pestle® motor, then centrifuged at 16,000×*g* for 10 min to remove tissue debris.
4. 300 μL of the supernatant was moved to a new 1.5-mL microcentrifuge tube, mixed with 300 μL of room-temperature isopropanol, and the DNA precipitated by several inversion of the tube. After centrifugation at 16,000×*g* for 10 min at room temperature, the DNA pellet was washed with 70% ethanol, resuspended in 50 μL of TE, extracted with phenol/chloroform, reprecipitated with ethanol, and washed with 70% ethanol according to standard procedures.
5. The dried DNA pellet was resuspended in 25 μL of TE, and ~4 μL (~100 ng) was used for PCR or sequencing analysis (Fig. 1j).

3.5. Site-Specific Integration in Plants (PEG Transformation of Protoplasts)

1. Fourteen healthy leaves about 3–4 cm long were placed in a sterile Petri dish and cut with a sterile scalpel perpendicular to the midrib into strips about 2 mm wide (see Note 3). Fourteen leaves typically yield 2×10^7 protoplasts for five samples of PEG transformation. The leaf strips were transferred as they were cut to another 100×25 mm Petri dish containing 20 mL of protoplasting enzyme solution.
2. Plates were sealed with Parafilm and agitated on a platform shaker at ~80 rpm for 10 min, followed by incubation of 14–18 h at room temperature in the dark (typically overnight). Afterward (day 2, Fig. 2a), the enzyme-treated leaf strips were agitated on a platform shaker at 80 rpm for 10 min.
3. To facilitate the release of protoplasts, the leaf debris were gently pipetted up and down several times with a 1-mL Gilson Pipetman®. Wide-bore tips were used (cut to 3 mm opening prior to autoclaving). The digestion mix was filtered through a sterile nylon mesh screen (Nylon Monofilament Filter Cloth, 100-μm mesh opening, 145 mesh screen, Cat No. CMN-0105-D, Small Parts, Inc., Miramar, FL) with a sterile glass funnel and collected in a sterile 50-mL Falcon® conical polystyrene tube (BD Biosciences, San Jose, CA). The nylon screen (which can be reused) and the funnel were sterilized by autoclaving.

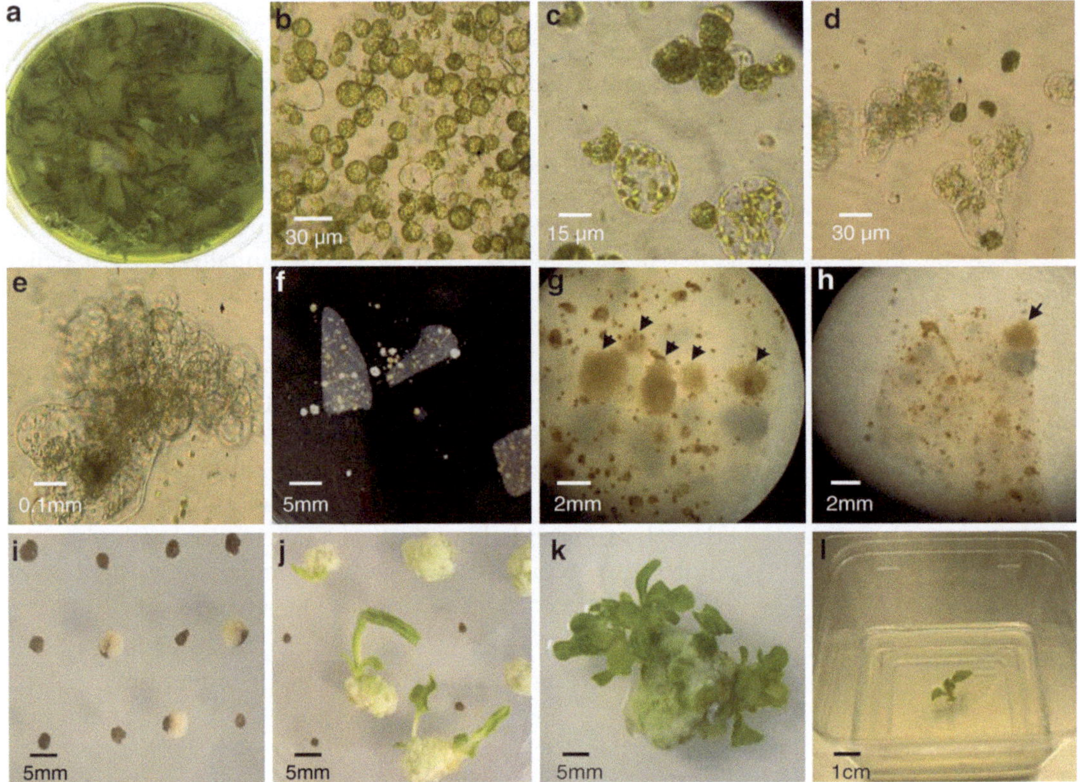

Fig. 2. PEG-mediated protoplast transformation. (**a**) Tobacco leaf strips treated with protoplasting enzyme solution. Protoplasts prior to PEG treatment (**b**), and recovery without selection for 1 (**c**), 2 (**d**), or 3 (**e**) weeks after PEG treatment, with newly formed cells *lighter green* in color. For selection of transformants, 2-week-old PEG-treated protoplasts are typically embedded in agarose for a week prior to antibiotic selection for another 6 weeks. Putative resistant calli in solution (**f**) or in embedded gel (**g**, **h**) transferred to plates (**i**) for shoot formation (**j**, **k**). Shoots excised to root in tissue culture boxes (**l**).

4. The filtered protoplast solution was centrifuged at 200×*g* for 20 min at room temperature in a swinging bucket rotor (Model TJ-6 centrifuge, Beckman, Fullerton, CA) to float the intact protoplasts. The floating protoplasts were transferred to a fresh 50-mL sterile Falcon® conical polystyrene tube with a 1-mL Gilson Pipetman®. This step requires patience and practice. To efficiently remove the floating protoplasts, touch the pipette tip (cut to 3 mm opening prior to autoclaving) to the floating protoplasts and gently draw in the protoplasts. Harvested volume should be around 7 mL. Make sure to keep the final volume less than 10 mL so that it can fit into a 50-mL Falcon® tube in the next step.
5. Harvested protoplasts were diluted with four volumes of W5 solution and mixed gently and centrifuged at 200×*g* for 5 min at room temperature. Supernatant was removed, and the protoplasts were resuspended and washed twice in 40 mL

of W5 solution, followed by centrifugation at 200×*g* for 5 min at room temperature.

6. After two washes, the protoplast pellet was resuspended in 10 mL of W5 solution and the protoplast count determined with a hemocytometer.
7. The protoplasts were centrifuged and resuspended in 10 mL of 1× MaMg solution, centrifuged again and resuspended again in 1× MaMg solution at a density of ~ 8×10^6 intact protoplasts per mL (Fig. 2b). Each PEG transformation sample used ~4×10^6 intact protoplasts (~0.5 mL). Transformation was performed in a sterile 50-mL Falcon® conical polystyrene tube.
8. DNA was mixed with the protoplasts. Typically, transformation DNA in a volume of less than 30 μL consists of 10 μg of a recombinase-expressing construct, 3 μg of an integration construct, and 50 μg of carrier DNA (see Note 4).
9. 0.6 mL of 40% PEG (in 1× MaMg) was gently added to the DNA–protoplast mixture, swirling gently, and the mixture left at room temperature for 25 min, occasionally swirling gently.
10. Protoplasts were diluted slowly and gently with 10 mL of K_3A medium by adding 2 mL dropwise while swirling the mixture, waiting for 10 min before another 2 mL was added for a total of 10 mL. This slow dilution is very tedious, but in some cases, it seems to have a major effect on protoplast regeneration.
11. Diluted protoplasts were transferred to 100 mm×25 mm Petri dishes, and another 10 mL K_3A medium was gently added to a total volume of 20 mL. Protoplasts may be observed on an inverted microscope. Many shrunken (evacuolated) or broken cells indicate problems.
12. Plates were sealed with Parafilm and placed in the dark (without shaking) at 26°C. After a week, protoplasts typically have undergone cell wall regeneration and several cell divisions (Fig. 2c–e). Two-week-old protoplasts were transferred to a sterile 50-mL Falcon® conical polystyrene tube and supplemented with an equal volume of the Wash solution, mixed gently, and centrifuged at 200×*g* for 7 min at room temperature.
13. Supernatant was removed except for ~0.5 mL, in which the protoplasts were gently resuspended and transferred to a new plate. 1% NuSieve® GTG low-melting agarose (Lonza, Rockland, ME) was prepared by adding NuSieve® GTG low-melting agarose to K_3A solution, autoclaving the agarose solution, and cooling to 37°C in a waterbath. 20 mL of 1% NuSieve® GTG low-melting agarose was added to the plates, swirling gently to distribute the protoplasts evenly. After gel formation, the agarose-embedded protoplasts were placed in

the dark at room temperature for a week before administering antibiotic selection. An inverted microscope can be used to monitor the growth of embedded protoplasts.

14. After 1 week, a sterile surgical blade was used to cut the disc of agarose-embedded protoplasts into eight pie slices. For a nonselected control to gauge protoplast viability, one slice was transferred to a fresh plate containing 30 mL K_3A. The other slices were transferred to 30 mL of K_3A containing a selection agent (such as 10 μg/mL hygromycin B) (see Note 5). A sterile spatula was used to loosen and float the agarose slices, and the plates were sealed with Parafilm and placed on a platform shaker, in the dark at 80 rpm, 26°C.
15. The medium was replaced every week with fresh medium. Many microcalli will loosen from the agarose and be free in the medium; care is needed to minimize losing these microcalli.
16. Resistant calli (approximately 1–2 mm) were visible by 3–6 weeks (Fig. 2f–h) and were transferred to solid shooting medium (MSNTS) with selection (such as hygromycin B at 10 μg/mL) and moved to a growth chamber with 16 h light at 26°C (Fig. 2i–k).
17. Shoots that generated from the calli were transferred to solid MSNT in Phytatray™ II boxes for rooting (Fig. 2l). A selection agent was not included in medium for rooting.
18. As plants grew, 1-cm square leaf explants were harvested from the putative transformants in a 1.5-mL microcentrifuge tube for genomic DNA isolation. To avoid contamination, plant tissues were collected in a sterile hood with a sterile surgical blade. Analysis of recombination junctions is described in Subheading 3.4 except that the phenol-chloroform-isoamyl alcohol extraction step was omitted (see Note 6). An extra phenol/chloroform extraction step and a more concentrated DNA solution are required for DNA isolated from the transient expression assay as only a small percentage of the cells take up DNA, resulting in lower signal-to-noise ratio.
19. After both recombination junctions of the putative integration lines were detected and confirmed by sequencing, some rooted plants were transferred to SUPER soil (fir/redwood bark, *Sphagnum* peat moss, sand, pH 5.5–6.5, Rod McLellan, Marysville, OH) in the greenhouse to set seed. Fertilizer [one teaspoon per 8″ pot of Sierra Slow Release Mix (15% (N)–9% (P)–12% (K) plus micronutrients)] was applied initially, followed weekly with supplemental fertilizer at 200 ppm N–P–K [Scott's General Purpose (20% N–P–K plus micronutrients)]. Plants were watered daily.
20. Southern hybridization analysis of the integration structure and the evaluation of the trait genes can be assessed on tissue-cultured or greenhouse-grown plants.

4. Notes

1. DNA oligonucleotides used:

   ϕC31 *attP* (P′P orientation):

   F1:5′-CCCTTGTGTCATGTCGGCGACCCTACGCCCC CAACTGAGAGAACTCAAAG-3′.
   R1: 5′-TCTCAGTTGGGGGCGTAGGGTCGCCGACATGAC ACAAGGGGGCC-3′(*Apa*I overhang underlined).
   F2:5′-GTTACCCCAGTTGGGGCACTACTCCCGAAAAC CGCTTCTGACGAATTCC-3′
   R2:5′-CCGGGGAATTCGTCAGAAGCGGTTTTCGGGA GTAGTGCCCCAACTGGGGTAACCTTTGAGTTC-3′(*XmaI* overhang, *Eco*RI site underlined).

   TP901 *attP* (P′P orientation):

   F1:5′-GATCCGCGAGGCGCGCCCTCCTTAAAAGGAGT TTTTTAGTTACCTTAATTGAAATAAACGAAATAAA-3′ (*Bam*HI overhang, *Asc*I site underlined).
   R1:5′-TATTTCAATTAAGGTAACTAAAAAACTCCTTTT AAGGAGGGCGCGCCTCGCG-3′.
   F2: 5′-AACTCGCAATTAAGCGAGTTGGAATTTAAATAT GATATCTGGGCC-3′ (*Apa*I overhang underlined).
   R2:5′-CAGATATCATATTTAAATTCCAACTCGCTTAAT TGCGAGTTTTTATTTCGTT-3′.

   U153 *attP* (PP′ orientation):

   F1:5′-GATCCAATTCCTAGGTAGCTTGTTTATTTAGAT TGTTTAGTTCCTCGTTTTCTC-3′
   (*Bam*HI overhang, *Avr*II site underlined).
   R1:5′-CGTCCAACGAGAGAAAACGAGGAACTAAACAA TCTAAATAAACAAGCTACCTAGGAATTG-3′
   F2:5′-TCGTTGGACGGAAACGAATCGAGAAACTAAAA TTATAAATAAAAAGTAACCTA-3′
   R2:5′-GATCTAGGTTACTTTTTATTTATAATTTTAGTT TCTCGATTCGTTTC-3′

   Bxb1 *attP* (P′P orientation):

   F1:5′-GATCCATGACTACCAGGGCTCTCATGGGTTTG TACCGTACACCA-3′ (*Bam*HI overhang underlined).
   R1:5′-GCGGTCTCAGTGGTGTACGGTACAAACCCAT GAGAGCCCTGGTAGTCATG-3′.
   F2:5′-CTGAGACCGCGGTGGTTGACCAGACAAACCA CGAAGACACAGGTCATCACGA-3′.
   R2:5′-GATCTCGTGATGACCTGTGTCTTCGTGGTTTG TCTGGTCAACCACC-3′ (GATC overhang underlined).

Bxb1 *att*B (B′B orientation):

F1:5′-GATCTGCCCGGATGATCCTGACGACGGAGACC GCCGTCGTCGACAAGCCGGCCGAGGGCCCTCGCG-3′(GATC overhang, *Apa*I site underlined).

R1:5′-GATCCGCGAGGGCCCTCGGCCGGCTTGTCG ACGACGGCGGTCTCCGTCGTCAGGATCATCCGGGCA-3′ (*Bam*HI overhang, *Apa*I site underlined).

U153 *att*B (B′B orientation):

F1:5′-CGGCTTATGGGTAACACCAATTAAGTGTTTAG TTCCCTCTTTGCGTCCTT-3′.

R1:5′-AGAGGGAACTAAACACTTAATTGGTGTTACCC ATAAGCCGGGCC-3′ (*Apa*I overhang underlined).

F2:5′-CATAGCTTGATCCGAAAAAGTTACAGCT GGTTTTAATTAGCTGGCGCGCCAATTG-3′ (*Asc*I site underlined).

R2:5′-GATCCAATTGGCGCGCCAGCTAATTAAAACC AGCTGTAACTTTTTCGGATCAAGCTATGAAGGACGC AA-3′ (GATC overhang, *Asc*I site underlined).

TP901 *att*B (B′B orientation):

F1:5′-GATCAAAATTCACGGAAGAAAGCTTTGGCAAA AAAAGCAAAAAG(GATC overhang underlined).

R1:5′-CAAGGTAAATGCTTTTTGCTTTTTTTGCCAAA GCTTTCTTCCGTGAATTTT-3′.

F2:5′-CATTTACCTTGATTGAGATGTTAATTGTG TTGGCAATTATCAGTATTTTAAGAATTCG-3′(*Eco*RI site underlined).

R2:5′-GATCCGAATTCTTAAAATACTGATAATT GCCAACACAATTAACATCTCAAT-3′ (*Bam*HI overhang underlined).

Primers for CinH site:

5′-AGTCGGCGCGCCAGATCTCACGTGCAGTGCCAAGCTTG CATG-3′(*Asc*I, *Bgl*II, *Pml*I sites underlined).

5′-AGTCGCGGCCGCAGTCTTAATTAAATCCATAACG CTAGCA-3′ (*Not*I, *Pac*I sites underlined).

Primers for ParA site:

5′-AGTCTTAATTAAAGTCGGATCCCCTTGGTCAAAT TGGGTA-3′ (*Pac*I, *Bam*HI sites underlined).

5′-AGTCGCGGCCGCACTAGTCACGTGATTAGCACATATGT GGGC-3′ (*Not*I, *Spe*I, *Pml*I sites underlined).

Primers for Tn1721 site:

5′-AGTCTTAATTAAAGTCGAGCTCTAGGTGCAAGCA AGTTAAGG-3′ (*Pac*I, *Sac*I sites underlined).

5′-AGTCGGATCCAGTCCTGCAGGTCTCCTTGGTTG AAGGCGGC-3′(*Bam*HI, *Pst*I sites underlined).

Primers for Tn5053 site:

5′-AGTCCTGCAGAGTCGACGTCGTCGTACCTGTCA CATATAC-3′ (*Pst*I, *Aat*II sites underlined).

5′-AGTCGGATCCCGACCCCGAATCCCTGG-3′ (*Bam*HI site underlined).

2. DMSO can be included in PCR amplifications to increase yield, specificity, and consistency (24).
3. All the steps described in tobacco protoplast isolation, PEG-mediated transformation, selection, and regeneration should be performed under aseptic conditions using sterile equipment and reagents. Leaf tissue and protoplasts should always be handled in a sterile laminar flow hood. Plant protoplasts are quite fragile and are sensitive to slight changes in osmoticum and to rough handling. All protoplast suspensions should be handled gently and transferred using wide-opening pipette tips.
4. Since 5–10 μg of each plasmid is needed for each experiment, we have used the Qiagen® Plasmid Midi Kit to isolate up to 100 μg of each plasmid. To keep plasmids sterile, after 70% ethanol wash and air-drying, DNA is resuspended in sterile water and stored in aliquots at –80°C.
5. We start with a lower level of selection. For example, hygromycin is used at 5 μg/mL the first 2 weeks, 10 μg/mL thereafter.
6. This protocol is designed for the rapid isolation of genomic DNA from a small amount of plant tissue (a leaf disc ~1 cm in diameter, or approximately 10 mg) in a microcentrifuge tube to detect recombinant junctions and to screen for the integration lines without phenol-chloroform extraction. Repeated freeze-and-thaw cycles degrade the DNA and affect PCR quality.

References

1. Terada, R., Urawa, H., Inagaki, Y., Tsugane, K. and Iida, S. (2002) Efficient gene targeting by homologous recombination in rice. *Nat. Biotechnol.* **20**, 1030–1034.
2. Shukla, V.K., Doyon, Y., Miller, J.C., DeKelver, R.C., Moehle, E.A., Worden, S.E., Mitchell, J.C., Arnold, N.L., Gopalan, S., Meng, X., et al. (2009) Precise genome modification in the crop species *Zea mays* using zinc-finger nucleases. *Nature* **459**, 437–441.
3. Townsend, J.A., Wright, D.A., Winfrey, R.J., Fu, F., Maeder, M.L., Joung, J.K. and Voytas, D.F. (2009) High-frequency modification of plant genes using engineered zinc-finger nucleases. *Nature* **459**, 442–445.
4. Ow, D.W. (2002) Recombinase-directed plant transformation for the post genomic era. *Plant Mol. Biol.* **48**, 183–200.
5. Dale, E.C. and Ow, D.W. (1990) Intra- and intermolecular site-specific recombination in plant cells mediated by bacteriophage P1 recombinase. *Gene* **91**, 79–85.
6. Albert, H., Dale, E.C., Lee, E. and Ow, D.W. (1995) Site-specific integration of DNA into wild-type and mutant *lox* sites placed in the plant genome. *Plant J.* **7**, 649–659.
7. Srivastava, V. and Ow, D.W. (2001) Biolistic-mediated site-specific integration in rice. *Mol. Breed.* **8**, 345–350.

8. Day, C.D., Lee, E., Kobayashi, J., Holappa, L.D., Albert, H. and Ow, D.W. (2000) Transgene integration into the same chromosome location can produce alleles that express at a predictable level, or alleles that are differentially silenced. *Genes Dev.* **14**, 2869–2880.
9. Srivastava, V., Ariza-Nieto, M. and Wilson, A.J. (2004) Cre-mediated site-specific gene integration for consistent transgene expression in rice. *Plant Biotechnol. J.* **2**, 169–179.
10. Chawla, R., Ariza-Nieto, M., Wilson, A.J., Moore, S.K. and Srivastava, V. (2006) Transgene expression produced by biolistic-mediated, site-specific gene integration is consistently inherited by the subsequent generations. *Plant Biotechnol. J.* **4**, 209–218.
11. Louwerse, J.D., van Lier, M.C., van der Steen, D.M., de Vlaam, C.M., Hooykaas, P.J. and Vergunst, A.C. (2007) Stable recombinase-mediated cassette exchange in Arabidopsis using *Agrobacterium tumefaciens*. *Plant Physiol.* **145**, 1282–1293.
12. Nanto, K., Yamada-Watanabe, K. and Ebinuma, H. (2005) *Agrobacterium*-mediated RMCE approach for gene replacement. *Plant Biotechnol. J.* **3**, 203–214.
13. Li, Z., Xing, A., Moon, B.P., McCardell, R.P., Mills, K. and Falco, S.C. (2009) Site-specific integration of transgenes in soybean via recombinase mediated DNA cassette exchange. *Plant Physiol.* **151**(3), 1087–1095.
14. Ow, D.W. (2005) Transgene management *via* multiple site-specific recombination systems. *In Vitro Cell Dev. Biol. Plant* **41**, 213–219.
15. Thomason, L.C., Calendar, R. and Ow, D.W. (2001) Gene insertion and replacement in *Schizosaccharomyces pombe* mediated by the *Streptomyces* bacteriophage phiC31 site-specific recombination system. *Mol. Genet. Genom.* **265**, 1031–1038.
16. Lutz, K.A., Corneille, S., Azhagiri, A.K., Svab, Z. and Maliga, P. (2004) A novel approach to plastid transformation utilizes the phiC31 phage integrase. *Plant J.* **37**, 906–913.
17. Marillonnet, S., Giritch, A., Gils, M., Kandzia, R., Klimyuk, V. and Gleba, Y. (2004) In planta engineering of viral RNA replicons: efficient assembly by recombination of DNA modules delivered by *Agrobacterium*. *Proc. Natl Acad. Sci. USA* **101**, 6852–6857.
18. Thomson, J.G. and Ow, D.W. (2006) Site-specific recombination systems for the genetic manipulation of eukaryotic genomes. *Genesis* **44**, 465–476.
19. Miller, E.M. and Nickoloff, J.A. (1995) *Escherichia coli* electrotransformation, in *Methods in Molecular Biology*, **47**, Nickoloff, J.A., ed., Humana Press, Totowa, NJ, 105.
20. Sambrook, J., Fristh, E.F. and Maniatis, T. (2001) *Molecular Cloning: A Laboratory Manual*, 3rd edn, Cold Spring Harbor Press Inc., New York.
21. Schenk, P.M., Sagi, L., Remans, T., Dietzgen, R.G., Bernard, M.J., Graham, M.W. and Manners, J.M. (1999) A promoter from sugarcane bacilliform badnavirus drives transgene expression in banana and other monocot and dicot plants. *Plant Mol. Biol.* **39**, 1221–1230.
22. Qin, M., Bayley, C., Stockton, T. and Ow, D.W. (1994) Cre recombinase-mediated site-specific recombination between plant chromosomes. *Proc. Natl Acad. Sci. USA* **91**, 1706–1710.
23. Horsch, R.B., Fry, J.E., Hoffman, N.L., Eichholz, D., Rogers, S.G. and Fraley, R.T. (1985) A simple and general method for transferring gene into plants. *Science* **227**, 1229–1231.
24. Varadaraj, K. and Skinner, D.M. (1994) Denaturants or cosolvents improve the specificity of PCR amplification of a G+C-rich DNA using genetically engineered DNA polymerase. *Gene* **140**, 1–5.

Chapter 9

Targeted Mutagenesis in *Arabidopsis* Using Zinc-Finger Nucleases

Feng Zhang and Daniel F. Voytas

Abstract

We report here an efficient method for making targeted mutations in *Arabidopsis thaliana* genes. The approach uses zinc-finger nucleases (ZFNs) – enzymes engineered to create DNA double-strand breaks at specific target loci. Imprecise repair of double-strand breaks by nonhomologous end-joining generates small insertions or deletions at the cleavage site. In this protocol, constructs encoding ZFNs for specific loci are transformed into *Arabidopsis* by *Agrobacterium*-mediated transformation. ZFN expression is induced during germination to initiate mutagenesis of the target locus. Typically, more than 20% of the primary transgenics segregate loss-of-function mutations in the next generation. ZFNs make it possible to expand the range of *Arabidopsis* mutants available for study and to create mutations in genes missed by random mutagenesis approaches, such as those using T-DNA, transposons, or chemical mutagens.

Key words: *Arabidopsis thaliana*, Mutagenesis, Zinc-finger nuclease, Nonhomologous end-joining

1. Introduction

Sequencing and annotation of the *Arabidopsis thaliana* genome revealed more than 30,000 genes (1). Significant progress has been made in determining the function of these genes, due in large part to powerful genetic and reverse genetic approaches available for this model plant (2). For example, RNAi-based gene silencing strategies and the manipulation of miRNAs make it possible to interfere with gene expression (3). A limitation to the use of such RNA-based methods, however, is that it is difficult to entirely knockout expression of a target gene. A more traditional approach for determining gene function is to study phenotypes conferred by knockout mutations. In several model plant species, including *Arabidopsis*, efficient mutagenesis approaches such as

James A. Birchler (ed.), *Plant Chromosome Engineering: Methods and Protocols*, Methods in Molecular Biology, vol. 701,
DOI 10.1007/978-1-61737-957-4_9,

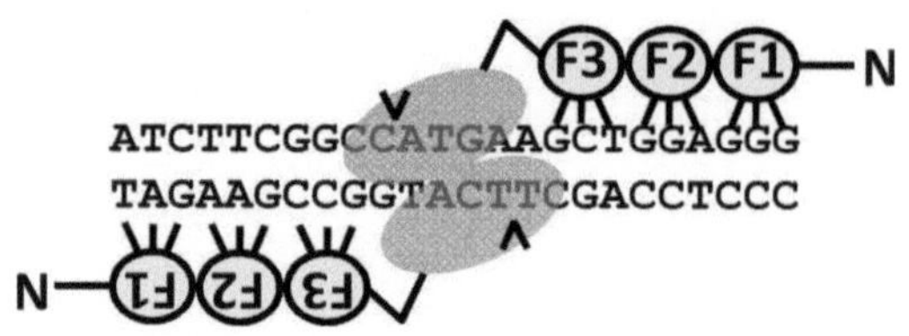

Fig. 1. Schematic of a zinc-finger nuclease.

TILLING, transposon tagging, or insertion of T-DNAs have allowed the identification of knockout mutations in most genes (4–6). All of these approaches, however, involve screening a large number of mutants for the desired knockout. Further, mutations in certain genes are sometimes not recovered, likely because their loss leads to lethality or because they are small in size and therefore poor targets for the mutagen. Genetic analysis is also frustrating because of the fact that more than 15% of plant genes are organized in tandem, and so, mutating one gene in the duplicated pair typically produces no measurable phenotype (7).

A particularly valuable tool for *Arabidopsis* functional genomics would be an efficient method for targeted mutagenesis that creates mutations either in genes missed by traditional approaches or in redundant or tandem genes. One promising approach for targeted mutagenesis is the use of zinc-finger nucleases (ZFNs) (8, 9). ZFNs are chimeric proteins with two functional domains (Fig. 1): a DNA recognition domain, composed of an array of Cys_2–His_2 zinc fingers, and the nonspecific DNA cleavage domain of the restriction endonuclease *Fok*I. Each zinc finger recognizes three nucleotides, and arrays of multiple fingers can be engineered that bind extended sequence targets with high affinity and specificity (10–12). ZFNs function as dimers and recognize two 9–12 bp half-sites in target genes. Binding of ZFNs to the target DNA allows the *Fok*I domains to dimerize and cleave within the spacer sequence (5–7 bp in length) that separates the two half-sites.

In the absence of a homologous DNA template, chromosome breaks created by ZFNs are repaired via nonhomologous end-joining (NHEJ) (13). Faithful repair by NHEJ restores DNA integrity by rejoining the broken chromosomes; however, imprecise repair by NHEJ introduces small insertions/deletions at ZFN cleavage sites that frequently knockout gene function (Fig. 2). Efficient methods of NHEJ-mediated targeted mutagenesis of endogenous loci have been developed for *Drosophila* (14), zebrafish (15–17), human cells (12, 18), and rat (19).

Here, we report a targeted mutagenesis protocol for *Arabidopsis* (summarized in Fig. 3) based on the imprecise repair of chromosome breaks introduced by ZFNs (20). The protocol requires ZFNs with high affinity and specificity for the desired target, and

```
CACGTATCTTCGGCCATGAAGCTGGAGGGTAAT     Wild type
CACGTATCTTCGGCCA---AGCTGGAGGGTAAT        -3
CACGTATCTTCGGC-----------AGGGTAAT        -11
CACGTATCTTCGGC--------TGGAGGGTAAT        -8
CACGTA-----------------TGGAGGGTAAT       -17
CACGTATCTTCGGCCAaTGAAGCTGGAGGGTAAT       +1
CACGTATCTTCGGCCAtaTGAAGCTGGAGGGTAAT      +2
```

Fig. 2. Examples of typical ZFN-induced mutations at a target site.

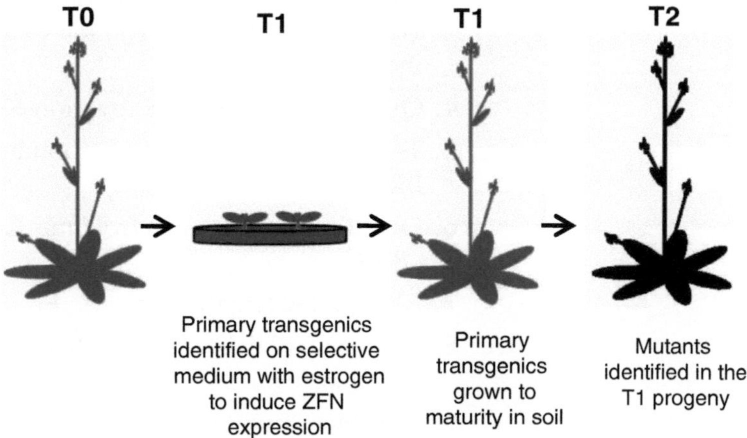

Fig. 3. Diagram of procedure for generating heritable ZFN-induced mutations.

it employs an estrogen-inducible expression system to provide a brief pulse of ZFNs early in *Arabidopsis* development (21). We recommend engineering zinc finger arrays (ZFAs) using the OPEN (Oligomerized Pool ENgineering) method, which was developed by the Zinc Finger Consortium (12, 22). Detailed protocols for OPEN have been published that describe: (1) the use of software tools for identifying ZFN target sites in genes of interest, (2) the construction and screening of libraries to identify ZFAs that bind the target gene, and (3) assays for testing ZFAs in bacterial two-hybrid assays to ensure that they have the requisite DNA affinity and specificity for targeted mutagenesis. Performing these steps yields functional ZFAs that recognize the target gene and provides a starting point for the three methods described in this protocol. The first method outlines the steps required to clone the ZFAs into a T-DNA vector and the use of *Agrobacterium* to transform *Arabidopsis*. The second method describes how to identify transformed plants and to test whether or not the ZFNs are creating mutations at target sites in somatic tissues. The final method describes procedures for recovering stably inherited mutations in the locus of interest.

2. Materials

2.1. Introducing ZFNs into Arabidopsis by Agrobacterium-Mediated Floral Transformation

1. Restriction enzymes *Xba*I, *Bam*HI, *Nhe*I, and *Bgl*II and their corresponding buffers (New England Biolabs).
2. Plasmid DNA for pMDC7 (24) (encoding components for estrogen-inducible transgene expression) and pFZ87 (20) (the Gateway entry vector that encodes the *Fok*I nuclease).
3. 1.5% agarose gel: prepare 1.5% agarose solution in 500 mL of ddH_2O. Microwave until completely melted. Add 10 μL of 10 mg/mL ethidium bromide solution, mix gently, and pour the gel.
4. Qiaex II Agarose Gel Extraction Kit (Qiagen).
5. T4 DNA ligase and corresponding buffer (New England Biolabs).
6. One Shot TOP10 Electrocomp Cells (Invitrogen). Store at -80°C.
7. LR Clonase (Invitrogen).
8. HotStar HiFidelity DNA Polymerase and corresponding buffer (Qiagen).
9. *Agrobacterium tumefaciens* strain GV3101 (23).
10. Kanamycin stock: dissolve 1 g of kanamycin in 20 mL of ddH_2O. Filter sterilize, aliquot out into Eppendorf tubes, and store at –20°C.
11. LB kanamycin plates: dissolve 10 g of tryptone, 5 g of yeast extract, 5 g of NaCl, and 15 g of bacteriological agar in 1 L of ddH_2O. Autoclave the medium, cool to 65°C, add 1 mL of kanamycin stock, and pour the plates.
12. LB liquid medium with kanamycin: dissolve 10 g of tryptone, 5 g of yeast extract, and 5 g of NaCl in 1 L of ddH_2O. Autoclave the medium, cool to 65°C, and add 1 mL of kanamycin stock.
13. Silwet L-77 (GE Silicones).
14. Approximately 4-week-old *Arabidopsis* plants grown in pots. The primary bolts should be clipped, and the secondary bolts allowed to grow to 2–10 cm in length.

2.2. Inducing ZFN Expression in Primary Transgenic Arabidopsis Plants

1. 50% bleach: mix equal volumes of ddH_2O and bleach (household chlorine bleach) and store at room temperature. Add Tween 20 to 0.05%.
2. 500 mL sterile ddH_2O.
3. 0.1% agarose: prepare 0.1% agarose in 500 mL of ddH_2O, autoclave, and store at room temperature.

4. MS selection/induction plates: Dissolve a package of MS salts (Caisson Laboratories) and 10 g of sucrose in 800 mL of ddH_2O; adjust the volume to 1 L. Add 7.5 g of phytagel (Sigma), autoclave, and cool to 65°C. Add hygromycin to a final concentration of 25 mg/L and 17β-estradiol to a final concentration of 10 μM. Mix and pour into Petri dishes. Store at 4°C for no more than 7 days. Caution: a chemical fume hood should be used when handling 17β-estradiol.
5. DNeasy Plant Mini Kit (Qiagen).
6. HotStar HiFidelity DNA Polymerase and corresponding buffer (Qiagen).
7. MinElute PCR Purification Kit (Qiagen).
8. TOPO TA Cloning Kit: pCR 2.1-TOPO Vector (Invitrogen).
9. QIAprep 96 Turbo Miniprep Kit (Qiagen).

2.3. Screening for Mutant Plants

1. DNeasy Plant Mini Kit (Qiagen).
2. HotStar HiFidelity DNA Polymerase and corresponding buffer (Qiagen).
3. MinElute PCR Purification Kit (Qiagen).
4. TOPO TA Cloning Kit: pCR 2.1-TOPO Vector (Invitrogen).
5. QIAprep 96 Turbo Miniprep Kit (Qiagen).

3. Methods

3.1. Introducing ZFNs into Arabidopsis by Agrobacterium-mediated Floral Transformation

1. DNA sequences encoding zinc-finger arrays (ZFAs) generated by OPEN (12) are released from their plasmid vector by digesting with restriction enzymes *Xba*I and *Bam*HI. Restriction digestions are typically performed with 1 μg of plasmid DNA.
2. The restriction digestion products are separated on a 1.5% agarose gel. DNA fragments encoding ZFAs are 264 bp in length. These fragments are excised from the gel and purified with a Qiagen Gel Extraction Kit.
3. Three sequential subcloning steps are used to transfer DNA encoding the left and right ZFA into the Gateway-compatible T-DNA destination vector pMDC7 (24) (see Note 1). The arrays are first subcloned into the gateway entry vector, pFZ87 (20), placing ZFA coding sequences in frame with the *Fok*I endonuclease and thereby creating the ZFN. In the first step, approximately 1 μg of pFZ87 is digested with *Xba*I and *Bam*HI.

4. The linearized backbone is separated on a 1.5% agarose gel, and the vector fragment is purified using a Qiagen Gel Extraction Kit.
5. The linearized vector is ligated to the *Xba*I/*Bam*HI fragment encoding the ZFA for the left half-site (from step 1).
6. Ligation reactions are transformed into *E. coli* and recombinant molecules identified.
7. The recombinant entry vector with the left ZFA is digested with *Nhe*I and *Bgl*II to produce DNA ends compatible with those generated by *Xba*I and *BamH*I, respectively. The linearized vector is ligated with the *Xba*I/*Bam*HI fragment encoding the ZFA for the right half site (from step 1).
8. Ligation reactions are transformed into *E. coli* and recombinant molecules identified.
9. The Gateway entry vector with both the left and right ZFNs is next recombined with the T-DNA expression vector, pMDC7, using LR Clonase from Invitrogen. The recombination reactions are transformed into *E. coli* and recombinant molecules identified.
10. The resulting T-DNA expression vector is transformed into *Agrobacterium tumefaciens* strain GV3101 by electroporation.
11. The transformed *Agrobacterium* cells are spread onto an LB plate with kanamycin (50 mg/L). The plate is incubated at 30°C for 2 days.
12. A single colony growing on the LB/kanamycin plate is inoculated into 5 mL of liquid LB medium with kanamycin (50 mg/L). The culture is shaken at 30°C, 220 rpm overnight.
13. The overnight culture is transferred to 250 mL of liquid LB medium with kanamycin (50 mg/L) in a 500-mL flask. The flask is shaken at 30°C, 220 rpm overnight.
14. The *Agrobacterium* culture is centrifuged at 4,000 × *g* for 5 min. The supernatant is removed, and the bacteria pellet is resuspended in 500 mL of 5% sucrose that contains 150 μL of Silwet L-77. The culture is transferred to a 1-L beaker.
15. *Arabidopsis* plants (primary bolts clipped, secondary bolts 2–10 cm in length) are inverted into the bacteria suspension with all above-ground tissues submerged. After 15 s, the plants are removed from the beaker and placed under a plastic cover to retain moisture. The plants are left in the dark overnight.
16. The next day the *Arabidopsis* plants are placed in a growth chamber. The plants are grown for an additional 3–4 weeks until the seed-bearing siliques are dry. The seeds are harvested and stored in microcentrifuge tubes.

3.2. Inducing ZFN Expression in Primary Transgenic Arabidopsis Plants

1. *Arabidopsis* seeds are surface sterilized in a 15-mL conical centrifuge tube by adding 10 mL of 50% bleach/0.05%Tween 20. The tube is shaken for 5 min at moderate speed on a rotary platform shaker.
2. The tube is centrifuged in a tabletop centrifuge at 100 × *g* for 2 min to collect the seeds at the bottom of the tube. The bleach solution is removed with a sterile pipette in a tissue culture hood. Steps 3 and 4 should also be performed in a tissue culture hood to ensure that the sterile seeds do not become contaminated.
3. Resuspend the seeds with 10 mL of sterile water, centrifuge as in step 2, and remove the water. Rinse the seeds two additional times, and after the third rinse, resuspend the seeds in a 0.1% agarose solution. The volume of the agarose solution should approximate 20 times the seed volume.
4. Spread the seeds on MS selection/induction plates with a sterile pipette (3–4 mL of seed/agarose mixture for each 150-mm plate; 1–2 mL of mixture for each 100 mm plate). The plates are wrapped in parafilm and placed at 4°C for 2 days. After stratification, the plates are placed in a growth chamber (16 h light/8 h dark; light intensity 75 μE/m/s).
5. Transgenic seedlings that grow on the selection/induction medium (see Note 2) are tested to determine whether the ZFNs have created mutations by NHEJ at their intended target sequences. Ten-, 7-day-old transgenic seedlings are pooled to make genomic DNA using the Qiagen DNeasy Plant Mini Kit.
6. Amplify the region encompassing the target site by PCR (see Note 3). Use approximately 100 ng of genomic DNA in a PCR reaction with HotStar HiFidelity DNA polymerase (see Note 4).
7. Purify the PCR product using a MinElute PCR Purification Kit.
8. The PCR products are cloned using a TOPO TA Cloning Kit and transformed into *E. coli.*
9. DNA is prepared in a 96-well plate from 96 independent colonies using QIAprep 96 Turbo Miniprep Kit.
10. DNA from the 96 independent colonies are sequenced. ZFN-induced mutations can be readily identified with characteristic of insertions, deletions, or nucleotide changes in the target sequences (Fig. 2). Somatic mutation rates can be estimated based on the number of independent polymorphic target sequences identified out of the 96 sequenced clones.

3.3. Screening for Mutant Plants

1. Somatic mutation rates provide an estimate of the number of primary transgenic plants that should be followed in the next generation to recover heritable mutations (see Note 5). Typically, 20–50 primary transgenic plants are moved from the selection/induction plates to soil and placed in a growth chamber. Seeds from individual plants are harvested and stored in microcentrifuge tubes.
2. If the mutant phenotype of the target gene is unknown, then proceed to step 3. However, if the phenotype has been previously characterized, then it should be apparent in plants derived from seeds of individual primary transgenics. Typically, 10–75% of the primary transgenic plants segregate mutations in the next generation (20).
3. If the mutant phenotype is not known or is difficult to score, then plants derived from seeds of individual primary transgenics should be screened molecularly for the presence of the mutation. Twenty seeds from each primary transgenic should be sown on soil. Leaves from individual, 2-week-old plants should be harvested and genomic DNA prepared from the pooled leaves using the Qiagen DNeasy Plant Mini Kit.
4. Amplify the region encompassing the target site by PCR (see Note 3). Use approximately 100 ng of genomic DNA in a PCR reaction with HotStar HiFidelity DNA polymerase (see Note 4).
5. Purify the PCR product using a MinElute PCR Purification Kit.
6. The PCR products are cloned using a TOPO TA Cloning Kit and transformed into *E. coli*.
7. Typically, ten to forty clones are sequenced to determine if a mutation is segregating in the progeny of a single primary transgenic (see Note 6). Mutations typically contain either insertions or deletions at the ZFN cleavage site (Fig. 2).
8. For those progeny populations that are segregating mutations, DNA is prepared from leaves of individual plants and analyzed molecularly as described in steps 4–7. Once individual plants are identified with the desired mutation, they can be subjected to genetic analysis (see Note 7).

4. Notes

1. The T-DNA vector, pMDC7, has several features that are critical for function (21, 24): (1) a hygromycin resistance gene that allows for selection of primary transgenics, (2) a gene for a constitutively expressed transcriptional activator with LexA

DNA binding sites that moves from the cytoplasm to the nucleus upon exposure to estrogen, (3) a Gateway cassette that facilitates cloning ZFNs into the expression construct, and (4) a promoter with LexA binding sites that is activated by the artificial transcriptional activator. We found that because some ZFNs have cellular toxicities, this expression system makes it possible to tightly control ZFN expression.

2. The MS selection medium contains hygromycin (25 mg/L) to select for transformed plants. The cotyledons of nontransformed plants are smaller than those of transformants, and they do not produce true leaves. The selection medium also contains 17β-estradiol (10 μM) to induce expression of the ZFNs in primary transgenic plants during germination. The estradiol is labile and only lasts a few days in the medium, thereby limiting ZFN expression. This short exposure to ZFN expression helps to minimize toxic effects of the ZFNs.
3. The PCR primers used to amplify the target locus should be carefully selected. They need to be close enough to the target site so that reliable DNA sequence can be obtained from the cloned PCR products. They are typically positioned about 150 bp on either side of the ZFN target site. Note that some ZFN-induced deletions will remove the primer binding site and that these mutations will not be amplified and scored.
4. It is important to use a high-fidelity polymerase for PCR because it is not possible to distinguish between PCR- and ZFN-induced point mutations at the ZFN cleavage site.
5. We typically observe somatic mutation frequencies in primary transgenics that range between 5 and 25%. Of these plants, 10–75% segregate mutations in the next generation. A few percent of the primary transgenics will sustain mutations in both alleles and produce homozygous mutant seed. Mutations in some genes may affect frequencies of germinal transmission.
6. We advocate DNA sequencing as the primary means to screen for ZFN-induced mutations. However, an easier assay can be employed for those targets with restriction endonuclease sites in the ZFN spacer sequence (20). Mutations created by NHEJ will typically destroy the restriction site, and so sequences with mutations are immune to digestion. This PCR screen is fast and easy and can also be used to screen individual plants. Another alternative genotyping approach uses the enzyme CEL-I (25, 26). In this strategy, PCR reactions are denatured and reannealed, and if ZFN-induced mutations are present, heteroduplexes form with unpaired bases at the cleavage site. The unpaired bases are recognized and cleaved by CEL-I, and the cleavage products can be visualized by electrophoresis.

7. There are several reports of ZFNs that recognize sites other than their intended targets (27). There is a possibility, therefore, that some ZFNs will introduce unwanted, secondary mutations. Because of this, we recommend crossing mutant plants to wild type to remove these unwanted mutations before exhaustive genetic analyses are performed.

References

1. Arabidopsis Genome Initiative (2000) Analysis of the genome sequence of the flowering plant *Arabidopsis thaliana*, *Nature* **408**, 796–815.
2. Alonso, J. M., and Ecker, J. R. (2006) Moving forward in reverse: genetic technologies to enable genome-wide phenomic screens in Arabidopsis, *Nat Rev Genet* 7, 524–536.
3. Ossowski, S., Schwab, R., and Weigel, D. (2008) Gene silencing in plants using artificial microRNAs and other small RNAs, *Plant J* **53**, 674–690.
4. Krysan, P. J., Young, J. C., and Sussman, M. R. (1999) T-DNA as an insertional mutagen in Arabidopsis, *Plant Cell* **11**, 2283–2290.
5. Parinov, S., and Sundaresan, V. (2000) Functional genomics in Arabidopsis: large-scale insertional mutagenesis complements the genome sequencing project, *Curr Opin Biotechnol* **11**, 157–161.
6. McCallum, C. M., Comai, L., Greene, E. A., and Henikoff, S. (2000) Targeting induced local lesions IN genomes (TILLING) for plant functional genomics, *Plant Physiol* **123**, 439–442.
7. Jander, G., and Barth, C. (2007) Tandem gene arrays: a challenge for functional genomics, *Trends Plant Sci* **12**, 203–210.
8. Cathomen, T., and Joung, J. K. (2008) Zinc-finger nucleases: the next generation emerges, *Mol Ther* **16**, 1200–1207.
9. Carroll, D. (2008) Progress and prospects: zinc-finger nucleases as gene therapy agents, *Gene Ther* **15**, 1463–1468.
10. Carroll, D., Morton, J. J., Beumer, K. J., and Segal, D. J. (2006) Design, construction and in vitro testing of zinc finger nucleases, *Nat Protoc* **1**, 1329–1341.
11. Wright, D. A., Thibodeau-Beganny, S., Sander, J. D., Winfrey, R. J., Hirsh, A. S., Eichtinger, M., Fu, F., Porteus, M. H., Dobbs, D., Voytas, D. F., and Joung, J. K. (2006) Standardized reagents and protocols for engineering zinc finger nucleases by modular assembly, *Nat Protoc* **1**, 1637–1652.
12. Maeder, M. L., Thibodeau-Beganny, S., Osiak, A., Wright, D. A., Anthony, R. M., Eichtinger, M., Jiang, T., Foley, J. E., Winfrey, R. J., Townsend, J. A., Unger-Wallace, E., Sander, J. D., Muller-Lerch, F., Fu, F., Pearlberg, J., Gobel, C., Dassie, J. P., Pruett-Miller, S. M., Porteus, M. H., Sgroi, D. C., Iafrate, A. J., Dobbs, D., McCray, P. B., Jr., Cathomen, T., Voytas, D. F., and Joung, J. K. (2008) Rapid "open-source" engineering of customized zinc-finger nucleases for highly efficient gene modification, *Mol Cell* **31**, 294–301.
13. Wyman, C., and Kanaar, R. (2006) DNA double-strand break repair: all's well that ends well, *Annu Rev Genet* **40**, 363–383.
14. Beumer, K. J., Trautman, J. K., Bozas, A., Liu, J. L., Rutter, J., Gall, J. G., and Carroll, D. (2008) Efficient gene targeting in Drosophila by direct embryo injection with zinc-finger nucleases, *Proc Natl Acad Sci USA* **105**, 19821–19826.
15. Meng, X., Noyes, M. B., Zhu, L. J., Lawson, N. D., and Wolfe, S. A. (2008) Targeted gene inactivation in zebrafish using engineered zinc-finger nucleases, *Nat Biotechnol* **26**, 695–701.
16. Doyon, Y., McCammon, J. M., Miller, J. C., Faraji, F., Ngo, C., Katibah, G. E., Amora, R., Hocking, T. D., Zhang, L., Rebar, E. J., Gregory, P. D., Urnov, F. D., and Amacher, S. L. (2008) Heritable targeted gene disruption in zebrafish using designed zinc-finger nucleases, *Nat Biotechnol* **26**, 702–708.
17. Foley, J. E., Yeh, J. R., Maeder, M. L., Reyon, D., Sander, J. D., Peterson, R. T., and Joung, J. K. (2009) Rapid mutation of endogenous zebrafish genes using zinc finger nucleases made by Oligomerized Pool ENgineering (OPEN), *PLoS One* **4**, e4348.

18. Perez, E. E., Wang, J., Miller, J. C., Jouvenot, Y., Kim, K. A., Liu, O., Wang, N., Lee, G., Bartsevich, V. V., Lee, Y. L., Guschin, D. Y., Rupniewski, I., Waite, A. J., Carpenito, C., Carroll, R. G., Orange, J. S., Urnov, F. D., Rebar, E. J., Ando, D., Gregory, P. D., Riley, J. L., Holmes, M. C., and June, C. H. (2008) Establishment of HIV-1 resistance in CD4+ T cells by genome editing using zinc-finger nucleases, *Nat Biotechnol* **26**, 808–816.
19. Geurts, A. M., Cost, G. J., Freyvert, Y., Zeitler, B., Miller, J. C., Choi, V. M., Jenkins, S.S., Wood, A., Cui, X., Meng, X., Vincent, A., Lam, S., Michalkiewicz, M., Schilling, R., Foeckler, J., Kalloway, S., Weiler, H., Menoret, S., Anegon, I., Davis, G. D., Zhang, L., Rebar, E. J., Gregory, P. D., Urnov, F. D., Jacob, H. J., and Buelow, R. (2009) Knockout rats via embryo microinjection of zinc-finger nucleases, *Science* **325**, 433.
20. Zhang, F., Maeder, M., Renyon, D., Unger-Wallace, E., Hoshaw, J., Pierick, C., Peterson, T., Dobbs, D., and Voytas, D. (2010) High frequency targeted mutagenesis of Arabidopsis genes using zinc finger nucleases, Proc Natl Acad Sci USA **107**(26), 12028–12033.
21. Zuo, J., Niu, Q. W., and Chua, N. H. (2000) Technical advance: an estrogen receptor-based transactivator XVE mediates highly inducible gene expression in transgenic plants, *Plant J* **24**, 265–273.
22. Maeder, M., Thibodeau-Beganny, S., Sander, J. D., Voytas, D. F., and Joung, J. K. (2009) Oligomerized Pool ENgineering (OPEN): an 'open-source' protocol for making customized zinc finger arrays, *Nat Protoc* **4**, 1471–1501.
23. Zhang, X., Henriques, R., Lin, S. S., Niu, Q. W., and Chua, N. H. (2006) Agrobacterium-mediated transformation of Arabidopsis thaliana using the floral dip method, *Nat Protoc* **1**, 641–646.
24. Curtis, M. D., and Grossniklaus, U. (2003) A gateway cloning vector set for high-throughput functional analysis of genes in planta, *Plant Physiol* **133**, 462–469.
25. Miller, J. C., Holmes, M. C., Wang, J., Guschin, D. Y., Lee, Y. L., Rupniewski, I., Beausejour, C. M., Waite, A. J., Wang, N. S., Kim, K. A., Gregory, P. D., Pabo, C. O., and Rebar, E. J. (2007) An improved zinc-finger nuclease architecture for highly specific genome editing, *Nat Biotechnol* **25**, 778–785.
26. Lombardo, A., Genovese, P., Beausejour, C. M., Colleoni, S., Lee, Y. L., Kim, K. A., Ando, D., Urnov, F. D., Galli, C., Gregory, P. D., Holmes, M. C., and Naldini, L. (2007) Gene editing in human stem cells using zinc finger nucleases and integrase-defective lentiviral vector delivery, *Nat Biotechnol* **25**, 1298–1306.
27. Townsend, J. A., Wright, D. A., Winfrey, R. J., Fu, F., Maeder, M. L., Joung, J. K., and Voytas, D. F. (2009) High-frequency modification of plant genes using engineered zinc-finger nucleases, *Nature* **459**, 442–445.

Chapter 10

Vectors and Methods for Hairpin RNA and Artificial microRNA-Mediated Gene Silencing in Plants

Andrew L. Eamens and Peter M. Waterhouse

Abstract

In plant cells, DICER-LIKE4 processes perfectly double-stranded RNA (dsRNA) into short interfering (si) RNAs, and DICER-LIKE1 generates micro (mi) RNAs from primary miRNA transcripts (pri-miRNA) that form fold-back structures of imperfectly dsRNA. Both si and miRNAs direct the endogenous endonuclease, ARGONAUTE1 to cleave complementary target single-stranded RNAs and either small RNA (sRNA)-directed pathway can be harnessed to silence genes in plants. A routine way of inducing and directing RNA silencing by siRNAs is to express self-complementary single-stranded hairpin RNA (hpRNA), in which the duplexed region has the same sequence as part of the target gene's mRNA. Artificial miRNA (amiRNA)-mediated silencing uses an endogenous pri-miRNA, in which the original miRNA/miRNA* sequence has been replaced with a sequence complementary to the new target gene. In this chapter, we describe the plasmid vector systems routinely used by our research group for the generation of either hpRNA-derived siRNAs or amiRNAs.

Key words: dsRNA, hpRNA, siRNA, miRNA, amiRNA, RNA silencing, Plant expression vectors pHellsgate12 and pBlueGreen

1. Introduction

Double-stranded RNA (dsRNA) is an effective trigger of RNA silencing in vertebrate, invertebrate, and plant systems (1, 2). This silencing operates by sequence-specific RNA degradation. In plants, a particularly effective method of silencing an endogenous gene is to transform the plant with a vector construct encoding a hairpin RNA (hpRNA) consisting of an inverted-repeat of a portion of the target gene sequence separated by a fragment of spacer material. Using an intron as the spacer fragment increases the frequency of obtaining silenced plants (3, 4). Based on this concept, we use the plasmid vector, pHellsgate12 (5), which allows for the simple and

James A. Birchler (ed.), *Plant Chromosome Engineering: Methods and Protocols*, Methods in Molecular Biology, vol. 701, DOI 10.1007/978-1-61737-957-4_10, © Springer Science+Business Media, LLC 2011

rapid generation of such vector constructs, taking advantage of the Gateway™ recombinase. An alternative method of directing RNA silencing in plants is to introduce a plant expression vector that produces a single artificial microRNA (amiRNA) silencing signal (6). For such an approach, we use our recently developed amiRNA plant expression vector, pBlueGreen (7), which is based on the *Arabidopsis* miR159b primary transcript (pre-miR159b).

Here we describe the PCR and bacterial cloning procedures for introducing either (1) long target sequences into the hpRNA vector, pHellsgate12, or (2) short target sequences into the amiRNA vector, pBlueGreen, as well as providing some simple design rules for their respective construction.

2. Materials

2.1. Hairpin RNA Construction Using pHellsgate12

2.1.1. hpRNA Target Selection and Design of Forward and Reverse PCR Primers

1. Personal computer with internet access.

2.1.2. Preparation of Template RNA

1. Plant material (100 mg fresh weight).
2. TRIzol reagent (Invitrogen).
3. 100% Isopropanol (ice-cold).
4. 75% Ethanol (RT).
5. Chloroform.
6. RNase-free deionised water (dH_2O).
7. Pestle and mortar (sterilised by autoclaving).
8. Liquid nitrogen (LN_2).
9. 1.5-ml Microfuge tubes (disposable).
10. Pipette tips (200 and 1,000 μL barrier-tipped).
11. Ice.
12. Benchtop microfuge (Eppendorf 5415D or similar) at 4°C and room temperature (RT).
13. Waterbath (at 65°C).
14. Freezer (at –20°C and –80°C).

2.1.3. Amplification of the Target Fragment by RT-PCR

1. Template RNA (~2 μg/μL).
2. Forward and reverse PCR primers (10 μM).
3. 6× Loading buffer (LB).

4. 10× Tris–borate–EDTA (TBE) buffer.
5. RNase-free dH_2O.
6. DNase-free dH_2O.
7. Ice.
8. 0.2-mL PCR tubes (disposable).
9. 1.5-mL Microfuge tubes (disposable).
10. QIAGEN OneStep RT-PCR kit (Qiagen).
11. QIAquick PCR Purification kit (Qiagen).
12. 1.0% w/v Agarose gel stained with ethidium bromide (EtBr).
13. 100 bp DNA ladder (MBI Fermentas).
14. Pipette tips (2, 20, and 200 μL).
15. Benchtop thermocycler (Bio-Rad MyCycler® or similar).
16. Benchtop microfuge (at RT).

2.1.4. Cloning the RT-PCR Fragment into pENTR™/D-TOPO®

1. PCR fragment from Subheading 3.2.3.
2. 6× LB.
3. 10× TBE buffer.
4. *Asc*I (10 U/μL, New England Biolabs).
5. *Not*I (10 U/μL, MBI Fermentas).
6. 10× Buffer 4 (New England Biolabs).
7. 10× Buffer O (MBI Fermentas).
8. Ice.
9. *pENTR™/D-TOPO®* cloning Kit (Invitrogen).
10. *Escherichia coli* One Shot Top10 chemically competent cells (Invitrogen).
11. Luria–Bertani (LB) liquid media.
12. LB agar plate (with 50 mg/mL kanamycin).
13. Bacterial cell spreader (sterilised).
14. 1.0% w/v Agarose gel (stained with EtBr).
15. 1 kb^+ DNA ladder (MBI Fermentas).
16. Pipette tips (2, 20, 200, and 1,000 μL).
17. 0.2-mL PCR tubes (disposable).
18. QIAprep Spin Miniprep kit (Qiagen).
19. QIAquick PCR Purification kit (Qiagen).
20. 1.5-mL Microfuge tubes (disposable).
21. Benchtop microfuge (at RT).
22. Incubated shaker (at 37°C).
23. Incubator (at 37°C).
24. Waterbath (at 42°C).

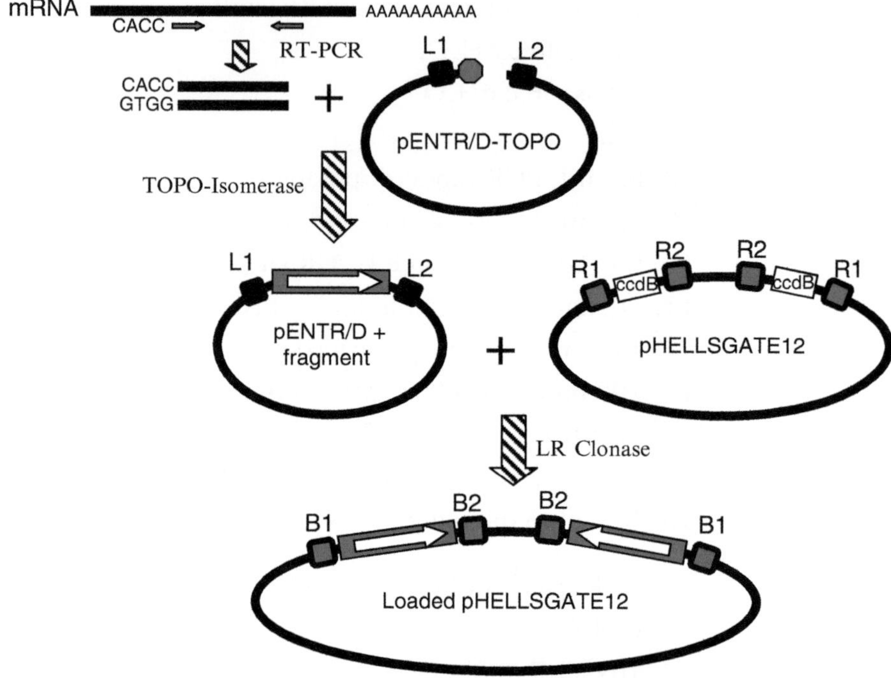

Fig. 1. Outline of the scheme to generate a pHELLSGATE hpRNA for expression in plants. Three steps are required (1) Reverse transcription and PCR amplification of a gene fragment, (2) cloning the fragment into pENTR™/D-TOPO® using TOPO-Isomerase, and (3) use of the LR Clonase® to make a double insertion of the fragment from pENTR/D into pHELLSGATE12 in an inverted-repeat orientation by the use of site-specific recombination sites. The *ccdB* negative selection gene in pHellsgate12 is replaced by the desired gene fragments in the LR reaction.

2.1.5. Cloning PCR Fragment from pENTR™/ D-TOPO® into pHellsgate12 (Fig. 1)

1. pHellsgate12 plant expression vector.
2. pENTR™/D-TOPO® plasmid preparation with target fragment (from Subheading 3.1.4).
3. 0.1 M Tris–EDTA (TE) buffer, pH 8.0.
4. LR Clonase® and 5× Clonase buffer (Invitrogen).
5. Proteinase K solution (Invitrogen).
6. *Xba*I (10 U/μL, MBI Fermentas).
7. *Xho*I (10 U/μL, MBI Fermentas).
8. 10× Buffer Tango™ (MBI Fermentas).
9. 10× Buffer R (MBI Fermentas).
10. Ice.
11. *E. coli* DH5α electro-competent cells.
12. LB liquid media.
13. LB agar plates (with 50 mg/mL spectinomycin).
14. Bacterial cell spreader (sterilised).
15. 1.5-mL Microfuge tubes (disposable).
16. 0.2-mL PCR tubes (disposable).
17. Pipette tips (2, 20, 200, and 1,000 μL).

18. 15-mL Capped centrifuge tubes.
19. QIAprep Spin Miniprep kit (Qiagen).
20. Electroporator (MicroPulser™ or similar) and cuvettes (Bio-Rad).
21. Benchtop microfuge (at RT).
22. Incubated shaker (at 37°C).
23. Incubator (at 37°C).
24. Waterbath (at 25 and 37°C).
25. Laminar flow cabinet.

2.2. Artificial miRNA Construction using pBlueGreen

2.2.1. Artificial miRNA Selection and Design of Forward and Reverse PCR Primers

1. Personal computer with internet access.
2. Template for amiRNA forward and reverse primers (Fig. 2b).

```
a  cDNA target          ...CAGCTGACTATGGTCAGTAGTAGCTGACGAGCTAGCTAGTG...
                                  |||||||||||||||||||||
   putative amiRNA                UACCAGUCAUCAUCGACUGCU

b  Syn-miR Primer-F(*)

   TATATGCTCTTCGAGAGG_._._._._._._._._._._.__.__._._._._._._._._GGAGGGTTTAGCAGGGTGAAGTAAAG
   TATATGCTCTTCGAGAGGa.a.g.a.g.c.t.c.c.t.t.g.a.a.g.t.t.c.a.a.tGGAGGGTTTAGCAGGGTGAAGTAAAG

   Syn-miR Primer-R

   TATATGCTCTTCGAGATG_._._._._._._._._._._.__.__._._._._._._._._GAAGAGTGAAGCCATTAAAGGG
   TATATGCTCTTCGAGATGa.a.g.a.g.c.t.c.c.c.t.t.c.a.a.t.c.c.a.a.aGAAGAGTGAAGCCATTAAAGGG

c  Syn-miR Primer-F(*);

   TATATGCTCTTCGAGAGGA.T.G.G.T.C.A.G.T.A.G.G.C.G.C.T.G.A.C.G.TGGAGGGTTTAGCAGGGTGAAGTAAAG
   TATATGCTCTTCGAGAGGa.a.g.a.g.c.t.c.c.t.t.g.a.a.g.t.t.c.a.a.tGGAGGGTTTAGCAGGGTGAAGTAAAG

   Order forward primer sequence;

   5'TATATGCTCTTCGAGAGGATGGTCAGTAGGCGCTGACGTGGAGGGTTTAGCAGGGTGAAGTAAAG 3'

   Syn-miR Primer-R;

   TATATGCTCTTCGAGATGA.T.G.G.T.C.A.G.T.A.G.T.A.G.C.T.G.A.C.G.AGAAGAGTGAAGCCATTAAAGGG
   TATATGCTCTTCGAGATGa.a.g.a.g.c.t.c.c.c.t.t.c.a.a.t.c.c.a.a.aGAAGAGTGAAGCCATTAAAGGG

   Order reverse primer sequence;

   5'TATATGCTCTTCGAGATGATGGTCAGTAGTAGCTGACGAGAAGAGTGAAGCCATTAAAGGG 3'
```

Fig. 2. Artificial miRNA forward and reverse PCR primer template. (**a**) Identify a putative amiRNA target sequence unique to the target gene to be silenced. (**b**) Forward and reverse PCR primer template sequences which maintain the endogenous shape of the *Arabidopsis* miR159b miRNA/miRNA* duplex. (**c**) Enter the exact 21-nt target sequence into the reverse primer template (this corresponds to your mature amiRNA 21-mer sequence) and also enter this sequence into the forward primer template but introduce dsRNA mismatched base pairings at positions 12, 13, and 21, respectively (this corresponds to the mature amiRNA* passenger 21-mer sequence).

2.2.2. Amplification of the Pre-amiRNA PCR Fragment

1. p*Ath-miR159b* template plasmid (~50 pg/μL).
2. amiRNA forward and reverse PCR primers (10 μM).
3. dNTPs (5 mM each; Fischer Biotech).
4. Expand Long Template Enzyme mix (5 U/μL, Roche).
5. 10× Expand Long Template buffer 1 (Roche).
6. DNase-free dH_2O.
7. 10× TBE buffer.
8. 6× LB.
9. 100 bp DNA ladder (MBI Fermentas).
10. 1.0% w/v Agarose gel (stained with EtBr).
11. Ice.
12. 0.2-mL PCR tubes (disposable).
13. Pipette tips (2, 20, and 200 μL).
14. 1.5-mL Microfuge tubes (disposable).
15. QIAquick PCR Purification kit (Qiagen).
16. Benchtop thermocycler.
17. Benchtop microfuge (at RT).

2.2.3. Cloning the Pre-amiRNA PCR Fragment into the pGEM®-T Easy Vector

1. Column-purified pre-amiRNA PCR fragment (from Subheading 3.2.2).
2. pGEM®-T Easy Vector (50 ng/μL, Promega).
3. 2× Rapid Ligation buffer (Promega).
4. T4 DNA ligase (3 U/μL, Promega).
5. 20% w/v 5-Bromo-4-chloro-3-indolyl-β-d-galactopyranoside (X-gal, Sigma-Aldrich).
6. 0.1 M Isopropyl β-d-1-thiogalactopyranoside (IPTG, Sigma-Aldrich).
7. LB liquid media.
8. LB agar plates (with 50 mg/mL ampicillin).
9. *E. coli* DH5α electro-competent cells.
10. Ice.
11. Pipette tips (2, 20, 200, and 1,000 μL).
12. 1.5-mL Microfuge tubes (disposable).
13. 15-mL Capped centrifuge tube (disposable).
14. Drawn-out glass pipette (with bulb).
15. Bacterial cell spreader (sterilised).
16. Bacterial loop (sterilised).
17. QIAprep Spin Miniprep kit (Qiagen).
18. Electroporator and cuvettes.

19. Benchtop microfuge (at RT).
20. Incubated shaker (at 37°C).
21. Incubator (at 37°C).
22. Waterbath (at 37 and 65°C).
23. Laminar flow cabinet.

2.2.4. Cloning of the Pre-amiRNA Restriction Fragment into the Plant Expression Vector pBlueGreen

1. pGEM-T::pre-amiRNA vector (from Subheading 3.2.3).
2. pBlueGreen plant expression vector.
3. pSoup helper plasmid.
4. *Lgu*I (5 U/μL, MBI Fermentas).
5. *Bam*HI (10 U/mL, MBI Fermentas).
6. T4 DNA ligase (5 U/μL, MBI Fermentas).
7. 10× Buffer Tango™ (MBI Fermentas).
8. 10× Buffer BamHI (MBI Fermentas).
9. 10× T4 DNA ligase buffer (MBI Fermentas).
10. 10× TBE buffer.
11. 6× LB.
12. DNase-free dH_2O.
13. Ice.
14. Vortex.
15. 20% w/v X-gal.
16. 0.1 M IPTG.
17. 1.0% w/v Agarose gel (stained with EtBr).
18. 100 bp DNA ladder (MBI Fermentas).
19. 1 kb$^+$ DNA ladder (MBI Fermentas).
20. LB liquid media.
21. LB agar plates (with 50 mg/mL kanamycin).
22. *E. coli* DH5α electro-competent cells.
23. Bacterial cell spreader (sterilised).
24. Bacterial loop (sterilised).
25. QIAquick PCR Purification kit (Qiagen).
26. Pipette tips (2, 20, 200, and 1,000 μL).
27. 1.5-mL Microfuge tubes (disposable).
28. Electroporator and cuvettes.
29. Benchtop mircofuge (RT).
30. Benchtop shaker (at 37°C).
31. Incubater (at 37°C).
32. Waterbath (at 37 and 65°C).
33. Laminar flow cabinet.

3. Methods

3.1. The Gateway® Compatible hpRNA Plant Expression Vector, pHellsgate12

The plant expression vector, pHellsgate12 (Fig. 3), was created to take advantage of the Gateway® unidirectional in vitro cloning system (Invitrogen, Carlsbad, CA) and is based on our findings that a PCR product flanked by *att*L1 and *att*L2 sites will recombine with a plasmid carrying two *att*R1–*att*R2 cassettes separated by an intron sequence when incubated with LR Clonase® to generate an inverted-repeat vector directing RNA silencing at high efficiencies (4, 8). The first step is the

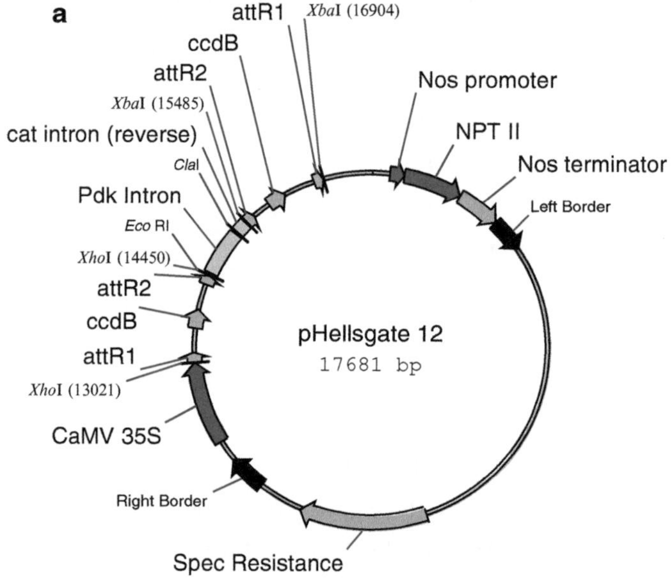

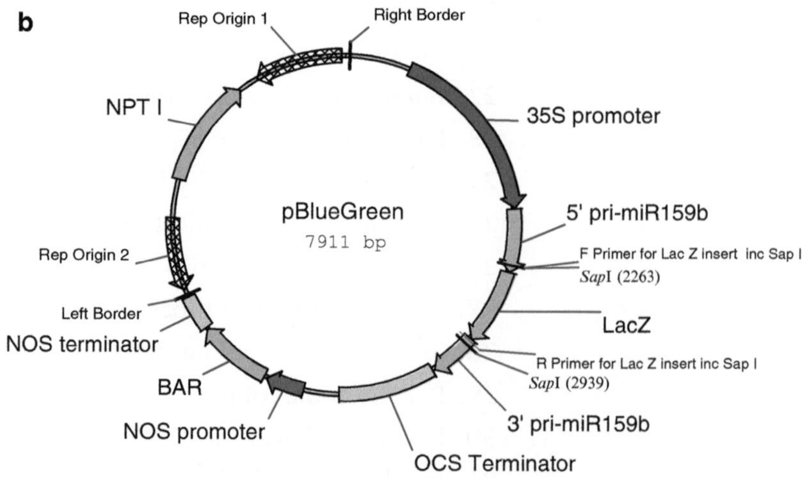

Fig. 3. Plasmid maps of the plant expression vectors (**a**) pHellsgate12 and (**b**) pBlueGreen.

amplification of a region of DNA from the target gene and to insert the amplified fragment it into the entry vector, pENTR™/D-TOPO®. This flanks the inserted PCR fragment with an *att*L1 site at one end and an *att*L2 site at the other, generating a donor fragment for an LR Clonase® reaction and entry into the plant expression vector pHellsgate12.

3.1.1. hpRNA Target Selection and Design of Forward and Reverse PCR Primers

1. For *Arabidopsis*, download the cDNA sequence of the gene to be targeted for hpRNA-directed RNA silencing from the TAIR website (http://www.arabidopsis.org/), using the Gene identifier search function (http://www.arabidopsis.org/servlets/Search?action=new_search&type=gene).
2. Select a region 300–800 nucleotide (nt) in length within the cDNA sequence and BLAST this against the *Arabidopsis* transcriptome, using the BLASTN search function of the TAIR database. Choose a sequence that shows no regions of perfect homology to another transcript encoded by *Arabidopsis* genome. If silencing is required in a plant species for which its genome has not been completely sequenced, use EST sequences to identify possible off-targets (see Note 1).
3. Order the forward (21 nt perfectly matching the target sequence plus four additional nucleotides – CACC at the 5′ end of the primer) and reverse (21 nt perfect match to target sequence) DNA oligonucleotides from your usual supplier to amplify the selected fragment (see Note 2).

3.1.2. Preparation of Template RNA

1. Collect 100 mg of plant leaf material and immediately freeze in liquid nitrogen (LN_2).
2. Using a sterilised mortar and pestle, grind the plant material into a fine powder under LN_2. Immediately transfer the powder to a pre-chilled (in LN_2) 1.5-mL microfuge tube, cap tube and invert 20×, then incubate at room temperature (RT) for 5 min.
3. Add 200 μL of chloroform, cap the tube and shake vigorously by hand for 15 s, then incubate at RT for 3 min.
4. Centrifuge at maximum speed (~16,000 *g*) for 10 min, at 4°C. Transfer the upper aqueous phase to a new 1.5-mL microfuge tube and repeat steps 3 and 4.
5. Add 500 μL of ice-cold 100% isopropanol, mix by inverting 20× and incubate at –20°C overnight (see Note 3).
6. Centrifuge at maximum speed for 20 min, at 4°C. A white pellet should be readily visible at the bottom of the 1.5-mL microfuge tube and using a drawn-out pipette tip, carefully pipette off and discard the supernatant.
7. Add 1 mL of 75% ethanol and wash pellet by inverting the tube 10× and centrifuging at 8,500 *g* for 5 min, at RT.

8. Pipette off supernatant and air dry the pellet (leave lid open) at RT for 5 min.
9. Add 50 μL of RNase-free dH_2O and incubate at 65°C for 10 min, then resuspend by pipetting (see Note 4). Either place the tube on ice and immediately proceed with the RT-PCR steps or store at −80°C until required.

3.1.3. Amplification of Target Fragment by RT-PCR

1. On ice, thaw components of the QIAGEN OneStep RT-PCR kit (including dNTP mix, 5× QIAGEN OneStep RT-PCR buffer, RNase-free dH_2O), the forward and reverse PCR primers (at working concentration of 10 μM), the QIAGEN OneStep RT-PCR Enzyme mix and template RNA from Subheading 3.1.2.
2. Add the above listed reaction components into a 0.2-mL PCR tube according to the manufacturers (QIAGEN) instructions for a 50 μL reaction, using 5 μL of the RNA template (~10 μg).
3. Pipette the reaction mixture up and down to ensure that all reaction components are thoroughly mixed, cap the 0.2-mL PCR tube and immediately transfer to a benchtop thermocycler. Amplify the fragment using the following program: 1×50°C/30 min (reverse transcription step); 1×95°C/15 min (HotStarTaq activation step), 30×94°C/30 s, 55°C/30 s, 72°C/60 s; and a final extension step of 1×72°C/10 min, then cool the reaction by incubating at 4°C/20 min.
4. Transfer a 10 μL aliquot to a labelled 1.5-mL microfuge tube containing 2 μL of 6× LB and vortex briefly to mix. Run on an ethidium bromide (EtBr)-stained 1.0% agarose gel in 1× TBE buffer along with a 10 μL aliquot of 100 bp DNA ladder to check for the correctly amplified product, migrating as a single band (see Note 5).
5. Purify the remaining PCR product using the QIAquick PCR Purification kit (Qiagen) according to the manufacturer's instructions and resuspend in 20 μL of DNase-free dH_2O. Store the amplified fragment at −20°C or use immediately.

3.1.4. Cloning the RT-PCR Fragment into pENTR™/D-TOPO®

1. On ice, thaw components of the pENTR™/Directional-TOPO® Cloning kit (salt solution, DNase-free dH_2O, topoisomerase-charged-pENTR™/D-TOPO® vector) and the RT-PCR product from Subheading 3.1.3.
2. Add reaction components into a 0.2-mL PCR tube according to manufacturers' instructions (Invitrogen) for a 6 μL reaction, using 1–4 μL of the RT-PCR product.
3. Mix the reaction gently by pipetting and incubate at RT for 5 min.

4. On ice, transform reaction into a thawed aliquot of One Shot TOP10 chemically competent *E. coli* cells (supplied with pENTR™/Directional-TOPO® Cloning kit) according to manufacturer's instructions. Spread the cell culture onto an LB agar plate (with 50 mg/mL kanamycin) and incubate overnight (16–24 h) in a 37°C incubator.
5. Analyse ten colonies by extracting their plasmid DNA using the QIAprep Spin Miniprep kit (Qiagen) according to manufacturer's instructions. Digest 10% of the plasmid preparation with the restriction endonucleases *Asc*I and *Not*I (two sequential digests are required and we recommend that the first digestion product is purified using the QIAquick PCR Purification kit prior to digestion with the second restriction endonuclease).
6. Add 4 μL of 6× LB to each reaction and run on an EtBr-stained 1.0% w/v agarose gel in 1× TBE buffer along with a 10 μL aliquot of 1 kb$^+$ DNA ladder and select a plasmid preparation that returns the appropriate sized restriction fragments (see Note 6).

3.1.5. Cloning PCR Fragment from pENTR™/D-TOPO® into pHellsgate12

1. On ice, thaw the 5× LR Clonase buffer, the selected plasmid preparation of the pENTR™/D-TOPO® clone from Subheading 3.1.4 and a plasmid preparation of the pHellsgate12 plant expression vector.
2. Once thawed and still on ice, add the above reaction components to a 0.2-mL tube. For a 10 μL reaction, add: 2 μL of 5× LR Clonase buffer, 2 μL of the selected pENTR™/D-TOPO®-PCR clone (~750 ng/μL), 2 μL of the pHellsgate12 plant expression vector (~250 ng/μL), 2 μL of TE buffer (pH 8.0), and 2 μL of the LR Clonase®.
3. Gently mix the reaction components by pipetting and incubate for 1–16 h in a 25°C waterbath (see Note 7). At the end of your selected incubation period, add 1 μL of proteinase K to the reaction tube, vortex to mix, spin down and incubate for 10 min, in a 37°C waterbath.
4. Transfer 2 μL of the LR Clonase® reaction to an electroporation cuvette and incubate on ice for 5 min. Thaw a 50 μL aliquot of highly efficient *E. coli* DH5α electrocompetent cells on ice and once thawed immediately transfer to the chilled cuvette containing the reaction mix.
5. Transfer the cuvette to an electroporator, electroporate, and immediately add 450 μL of LB liquid media to the cuvette. Transfer the cellular mixture to a labelled 1.5-mL microfuge tube using a drawn-out glass pipette, cap the tube, and incubate at 37°C for 60 min, in a benchtop shaker (at 200 rpm).

6. In a laminar flow cabinet, transfer a 100 μL aliquot of the above cell culture onto an LB agar plate (containing 50 mg/mL spectinomycin) and evenly spread the mixture over the entire surface of the plate with a sterilised bacterial cell spreader. Allow the LB agar plate to dry in the laminar flow cabinet for 10 min. Place the lid back onto the agar plate and incubate for 16–24 h in a 37°C incubator.
7. Analyse ten colonies by extracting their plasmid DNA using the QIAprep Spin Miniprep kit (Qiagen) according to manufacturer's instructions. Digest 10% of the plasmid preparation with the restriction endonucleases *Xho*I or *Xba*I (see Note 8).
8. Select a plasmid preparation that returns the desired restriction pattern (Fig. 3). This plasmid preparation is used for *Agrobacterium tumefaciens* (*Agrobacterium*)-mediated transformation of wild-type *Arabidopsis* plants.

3.2. The amiRNA Plant Expression Vector pBlueGreen

The pBlueGreen vector (Fig. 3) was constructed to generate a plant expression vector that produces a single 21-nt silencing signal. Long segments of perfectly dsRNA are processed along their entire length by DICER-LIKE4 into many heterologous 21-nt sRNAs that can potentially direct silencing of other mRNAs in addition to the targeted transcript. Artificial miRNA technology allows for the removal of such "off-target" effects due to the fact that a single homozygous 21-nt silencing signal is processed by DICER-LIKE1 from the modified endogenous miRNA precursor transcripts on which amiRNA plant expression vectors are based. More recently, this technology has been widely applied in several plant species and has been demonstrated to direct highly efficient and specific RNA silencing of targeted genes (6, 9–11). Here we outline the steps required for the PCR amplification of a pre-amiRNA insertion fragment, flanked by *Lgu*I restriction sites for convenient insertion into our recently developed amiRNA plant expression vector, pBlueGreen.

3.2.1. Artificial miRNA Selection and Design of Forward and Reverse PCR Primers

1. For *Arabidopsis* amiRNA design, download the cDNA sequence of the gene to be targeted for amiRNA-directed RNA silencing from the TAIR website, using the Gene Search function (http://www.arabidopsis.org/servlets/Search?action=new_search&type=gene).
2. Identify a 19-nt sequence within the cDNA sequence containing either a cytosine (C) or guanine (G) at position 1, a thymine residue (T) at position 10 and an adenine (A) at position 19 (see Note 9), as shown in Fig. 2a.
3. Add two additional nucleotides 5′ (upstream) of position 1 of the putative amiRNA target sequence and BLAST the complementary 21-nt sequence (putative mature amiRNA

21-mer silencing signal), against the *Arabidopsis* transcriptome using the BLASTN search function of the TAIR database (see Note 10).

4. If the putative 21-nt amiRNA sequence only shows complementarity to the cDNA sequence to be targeted for amiRNA-directed RNA silencing, enter the exact 21-nt target sequence (of your cDNA target sequence, 5′–3′) into the amiRNA reverse primer template (Fig. 2b) as shown in Fig. 2c.
5. If the putative 21-nt amiRNA sequence shows complementarity to another *Arabidopsis* transcript, which is not to be targeted for amiRNA-directed RNA silencing, discard the sequence and repeat steps 2 and 3 in Subheading 3.2.1 until a 21-nt sequence only showing complementarity to the target transcript is identified.
6. Enter the 21-nt target sequence (again enter the cDNA target sequence in the 5′–3′ direction) into the amiRNA forward primer template (Fig. 2b), but introduce three mismatched nucleotides at positions 12, 13, and 21, respectively (see Note 11), as shown in Fig. 2c.
7. Order the amiRNA forward (65 nt) and reverse (61 nt) DNA oligonucleotides from your usual supplier (see Note 2).

3.2.2. Amplification of the Pre-amiRNA PCR Fragment

1. On ice, add reaction components in the following order to a 0.2-mL PCR tube; 37.5 μL of DNase-free dH_2O, 1.0 μL of the p*Ath-mR159b* template vector (~50 pg/μL), 2.5 μL of 10 μM amiRNA forward primer, 2.5 μL of 10 μM amiRNA reverse primer, 1.0 μL of 5 mM dNTPs, 5.0 μL of 10× Expand Long Template buffer 1 and 0.5 μL of Expand Long Template Enzyme mix (see Note 12).
2. Pipette the reaction mixture up and down to ensure that all reaction components are thoroughly mixed, cap the PCR tube, and immediately transfer to a benchtop thermocycler.
3. Amplify the pre-amiRNA fragment from the p*Ath*-miR159b template vector using the following program; 1×95°C/3 min; 28×94°C/20 s, 56°C/30 s, 72°C/60 s; 1×72°C/7 min, 4°C/20 min.
4. Transfer a 10 μL aliquot of the PCR product to a labelled 1.5-mL microfuge tube containing 2 μL of 6× LB and vortex to mix. Run the PCR product on an EtBr-stained 1.0% w/v agarose gel in 1× TBE buffer along with a 10 μL aliquot of the 100 bp DNA ladder to check that the 224-nt pre-amiRNA PCR product was amplified successfully.
5. Purify the remaining PCR product using the QIAquick PCR Purification kit according to the manufacturers instructions and resuspend in 20 μL of DNase-free dH_2O.

3.2.3. Cloning the Pre-amiRNA PCR Fragment into the pGEM®-T Easy Vector

1. Add 4 μL of the column-purified pre-amiRNA PCR product (from Subheading 3.2.2) and 0.5 μL of the pGEM®-T Easy vector to a labelled 1.5-mL microfuge tube, cap the tube, briefly vortex to mix, and incubate in a 65°C waterbath for 5 min, then immediately transfer the tube to ice and incubate for an additional 5 min (see Note 13).
2. On ice, add 5 μL of 2× Rapid Ligation buffer (Promega) and 0.5 μL of T4 DNA ligase (Promega), gently mix reaction components by pipetting, cap the tube, and incubate in a 37°C waterbath for 60 min.
3. Transfer 2 μL of the ligation reaction to an electroporation cuvette and incubate on ice for 5 min. At the same time, thaw a 50 μL aliquot of *E. coli* DH5α electrocompetent cells on ice and once thawed immediately transfer to the chilled cuvette containing the ligation reaction.
4. Transfer the cuvette to an electroporator, electroporate, and immediately add 450 μL of LB liquid media to the cuvette. Transfer the cellular mixture to a labelled 1.5-mL microfuge tube using a drawn-out glass pipette, cap the tube, and incubate at 37°C for 60 min, in a benchtop shaker (at 200 rpm).
5. In a laminar flow cabinet, transfer a 100 μL aliquot of the above cell culture to an LB agar plate (containing 50 mg/mL ampicillin) and evenly spread the mixture over the entire surface of the plate with a sterilised bacterial cell spreader (see Note 14). Allow the LB agar plate to dry in the laminar flow cabinet for 10 min. Place the lid back onto the agar plate and incubate for 16–24 h in a 37°C incubator.
6. Using a sterilised loop, pick off a single white-coloured colony (see Note 14) and use to inoculate a 5 mL culture of LB liquid media (containing 50 mg/mL ampicillin). Cap the 15-mL centrifuge tube and incubate at 37°C for 16–24 h in a benchtop shaker (at 200 rpm).
7. Isolate plasmid DNA from the overnight culture using the QIAprep Spin Miniprep kit (Qiagen) according to the manufacturer's instructions. The plasmid preparation, pGEM-T®::pre-amiRNA, contains the modified precursor transcript of *Ath*-miR159b, where the endogenous miR159/miR159* sequences have been replaced with amiRNA guide and passenger strand sequences by PCR.

3.2.4. Cloning of the Pre-amiRNA Restriction Fragment into the Plant Expression Vector pBlueGreen

1. On ice and in two separately labelled 1.5-mL microfuge tubes, add reaction components in the following order; 7 μL of DNase-free dH_2O, 2 μL of 10× Buffer Tango™ and 1 μL of 5 U/μL *Lgu*I.
2. Add 10 μL of each plasmid preparation, pGEM-T::pre-amiRNA or pBlueGreen, to the appropriately labelled 1.5-mL

microfuge tube. Briefly vortex each tube to thoroughly mix reaction components, spin down in a microfuge and incubate overnight (16–24 h) in a 37°C waterbath.

3. Transfer a 5 μL aliquot of each digested plasmid to a new labelled 1.5-mL microfuge tube containing 1 μL of 6× LB and run on an EtBr-stained 1.0% w/v agarose gel in 1× TBE buffer along with 10 μL of 1 kb+ DNA ladder to determine whether *Lgu*I digestion is complete (see Note 15).
4. Purify digestion products with the QIAquick PCR Purification kit according to the manufacturer's instructions and resuspend each *Lgu*I digestion product in 20 μL of DNase-free dH_2O.
5. To an appropriately labelled 1.5-mL microfuge tube, add 10 μL of the pGEM-T::pre-amiRNA/*Lgu*I insertion fragment, 0.5 μL of the *Lgu*I-digested pBlueGreen plant expression vector and 7 μL of DNase-free dH_2O. Mix thoroughly by pipetting and incubate at 65°C for 5 min in a waterbath, then immediately transfer the reaction tube to ice and incubate for 5 min (see Note 13).
6. On ice, add 2 μL of 10× T4 DNA Ligase buffer (MBI Fermentas) and 0.5 μL of T4 DNA ligase (MBI Fermentas), mix by pipetting and incubate for 60 min, in a 37°C waterbath.
7. Transfer 2 μL of the above ligation reaction to an electroporation cuvette and incubate on ice for 5 min. At the same time, thaw a 50 μL aliquot of *E. coli* DH5α electrocompetent cells on ice and once thawed immediately transfer to the chilled cuvette containing the ligation reaction.
8. Transfer the cuvette to an electroporator, electroporate, and immediately add 450 μL of LB liquid media to the cuvette. Transfer the cellular mixture to a labelled 1.5-mL microfuge tube using a glass drawn-out pipette, cap the tube and incubate at 37°C for 60 min, in a benchtop shaker (at 200 rpm).
9. In a laminar flow cabinet, pipette a 100 μL aliquot of the above cellular culture onto an LB agar plate (containing 50 mg/mL kanamycin) and evenly spread over the entire surface of the plate with a sterilised bacterial cell spreader (see Note 14). Allow the LB agar plate to dry in the laminar flow cabinet for 10 min, place the lid back onto the plate, and incubate for 16–24 h in a 37°C incubator.
10. Using a sterilised loop, pick off a single white-coloured colony (see Note 14) and use to inoculate a 5 mL culture of LB liquid media (containing 50 mg/mL kanamycin). Cap the 15-mL centrifuge tube and incubate at 37°C for 16–24 h in a benchtop shaker (at 200 rpm).
11. Isolate plasmid DNA from the overnight culture using the QIAprep Spin Miniprep kit (Qiagen) according to the manufacturer's instructions.

12. Using the above plasmid preparation, set-up a 20 μL *Bam*HI digestion in a labelled 1.5-mL microfuge tube as follows; 12 μL of DNase-free dH_2O, 5 μL of plasmid preparation, 2 μL of 10× Buffer BamHI, and 1 μL of 10 U/μL *Bam*HI. Mix by pipetting, spin reaction contents down in a benchtop microfuge and incubate for 60 min, in a 37°C waterbath.
13. Add 4 μL of 6× LB to the digestion reaction, mix by pipetting and run on an EtBr-stained 1.0% w/v agarose gel in 1× TBE buffer along with a 10 μL aliquot of the 100 bp DNA ladder. pBlueGreen plasmid preparations containing the pre-amiRNA/*Lgu*I insertion fragment in the desired sense orientation (5′–3′) will return a 440 bp *Bam*HI-restriction fragment (see Note 16).
14. For plant transformation, 1 μL of the selected pBlueGreen plant expression vector (containing the pre-amiRNA insert in the sense orientation) is mixed with 1 μL of the helper plasmid, pSoup, prior to *A. tumefaciens* transformation of *Arabidopsis* wild-type plants (see Note 17).

4. Notes

1. In practice, we have found little evidence of cross-silencing effects except for closely related gene family members and, if little or no genomic sequence is available, we recommend using the 3′ terminal 300–500 nt from the target gene mRNA sequence as higher levels of sequence variability occur in this region.
2. DNA oligonucleotides (amiRNA forward and reverse primers) are ordered without requesting any additional modifications.
3. Overnight incubation is very important, do not reduce this incubation period. Reduction of this incubation period will result in severe losses to overall yields of precipitated nucleic acids.
4. The quality of extracted RNA can be checked by running a 5 μL aliquot with 1 μL of 6× LB on a EtBr-stained 1.0% agarose gel (in 1× TBE buffer).
5. If multiple bands are amplified, including a major band of the correct size, it is advisable to gel purify this band or to refine the RT-PCR conditions (i.e., optimise the annealing temperature) and repeat until a single PCR fragment of the correct size is amplified.
6. We often screen the kanamycin-resistant colonies directly for inserted fragments in the desired orientation, using the cycles, conditions, and primers made for the RT-PCR step

(but omitting the RT step), rather than double-digestion with *Asc*I and *Not*I, which is cumbersome, requiring changing of digestion buffers, purification of the initial digestion product and a subsequent digestion step.

7. The LR Clonase® reaction into the pHellsgate12 plant expression vector inserts the PCR fragment from the pENTR™/D-TOPO® vector into the two *att*R cassettes forming both arms of the hairpin. As the insertion into two *att*R cassettes is less efficient than into a single cassette, we recommend extending the incubation time for the recombination reaction. In addition, we recommend the use of highly efficient DH5α *E. coli* electrocompetent cells for this bacterial transformation step.
8. The pHellsgate12 vector contains convenient restriction sites that can be used to check (1) the presence of an inserted sequence and (2) diagnose the orientation of the intron spacer fragment. Incubation with *Xho*I will excise the first arm of the inverted-repeat and incubation with *Xba*I will excise the second arm, unless the intron has flipped in orientation during the recombination process. In this case, each arm will be excised along with the intron itself. However, in pHellsgate12, this flipped orientation is permissible as the spacer region contains two introns (in opposite orientations) so that one intron is processed *in planta*, irrespective of the orientation. An alternative way of screening is by PCR using primers flanking the insertion sites. It is advisable to sequence the selected clone. This can be carried out by using primers based on the promoter and terminator sequences of the inverted-repeat cassette. However, before performing the sequencing reactions, the plasmid DNA should be linearised with *Cla*I (cleaves within the intron). This alleviates the reduction in sequence trace signal that is often seen for the duplexed vector regions (caused by the inverted-repeat nature of the vector). Please also note that sequencing through *att*R, *att*P, or *att*L sites can be troublesome, probably because of the secondary structures formed in these sites.
9. Putative amiRNA target sequences are identified in this way due to the fact that target sequence positions 1, 10, and 19 correspond to mature amiRNA positions 19, 10, and 1, respectively. The majority of endogenous plant miRNAs, including *Ath*-miR159b, possess uracil (U) at their 5′ terminus and sRNAs expressing a U residue at this position preferentially associate with AGO1 (12). Similarly, the endonucleolytic Slicer activity of the AGO1 protein appears to preferentially cleave mRNA targets next to adenine residues (6), and a more-stable dsRNA base pairing is preferred at amiRNA position 19 to ensure preferential loading of the amiRNA guide strand over the

amiRNA* passenger onto the AGO1-catalysed RISC complex (13–15). These features are therefore retained in the design of amiRNAs to ensure efficient amiRNA-directed RNA silencing by endogenous miRNA biogenesis pathway machinery.

10. Two additional bases are added to the putative amiRNA target sequence as this 21-nt cDNA sequence corresponds to the full length of mature amiRNA. BLAST search the putative amiRNA 21-nt sequence against transcript sequences only (TAIR9 Transcripts), which is a selectable option on a drop-down menu within the TAIR website's BLASTN tool.
11. The endogenous *Arabidopsis* miR159b precursor transcript (pre-miR159b) contains mismatched dsRNA base pairings at the indicated positions within the miR159/miR159* duplex. These mismatches are maintained in the design of the amiRNA* passenger strand (corresponding to the amiRNA forward primer) to retain the endogenous pre-miR159b shape of the introduced amiRNA/amiRNA* duplex within the pri-miR159b transcript.
12. The Expand Long Template Enzyme mix is used to amplify the pre-amiRNA fragment as this system contains a mixture of two Taq DNA polymerases to allow for (1) proofreading of amplified PCR products and (2) A-tailing of the amplified product. A-tailing enables ligation between the pre-amiRNA PCR product and the pGEM®-T Easy vector.
13. Incubating linearised DNA fragments at 65°C for 5 min, removes secondary structure and immediately transferring the mixture to ice and incubating for an additional 5 min, maintains these molecules in a denatured state.
14. Ten minutes prior to spreading the bacterial culture onto the LB agar plate, and using a sterilised bacterial cell spreader, evenly spread 40 μL each of 0.1 M IPTG and 20% w/v X-gal over the agar plate. This allows for blue–white selection of insert positive bacterial colonies, which are selected for subsequent plasmid preparations.
15. A completely digested pGEM-T::pre-amiRNA returns *Lgu*I restriction fragments of 250, 378, and 3,000 nt, respectively. *Lgu*I digestion of the pBlueGreen plant expression vector will return two restriction fragments of 700 and >10 kb, respectively. Only use plasmid preparations returning these respective *Lgu*I banding patterns for subsequent molecular manipulations.
16. The pre-amiRNA/*Lgu*I restriction fragment can insert into the similarly digested plant expression vector, pBlueGreen, in either orientation (5′–3′ or 3′–5′). If the selected plasmid preparation contains the pre-amiRNA insert in the antisense orientation (376 bp *Bam*HI restriction fragment), discard

this preparation and screen additional kanamycin-resistant white-coloured colonies until a plasmid preparation containing the insert in the desired sense orientation is identified.

17. The pBlueGreen plant expression vector has been modified to contain a pri-miR159b sequence where the sequences corresponding to the pre-miR159b have been replaced with the *LacZ* gene (for blue/white colony selection) which itself is flanked by *Lgu*I restriction sites. This sequence was flanked with the 35S promoter and OCS terminator and placed between the right and left borders of the plant expression vector, pGreen (16). This vector requires the helper plasmid, pSoup for replication in *Agrobacterium*.

References

1. Sharp, P. A. (2001). RNA interference. *Genes Dev.* **15,** 485–490.
2. Waterhouse, P. M., Wang, M-B., and Lough, T. (2001). Gene silencing as an adaptive defence against viruses. *Nature* **411,** 834–842.
3. Smith, N. A., Singh, S. P., Wang, M-B., Stoutjesdijk, P. A., Green, A. G., and Waterhouse, P. M. (2000). Total silencing by intron-spliced hairpin. *Nature* **407,** 319–320.
4. Wesley, S. V., Helliwell, C. A., Smith, N. A., Wang, M-B., Rouse, D. T., Liu, Q., Gooding, P. S., Singh, S. P., Abbott, D., Stoutjesdijk, P. A., Robinson, S. P., Gleave, A. P., and Waterhouse, P. M. (2001). Constructs for efficient, effective and high throughput gene silencing in plants. *Plant J.* **27,** 581–590.
5. Wielopolska, A., Townley, H., Moore, I., Waterhouse, P., and Helliwell, C. (2005). A high-throughput inducible RNAi vector for plants. *Plant Biotechnol. J.* **3,** 583–590.
6. Schwab, R., Ossowski, S., Riester, M., Warthmann, N., and Weigel, D. (2006). Highly specific gene silencing by artificial microRNAs in *Arabidopsis. Plant Cell* **18,** 1121–1133.
7. Eamens, A. L., Smith, N. A., Curtin, S. J., Wang, M-B., and Waterhouse, P. M. (2009). The double-stranded RNA binding domain protein DRB1 is involved in guide strand selection from microRNA duplexes in *Arabidopsis thaliana. RNA* **15,** 2219–2235.
8. Helliwell, C. A., Wesley, S. V., Wielopolska, A. J., and Waterhouse, P. M. (2002). High throughput vectors for efficient gene silencing in plants. *Funct. Plant Biol.* **29,** 1217–1225.
9. Niu, Q. W., Lin, S. S., Reyes, J. L., Chen, K. C., Wu, H. W., Yeh, S. D., and Chua, N. H. (2006). Expression of artificial microRNAs in transgenic *Arabidopsis thaliana* confers virus resistance. *Nat. Biotechnol.* **24,** 1420–1428.
10. Warthmann, N., Chen, H., Ossowski, S., Weigel, D., and Hervé, P. (2008). Highly specific gene silencing by artificial miRNAs in rice. *PLoS One* **3,** e1829.
11. Park, W., Zhai, J., and Lee, J.Y. (2009). Highly efficient gene silencing using perfect complementary artificial miRNA targeting AP1 or heteromeric artificial miRNA AP1 and CAL genes. *Plant Cell Rep.* **28,** 469–480.
12. Mi, S., Tao, C., Hu, Y., Chen, Y., Hodges, E., Ni, F., Wu, L., Li, S., Zhou, H., Long, C., Chen, S., Hannon, G., and Qi, Y. (2008). Sorting of small RNAs into *Arabidopsis* argonaute complexes is directed by the 5′ terminal nucleotide. *Cell* **133,** 1–12.
13. Matranga, C., Tomari, Y., Shin, C., Bartel, D. P., and Zamore, P. D. (2005). Passenger-strand cleavage facilitates assembly of siRNA into Ago2-containing RNAi enzyme complexes. *Cell* **123,** 607–620.
14. Rand, T. A., Petersen, S., Du, F., and Wang, X. (2005). Argonaute2 cleaves the anti-guide strand of siRNA during RISC activation. *Cell* **123,** 621–629.
15. Tomari, Y., Matranga, C., Haley, B., Martinez, N., and Zamore, P. D. (2004). A protein sensor for siRNA asymmetry. *Science* **306,** 1377–1380.
16. Helen, R. P., Edwards, E. A., Leyland, N. R., Bean, S., and Mullineaux, P. M. (2000). pGreen: a versatile and flexible binary Ti vector for *Agrobacterium*-mediated plant transformation. *Plant Mol. Biol.* **42,** 819–832.

Chapter 11

Recombinant Protein Expression in *Nicotiana*

Nobuyuki Matoba, Keith R. Davis, and Kenneth E. Palmer

Abstract

Recombinant protein pharmaceuticals are now widely used in treatment of chronic diseases, and several recombinant protein subunit vaccines are approved for human and veterinary use. With growing demand for complex protein pharmaceuticals, such as monoclonal antibodies, manufacturing capacity is becoming limited. There is increasing need for safe, scalable, and economical alternatives to mammalian cell culture-based manufacturing systems, which require substantial capital investment for new manufacturing facilities. Since a seminal paper reporting immunoglobulin expression in transgenic plants was published in 1989, there have been many technological advances in plant expression systems to the present time where production of proteins in leaf tissues of nonfood crops such as *Nicotiana* species is considered a viable alternative. In particular, transient expression systems derived from recombinant plant viral vectors offer opportunities for rapid expression screening, construct optimization, and expression scale-up. Extraction of recombinant proteins from *Nicotiana* leaf tissues can be achieved by collection of secreted protein fractions, or from a total protein extract after grinding the leaves with buffer. After separation from solids, the major purification challenge is contamination with elements of the photosynthetic complex, which can be solved by application of a variety of facile and proven strategies. In conclusion, the technologies required for safe, efficient, scalable manufacture of recombinant proteins in *Nicotiana* leaf tissues have matured to the point where several products have already been tested in phase I clinical trials and will soon be followed by a rich pipeline of recombinant vaccines, microbicides, and therapeutic proteins.

Key words: *Nicotiana benthamiana*, Recombinant protein expression, Transient expression, Molecular farming, Plant-made pharmaceutical

1. Introduction to Plant-Made Pharmaceuticals

Since the ability to genetically engineer plants was established in the early 1980s (1), researchers have endeavored to utilize plant-produced proteins for a variety of applications, including the production of recombinant pharmaceutical proteins. The potential advantages of using plant-based expression systems include the

James A. Birchler (ed.), *Plant Chromosome Engineering: Methods and Protocols*, Methods in Molecular Biology, vol. 701,
DOI 10.1007/978-1-61737-957-4_11,

ability to produce complex proteins that require posttranslational modifications, excluding the possibility of introducing human pathogens during the manufacturing process, and the ability to scale up production efficiently and cost-effectively (2–6). The field of plant-made pharmaceuticals (PMPs) has steadily evolved since the expression of a functional mouse IgG in tobacco was published in 1989 (7) to the point where several plant-produced proteins have been used in clinical trials (8–10). Having several PMP therapeutics in advanced clinical trials is a major breakthrough and has initiated more focused efforts on developing a regulatory process for approving plant-made biologics (11).

A number of plant-based expression systems have been developed to express recombinant proteins. These expression systems include plant cell cultures and intact plants, and the use of both stable transformation and transient expression systems (12). Plant cell culture systems derived from plants that have been used to produce biopharmaceuticals include tobacco, tomato, soybean, rice, carrot cells, and *Arabidopsis thaliana* (13–15). Other cell-based systems include algae and moss bioreactors (16–18). Advantages that cell-based systems may provide compared to whole plants include the ability to produce proteins under more controlled conditions, which is important for good manufacturing practice (GMP) protein production, shorter turnaround times for production runs, and in some cases, protein purification can be less expensive due to streamlined downstream processing (15, 16).

A wide variety of plant species have been tested for their ability to produce recombinant pharmaceutical proteins, including *Nicotiana* species, safflower, tomato, potato, soybean, alfalfa, spinach, *A. thaliana*, maize, and rice (4, 12). The major strategies for expressing proteins in whole plants are transient expression with viral vectors and stable transformation where transgenes are targeted to either the nuclear or chloroplast genome (3, 12, 19–23). Stable transformation offers the advantage that it is scalable to large field production methods. However, this can be offset in some cases by the lower expression level often obtained in transgenic plants compared to transient expression and the length of time required to create transgenic lines that have high expression levels that are stable across multiple generations.

Nicotiana species have long been model systems in plant biology and became particularly prominent as the field of plant molecular biology and plant transformation began. *Nicotiana benthamiana*, in particular, has been a workhorse for studies of plant–virus interactions due to its susceptibility to a wide variety of viruses and ease of use in laboratory settings (24). Based on the detailed understanding of viral replication and protein expression obtained from decades of research from many groups, transient expression using viral vectors has become a major strategy for expressing proteins in plants, with the majority of current efforts focused on *N. benthamiana* and tobacco mosaic virus (TMV)-based vector systems (19, 20, 25, 26). Transient expression in *N. benthamiana* offers the advantages of high expression levels, relatively short production times of days to several weeks, and ease of use in controlled growth conditions where optimal parameters for biomass production under GMP can be obtained. In addition, since *N. benthamiana* is not a food crop, there is less concern about contaminating food supplies. A limitation of the infectious TMV vector systems is that the expression of proteins larger than 50 kDa becomes difficult. This has been largely solved with the introduction of "deconstructed" viral vectors that can be delivered to plant cells by *Agrobacterium tumefaciens*, resulting in high level expression of larger proteins (20, 27).

A number of different recombinant proteins have been expressed in these various plant systems including numerous antigens for vaccine production, hormones, growth factors, blood proteins, cytokines, enzymes, and antibodies (2–4, 28, 29). In many cases, these proteins retained biological activity, and at least, a few were expressed at levels sufficient to support commercial production (30). As has been observed in all prokaryotic and eukaryotic expression systems, there is significant variability in plant expression levels of specific proteins and not all proteins express well in plants. Moreover, within the arena of plant-based expression, a given protein may express much better in one system versus another (12). However, one clear conclusion that can be drawn from these studies is that plant-based expression is a viable alternative for the production of some pharmaceutical proteins. It is very likely that within the next several years the first plant-produced therapeutic protein will be approved and made available to patients. This milestone will usher in a new wave of development in PMP research and help drive further improvements in plant-based expression technologies and downstream processing of plant-derived proteins.

2. Protein Expression Methods in *Nicotiana* Species: Historical Overview and Progress

2.1. Stable Expression Systems

Nuclear genome transformation: Historically, transgenic plants obtained via nuclear genome transformation have been most widely used for the expression of recombinant proteins. Nuclear genome transformation of *Nicotiana* plants is easily accomplished by methods exploiting *A. tumefaciens*. A specific tissue culture procedure for this purpose has been well established, which usually takes about 3 months (31). Once transformants are established, the transferred gene is stably maintained and inherited as part of the plant chromosome in a Mendelian fashion. However, varying degrees of target gene expression are commonly observed for different transgenic lines (e.g., see (32)). This is due to random transgene integration into the chromosome in *Agrobacterium*-mediated transformation and is commonly referred to as "position effect." Therefore, screening multiple transgenic lines (a few dozen or more) is necessary to identify good expressors. Such interline expression variance can also occur within different progenies from the same parental line albeit with stable transgene heredity. While homozygous lines may show higher expression than heterozygous plants (33), suppressed expression is often observed due to transgene silencing (34). Furthermore, even within the same plant, there can be considerable degrees of diversity in transgene expression levels, depending on growth stage and physiological conditions; typically, young and growing tissue expresses higher levels of recombinant proteins. Therefore, even though the transgene stably exists in the system, the reality is that careful monitoring of expression along with good agricultural practice is essential to obtain a sustainable supply of the protein of interest. Typical expression levels in nuclear genome transformants usually do not exceed 1–2% of total proteins in leaf tissue (less than 100 mg/kg of fresh biomass) and are largely dependent on the transgene construct (promoter, 5′ and 3′ untranslated regions (UTRs), transcript stability, codon usage, etc.) and the nature of expressed proteins (size, solubility, stability, toxicity, accumulation sites, etc.).

Transplastomic plants: The plastid genome can also be transformed via biolistic transgene delivery, generating a different type of transgenic plants called "transplastomic" plants (35). Although the procedure to develop transplastomic *Nicotiana* plants is not as straightforward as *Agrobacterium*-mediated nuclear genome transformation, as it takes 7–12 months (36), chloroplasts have been shown to be capable of expressing very high levels of recombinant proteins, often reaching an order of magnitude higher than those obtained with best expressors via nuclear transformation (23). The advantages of plastid-based recombinant protein expression are multifold. These include the high ploidy of the organelle, lack of gene silencing, and the availability of site-directed transgene integration through homologous recombination. These features provide higher, more controlled, and uniform transgene expression than conventional transgenic plants based on nuclear transformation. Concurrent expression of multiple genes can be readily achieved by utilizing polycistronic operons (37–39). Furthermore, environmental concerns are alleviated by the strict maternal inheritance of the plastid in most species, including *Nicotiana* plants (23); thus, transgenes will not disseminate into nearby plants through pollen outflow.

While many of the unique advantages of protein expression in this organelle listed above are associated with the prokaryotic nature of plastids, the same is also true for the system's shortcomings. Among the major limitations is the plastid's inability to perform posttranslational modifications such as N-glycosylation, which plays important roles in the bioactivity and pharmacokinetics of many therapeutic proteins, including monoclonal antibodies (40).

Plant cell/tissue culture: Transgenic cell and tissue cultures derived from *Nicotiana* plants have also been explored as platforms for recombinant protein expression (41). Historically, plant cells and hairy roots have been developed for the production of various plant secondary metabolites, including food additives, nutraceuticals, and pharmaceuticals (42, 43). For cell culture-based recombinant protein production, tobacco-derived Bright Yellow-2 (BY-2) and *Nicotiana tabacum*-1 (NT-1) cells are frequently chosen for protein expression due to well-established transformation and propagation procedures and favorable growth characteristics (13).

Meanwhile, hairy root culture represents a type of tissue culture-based protein expression (44, 45). Neoplastic, clonally expanding roots are readily developed by transformation of preestablished transgenic plants expressing a protein of interest with *Agrobacterium rhizogenes*. The advantages of these in vitro systems over whole-plant systems are mainly attributed to their greater amenability to controlled production conditioning for consistent product quality and yield, similar to that obtained in microbial fermentations and mammalian cell culture systems. From a product development standpoint, such features might make regulatory approval easier to obtain for plant culture systems than whole-plant systems. Indeed, the world's first licensed, plant-made protein pharmaceutical is a vaccine antigen for poultry Newcastle disease expressed in tobacco cell culture; the first product for human use will likely be glucocerebrosidase, which is produced in carrot suspension cells for Gaucher's disease (46). However, tight regulation in a closed system in turn limits production scalability, which is one of the most attractive advantages of whole-plant-based systems. Obvious challenges for plant culture systems would, therefore, be the efficiencies of protein expression and recovery. In this context, a critical factor determining the feasibility of the culture systems would be whether the protein of interest can be efficiently secreted in culture medium or not because this will significantly impact the complexity and cost of downstream processing (41).

2.2. Transient Expression Systems

Each of the above transgenic-based expression systems has its own unique advantages. Nevertheless, transgenic approaches require long development times, which have been a major bottleneck not only for their progress but also for technical improvements. As a result, a general concept for plant-based protein expression has been "low and slow," a few successful cases aside.

Meanwhile, the recent advent of highly efficient transient expression systems has completely changed the concept and revolutionized plant-made pharmaceutical research.

Whole recombinant virus-based expression: Transient expression of recombinant proteins in *Nicotiana* plants is currently performed by the use of engineered infectious plant viruses or *Agrobacterium*-mediated DNA transfer (agroinfiltration). Over the last decade, many plant virus vectors have been engineered for the expression of foreign proteins in *Nicotiana* plants. Most of these are based on RNA viruses such as tobacco mosaic virus (TMV) and potato virus X (PVX) (19). The first generation plant virus vectors utilize infection-competent viruses, represented by the modified TMV-based Geneware® system (Kentucky BioProcessing, LLC, Owensboro, KY). In essence, such a vector is comprised of the viral cDNA harboring a gene of interest either as a fusion to viral coat proteins (CPs), mainly for epitope presentation as vaccine antigen (47), or placed downstream of an additional subgenomic promoter (19, 48). Viruses are inoculated into the leaf initially as infectious RNA, which is created from the vector either through in vitro transcription or agroinfiltration followed by *in planta* transcription (48). Thus, the protein of interest is coexpressed along with systemic viral spread and replication, with maximal expression usually obtained within 2–3 weeks postinfection. A recent notable example of recombinant proteins expressed by infectious virus-based systems is the antiviral lectin Griffithsin. Using the Geneware® system, functional Griffithsin was expressed in *N. benthamiana* at a very high level, reaching as high as 5 g of the protein per kg of leaf biomass (30). Such high levels of expression with this type of virus vectors are, however, usually limited to small proteins whose coding sequences are less than 1.5 kb. This is due to the increased genetic instability of recombinant viruses carrying a larger foreign sequence (20, 49). Despite this limitation, this method offers a viable option for the mass production of small proteins such as antiviral lectins and monoclonal antibody single-chain variable fragments (scFvs).

Deconstructed virus-based expression: A major breakthrough in viral expression strategies was facilitated by the recent advent of deconstructed virus vectors, originally reported for the TMV-based magnICON® system, developed by ICON Genetics GmbH (Halle, Germany) (50). The essence of improvements in this system from the first generation viral vectors are: (1) deletion of the viral CP gene to enhance the stability and size compatibility of a transgene, (2) viral cDNA modifications facilitating *in planta* RNA replicon recovery upon *Agrobacterium*-mediated DNA transfer, and (3) efficient whole-plant vector delivery by vacuum-based agroinfiltration ("magnifection") to compensate for defective systemic movement due to CP deletion (27, 51). These improvements allowed the uniform and high-level (gram per kg biomass) expression of larger proteins in *N. benthamiana* plants within 10 days. ICON Genetics further developed a similar deconstructed viral vector system based on PVX. Taking advantage of the fact that TMV and PVX do not compete during replication, fully assembled immunoglobulin (Ig)G molecules were expressed at up to 0.5 g per kg of leaf by codelivering deconstructed TMV and PVX vectors (each encoding a gene for Ab heavy or light chains) (52). This technology may provide the most rapid means among all currently available recombinant expression systems for the production of full length monoclonal antibodies from genes in various production scales ranging from bench to commercialization (53). A potential limitation of the magnifection method is that it is technically challenging to scale up; however, this impediment has recently been solved by development of a robotic magnifection system by Kentucky BioProcessing, LLC.

Nonviral vector-based expression: Agroinfiltration with conventional nonviral binary vectors, on the other hand, had been primarily used for analytical purposes before constructing transgenic plants (e.g., see (54)). However, progress made in recent years now allows even these vectors to express proteins at higher levels with agroinfiltration compared to transgenic plants. One of the key factors for high expression with agroinfiltration-delivered nonviral vectors appears to be the coexpression of a posttranscriptional gene silencing (PTGS) inhibitor such as tomato busy stunt virus-derived p19 and potyviral helper component proteinase (HC-Pro) (55–60). Other important factors of successful *Agrobacterium*-based transient expression include the strain of *Agrobacterium*, density of the bacteria, infiltration media, infiltration condition, and the plant's physiological condition (61, 62), along with vector and transgene design. Detailed protocols of agroinfiltration-based protein expression have been recently published (63, 64). A notable example of agroinfiltration-based nonviral transient protein expression can be seen in a recent report by Vezina et al. (55). Up to 1.5 g of full-size, assembled IgG was expressed in 1 kg of *N. benthamiana* leaf in 4–6 days by coexpressing the heavy and light chains, HC-Pro, and a chimeric human β1,4-galactosyltransferase (GT) by infiltrating a mixture of four *Agrobacterium* strains, each delivering either of four constructs. Notably, GT coexpression led to not only the addition of the terminal β1,4-galactose residue but also the efficient reduction of the plant-specific glycotope structure (α1,3-fucose and β1,2-xylose) on the IgG N-glycans, making them more "human-like" (55). Very recently, a series of new highly efficient agroinfiltration expression vectors (pEAQ vectors) has been constructed based on a conventional binary vector containing cauliflower mosaic virus 35S promoter, and modified 5′-UTR and the 3′-UTR from Cowpea mosaic virus RNA-2 within the T-DNA region (65). These vectors were shown to express multiple polypeptides along with P19 from a single plasmid at a high level within a few days.

As exemplified above, agroinfiltration currently provides a key technique for efficient transient expression of recombinant proteins in *Nicotiana* plants. The above-mentioned magnifection method allows simple scale-up of protein production based on these transient systems even to a commercial scale. In theory, grams of recombinant proteins can be obtained within the shortest period of time (days) among all currently available recombinant production systems (53). Such a rapid turnaround from DNA to product may provide a powerful tool for those proteins that need to be produced in a very short time, such as influenza vaccines, biodefense agents, and individualized cancer vaccines (20). Combined with other unique advantages of plant expression systems (i.e., cost-effectiveness, safety, eukaryotic protein modification), transient plant expression can now offer a robust and efficient method to produce valuable proteins.

Given all the marked progress in transient expression systems illustrated above, the main stream of recombinant protein expression in *Nicotiana* plants has almost entirely moved away from constructing transgenic plants. However, transgenic plants still provide important resources to recombinant protein production. One particularly important application would be the engineering of expression hosts, thereby providing optimal features for protein production in terms of quantity and quality.

2.3. Host Engineering for Optimal Protein Expression

There is no perfect biological system for recombinant protein expression. In other expression systems available today, it has been a common practice to engineer hosts for optimal protein production, along with expression vector design (66–69). Efforts in this regard have just begun in the field of plant-based protein expression. Below are some of the plant-host factors targeted for improvement to enable improved protein expression.

Host species: An ideal host should support high-level foreign protein expression while providing large biomass with easy and rapid growth profiles in both the greenhouse and field. Currently, *N. benthamiana* is used as a primary host for recombinant virus and agroinfiltration-based transient expression. The plant exhibits high susceptibility to many plant pathogens (24), which may be contributing to efficient expression with transient systems. However, *N. benthamiana* has a relatively small biomass and does not grow well in the field compared to other *Nicotiana* species. Large Scale Biology Company created a hybrid line between *Nicotiana excelsior* and *N. benthamiana* (*N. excelsiana*), which was shown to support high-level expression with the Geneware® system while maintaining good field growth properties (47, 70). Sheludko et al. recently tested several

Nicotiana species (*N. benthamiana, Nicotiana debneyi, N. excelsior, Nicotiana exigua, Nicotiana maritima,* and *Nicotiana simulans*) for *Agrobacterium*-mediated transient expression of a model protein, i.e., green fluorescent protein (GFP) with the magnICON® deconstructed TMV vector and a nonviral construct based on cauliflower mosaic virus 35S promoter (71). The authors concluded that *N. excelsior* has the best characteristics in terms of biomass and GFP accumulation level for both types of the expression vectors.

PTGS: Transgene-specific PTGS often leads to poor expression in plants. As described above, coexpression of a PTGS inhibitor often improves levels of protein expression. Creating a host line expressing such an inhibitor would theoretically circumvent the need of codelivering vectors for the PTGS inhibitor and a target protein upon transient protein expression. However, P19-expressing transgenic plants exhibit altered morphology because PTGS is actively involved in plant development (72, 73). A potential solution may be the control of PTGS inhibitor expression, for example, by the use of an inducible promoter (74).

Glycosylation: As described above, glycans often found on human endogenous bioactive proteins have important roles in their bioactivities and pharmacokinetics, constituting an important factor for their recombinant production in heterologous hosts. For example, in the case of N-glycans on the Fc fragment of IgG, the presence and absence of terminal sugars such as sialic acid, core fucose, bisecting N-acetylglucosamine, and mannose residues impact the longevity and effector functions such as complement-dependent cytotoxicity and Ab-dependent cell-mediated cytotoxicity (75). Glycoengineering is therefore a major focus of protein expression systems with virtually all organisms, as none of the currently available systems (including mammalian cells) can provide a uniform glycoform that is completely human-like that possesses desirable biological properties (40). For plant-expressed therapeutic proteins, a concern has been the presence of the plant-specific α1,3-fucose and β1,2-xylose structure and the lack of terminal β1,4-galactose and sialic acid on complex N-glycans (76). In particular, the plant-specific glycan structure can be immunogenic and reduce bioavailability upon human use (77, 78). Thus, many efforts have been centered on reducing plant-specific glycotopes by targeting α1,3-fucosyltransferase (FT) and β1,2-xylosyltransferase (XT) and further humanizing by adding the terminal β1,4-galactose via expressing GT (76). Recently, Strasser et al. have used RNA interference (RNAi) technology to construct transgenic *N. benthamiana* that possesses downregulated FT and XT activities (ΔFT/XT). A monoclonal antibody transiently expressed in this plant with a nonviral expression vector via agroinfiltration was shown to contain an almost homogeneous *N*-glycan without detectable β1,2-xylose and α1,3-fucose residues (79). More recently, the same group has overexpressed GT in the ΔFT/XT either stably (via nuclear genome transformation) or transiently (via agroinfiltration). A monoclonal antibody expressed in these plants was shown to contain terminal β1,4-galactose and lack β1,2-xylose and α1, 3-fucose residues in the N-glycan structure (80).

Other plant-derived factors influencing protein production: Plants contain numerous endogenous factors (e.g., phenolics and proteases) that potentially affect the yields, homogeneity, and quality of recombinant proteins before or after extraction (81). Particularly, proteolysis often accounts for low expression and recovery of recombinant proteins. As has been done with *E. coli* and yeast (66, 82, 83), constructing protease-deficient or protease-deactivated hosts might be a potential approach to this issue. A built-in protease inhibitor system was reported with transgenic potato plants constitutively expressing tomato cathepsin D inhibitor or bovine aprotinin, which was shown to stabilize a foreign protein (neomycin phosphotransferase II) in leaf tissue (84). Protease activity might be downregulated by RNAi. However, such approaches may not be as simple as would be in monocellular organisms like *E. coli* and yeast, given the higher complexity of proteases and their roles in plant development, growth, and homeostasis (81). Alternatively, protease inhibitors or silencing constructs may be coexpressed with the protein of interest using transient expression methods described above. Such a strategy has been proposed (85). Other possible targets of host engineering for enhanced protein expression may be the coexpression of molecular chaperones (21, 86). These would thereby promote correct protein folding and help avoid the stress response associated with misfolded protein accumulation (87).

3. Tips for Successful Protein Expression in *Nicotiana* Plants

As described above, significant progress has been made over the last few years with respect to *in planta* expression vectors and methodologies. Plant expression technologies have reached a point where many proteins can be expressed at levels within days that approach ~5 g/kg leaf biomass. With such robust expression vectors becoming available, researchers will soon only need to clone a gene of interest into an appropriate expression vector and follow established procedures to obtain complex bioactive proteins of interest. However, even with the cutting-edge expression vectors, high-level expression is not guaranteed for all proteins, as is the case with any recombinant expression system. Protein expression levels would be reflected not only by transcription efficiency but also by transcript stability, translation efficiency as well as protein stability. The nature of a transgene can have a significant impact on transcript stability and translation efficiency. Meanwhile, poor expression may be due to the protein's high susceptibility to proteolysis, potential toxicity, or physiological incompatibility with plant cells. Last but not the least, there are important factors to consider for the extraction and purification of proteins expressed in plants. These are discussed in this section.

3.1. Transgene Design

Recombinant protein expression in heterologous hosts is often inefficient due to nonoptimal transgene construction, which could compromise mRNA stability and/or protein translation efficiency (66, 88). Given the current availability of quick and cost-effective gene synthesis services, it is preferable whenever possible to redesign the entire coding sequence to maximize the likelihood of high protein expression. Several companies offer gene synthesis services using proprietary algorithms to optimize the factors described below toward efficient transgene expression in a given organism.

One of the most important factors is codon usage. Different organisms tend to have a particular preference for one or a small set of many synonymous codons for a given amino acid (89). Codon usage reflects the availability of cognate amino-acylated tRNAs and thus can correlate with translation efficiency (89, 90). Codon frequencies in *Nicotiana* plants can be found at http://www.kazusa.or.jp/codon/cgi-bin/spsearch.cgi?species=nicotiana&c=i. For example, among the four alanine codons, GCG occurs at the lowest frequency (9%) in *N. benthamiana*, according to the database, and is therefore considered "nonpreferred." Avoiding such rare codons would be theoretically important for codon optimization. However, a caveat for codon optimization across the whole coding sequence is that rare codons at certain positions may play an important role in proper protein folding by regulating local translational efficiency (91–93).

Meanwhile, it is well known that the sequence surrounding the AUG start codon constitutes an important factor in the control of mRNA translation efficiency in eukaryotes. In addition to the plant-specific Kozak-like consensus sequence (94, 95), the sequence and secondary structure downstream of the initiation codon may also influence the transcript's translational efficiency and stability (91, 96–98). Removing potential mRNA modification sequences constitutes another factor for consideration toward transcript stability. For example, bacterial genes tend to be AT-rich, whereas many plant genes have a higher GC content (88). AT-rich sequences may contain potential polyadenylation signals such as AATAAA and RNA destabilizing elements such as ATTTA (88). These could lead to poor accumulation of intact transcripts upon expression in plant cells, although they are not always functionally active (91). Information about some of these cryptic sequences is available (88). Usually, codon optimization helps remove many of such cryptic sites. However, the list is not comprehensive, and unidentified deleterious sequences may still exist, thus requiring a detailed analysis if transcript quality is suspected for poor protein expression. Finally, mRNA's secondary structure may also need attention in terms of a potential target for silencing (99); however, this may be addressed by the coexpression of a silencing inhibitor.

3.2. Strategies to Improve Protein Accumulation

Poor *in planta* protein accumulation due to stability or toxicity issues can be sometimes addressed by targeting to a specific organelle such as the apoplast, endoplasmic reticulum (ER), chloroplast, or vacuole through specific signal peptides attached to the protein of interest. Detailed discussion on the effects of subcellular targeting is provided in recent review articles (100, 101). Each cellular compartment has a different influence on the folding, assembly, posttranscriptional modification, and metabolism for proteins. Therefore, an optimal accumulation site for a given protein has to be determined experimentally. However, proteins that require special posttranslational modification such as glycosylation must be targeted to the endomembrane systems, thus limiting potential options. For such proteins, the ER has been shown to be able to support higher accumulation levels (100, 101), although targeting expression to this cellular compartment will limit N-glycosylation primarily to high-mannose types (e.g., (32)).

In conventional recombinant protein expression systems, a wide range of translational fusion partners has been developed to enhance expression, solubility, and stability of the passenger protein (66). Although not yet extensively investigated, such a strategy may also improve the protein accumulation in plants. A unique approach has been recently proposed, in which a recombinant protein is designed to be accumulated within a highly stable ER-derived protein body in leaf tissue by fusing a unique sequence repeat of elastin-like polypeptides (102) or a sequence derived from prolamin seed storage proteins (103, 104). A potential challenge for these fusion strategies is the requirement for the cleavage of the fusion partner upon purification, which could significantly increase the downstream processing costs.

3.3. Extraction and Purification

Extraction of recombinant proteins from *Nicotiana* leaves poses a number of challenges that originate in the physical and biochemical environment of the leaf. Plant cell walls and extensive endomembrane structures provide the first purification challenge. They encapsulate a cytoplasm rich in alkaloids and polyphenolic compounds such as flavonoids and tannins, and a wide array of protease activities that can complicate downstream purification of the desired recombinant protein products. Obviously, the level of accumulation of the protein of interest is critically important, with higher accumulation levels reducing the purification burden.

Several groups have explored methods that avoid disruption of the cellular structure to harvest the secreted proteome that contains the target protein. Julian Ma and colleagues have recently described methods for recovery of monoclonal antibodies and an antiviral lectin (cyanovirin-N) from root exudates of plants grown hydroponically (105). This approach has significant appeal, since it allows continuous production and harvest from live plants, and given that the complexity of the root exudate proteome is low, a protein that is expressed at relatively high levels could be purified by conventional filtration and chromatography methods with relative ease.

Proteins that are targeted into the secretory pathway in *Nicotiana* leaves usually accumulate in an appropriately folded form in the interstitial space between plant cells. The soluble protein complement in the interstitial space is comparatively simple, and the secreted recombinant protein can be one of the most abundant proteins present (e.g., see (10, 106–108)). Extraction of the protein component of the interstitial space relies on gentle treatment of the leaf material to limit cell breakage: intact leaves are harvested from plants expressing the secreted recombinant protein of interest and infiltrated with extraction buffer under vacuum. The secreted proteome, hopefully including the recombinant protein, is then recovered in the interstitial fluid (IF) by centrifugation under low gravitational forces that limit tissue damage. This method has been used to recover a wide array of secreted proteins from *Nicotiana* leaf tissues, from small proteins such as scFvs, lysozyme, and aprotinin, through molecules as large as monoclonal antibodies. Key examples are published by McCormick et al. (10, 106, 107) and by Du et al. (108). In our experience, the target protein may represent at least 10% and frequently more than half of the secreted protein component recovered from the IF extract, greatly reducing the purification burden and facilitating purification by conventional affinity, ion exchange, or hydrophobic interaction chromatography methods. The methods that describe the IF purification procedure are described in the United States Patent literature in several US Patents entitled "Method for recovering proteins from interstitial fluid of plant tissues"; these are available at http://www.uspto.gov. Kentucky Bioprocessing LLC has invented an apparatus that is capable of processing multikilogram amounts of *Nicotiana* leaf tissue for recovery of secreted proteins, which facilitates the IF process at industrial scale (see US Patent 6,284,875 entitled "Centrifuge for extracting interstitial fluid").

For proteins targeted to the cytoplasm or retained in the ER, and in instances where IF extraction of secreted proteins is economically unfeasible, it is necessary to extract the whole soluble protein component. When designing an optimal extraction and purification strategy of a target recombinant protein, it is important to investigate as thoroughly as possible the biochemical and biophysical properties of the native protein. Answers to questions such as (1) what is the protein's isoelectric point (pI), (2) does it contain disulphide bonds, (3) does it prefer a certain ionic strength and pH of buffer to maintain solubility, and (4) does it tolerate heat will help guide the design of the initial extraction process. Some general rules about plant protein extraction should be considered. In the first instance, plant cells are replete with proteases and phenolics that will hamper downstream purification and product quality if not addressed early in the extraction procedure. It is generally unhelpful to use reagents that are costly, overly toxic, or which present a complicated disposal problem, in the initial laboratory extraction procedure, since this will render the protocol difficult to scale-up. The protein extract is derived by grinding leaf material with buffer and must be separated from the fibrous material by facile filtration processes (at bench scale) or by continuous flow centrifugation (at industrial scale). For process scale-up, it is important to keep extract volumes as low as possible, so we often strive to use buffer to leaf ratios as low as 0.5:1.0, and preferably not greater than 1:1. It is also desirable to keep the pH of the extraction buffer below 5.0 because the majority of plant proteins are insoluble at this pH, in particular the proteins involved in the photosynthetic complex such as the large and small subunits of RUBISCO, which can be challenging to eliminate during purification of the target proteins. Oxidized phenolics and RUBISCO-associated membranous structures can very rapidly foul filtration membranes and chromatography columns, so it is vital that green (RUBISCO-associated membranes) and brown (polyphenols) color is eliminated from the protein extract. A low pH buffer (<5.0) containing a strong reducing agent such as 10 mM sodium metabisulphite that will inhibit polyphenol oxidase activity, and EDTA to inhibit metalloproteinases is, in our experience, ideal. We adjust the initial extract to <pH 5.0 and remove the majority of the insoluble proteins (and green pigment) by centrifugation. If the target protein tolerates heat, the extract may be heated to around 48°C for 10 min to further remove contaminating proteins. In our experience, this procedure, with slight variation in buffer concentration, pH, and ionic strength, yields straw-colored extracts highly enriched for the target protein, which can subsequently be purified using standard protein purification tools.

Of course, there are many proteins that do not tolerate extraction at acidic pH, or which are susceptible to proteolysis at acidic pH. Proteolytic activity can sometimes be reduced by addition of specific inhibitors such as phenylmethylsulfonyl fluoride (PMSF). A list of appropriate protease inhibitors is given by Gegenheimer (109); once again, it is important to consider issues that might limit process scale-up in deciding on appropriate methods to mitigate proteolysis. For proteins intolerant of pH below 5.0 where the RUBISCO complex is insoluble, elimination of the inevitable green pigmentation in the extract is vital. Filtration of the extract through a cake of diatomaceous earth or similar material can sometimes be helpful; this is a scalable process with precedence for use in plant virus purification. When using recombinant plant viral expression systems, the plant virus itself frequently becomes a major contaminant that can be removed by size exclusion chromatography, treatment of the extract with high-molecular-mass polyethylene glycol to precipitate the macromolecules, or by filtration. We showed that the TMV coat protein could be very efficiently removed from an initial extract by filtration through a ceramic membrane (30).

Phospholipids containing the photosynthetic complex can sometimes be eliminated by precipitation with the synthetic polycation polyethyleneimine (PEI) (109), a highly branched polymer with a relatively high molecular weight. The plant extract is treated with PEI with gentle stirring, followed by centrifugation to separate the membranous material from the supernatant fraction that contains the target protein. Aqueous two-phase partitioning systems methods that employ polyethylene glycol and phosphate have been used to impressive effect in separation of monoclonal antibodies from contaminating plant proteins (110–112). On occasion, we have used the Triton X114 detergent-based phase-separation technique of Bordier (113) to separate soluble, hydrophilic proteins of interest from membrane-associated proteins of the photosynthetic complex. Clearly, proteins that do not tolerate low pH extraction present additional complexities for purification from plants, but these issues are not insurmountable. Once the protein extract is free of contaminating RUBISCO and polyphenols, the next step is to purify the protein using the standard battery of filtration and chromatography methods (e.g., size exclusion, ion exchange, and hydrophobic interaction chromatography) that are available to protein biochemists. The ease of use of immobilized metal affinity chromatography (IMAC) with polyhistidine-tagged proteins has made this technique highly popular amongst academic groups as well as some industrial players, and protein A or protein G affinity chromatography remains the gold standard for antibody purification by plant biotechnologists, as it does in industries that use mammalian cell production methods. An innovative method that uses affinity nanoparticles based on tobamoviruses displaying domains of protein A may allow application of a unique plant biotechnology product to industrial purifications (114).

At this point, only a handful of publications show purification of recombinant proteins from *Nicotiana* leaf tissues with sufficient purity, scale, and efficiency to facilitate serious preclinical and clinical evaluation, e.g., (10, 30, 106–108), but a number of other candidate products are in the pipeline.

4. Conclusion

The development of plant-based expression systems has begun to mature, and there are currently several promising technologies available that will support commercial-scale production of recombinant proteins in plants. These plant-based expression systems provide a robust and economical method to produce recombinant proteins for use as therapeutics, diagnostic reagents, and industrial enzymes. In addition, plant-based systems have been shown to be particularly attractive for producing complex proteins such as antibodies that are often difficult to produce in other expression systems. With an increased focus on engineering host plants to improve protein expression levels and biological activity, and the further development of robust downstream processing methods, it is likely that plant-based expression will become more prominent over the coming years and become a valuable tool in the production of biopharmaceuticals.

Acknowledgement

We thank Andrew Marsh for editorial assistance.

References

1. Horsch, R. B., Rogers, S. G., and Fraley, R. T. (1985) Transgenic plants, *Cold Spring Harb Symp Quant Biol* 50, 433–437.
2. Karg, S. R. and Kallio, P. T. (2009) The production of biopharmaceuticals in plant systems, *Biotechnol Adv* 27(6), 879–894.
3. Mett, V., Farrance, C. E., Green, B. J., and Yusibov, V. (2008) Plants as biofactories, *Biologicals* 36, 354–358.
4. Lienard, D., Sourrouille, C., Gomord, V., and Faye, L. (2007) Pharming and transgenic plants, *Biotechnol Annu Rev* 13, 115–147.
5. Boehm, R. (2007) Bioproduction of therapeutic proteins in the 21st century and the role of plants and plant cells as production platforms, *Ann NY Acad Sci* 1102, 121–134.
6. Ma, J. K., Chikwamba, R., Sparrow, P., Fischer, R., Mahoney, R., and Twyman, R. M. (2005) Plant-derived pharmaceuticals – the road forward, *Trends Plant Sci* 10, 580–585.
7. Hiatt, A., Cafferkey, R., and Bowdish, K. (1989) Production of antibodies in transgenic plants, *Nature* 342, 76–78.
8. Aviezer, D., Brill-Almon, E., Shaaltiel, Y., Hashmueli, S., Bartfeld, D., Mizrachi, S., Liberman, Y., Freeman, A., Zimran, A., and Galun, E. (2009) A plant-derived recombinant human glucocerebrosidase enzyme – a preclinical and phase I investigation, *PLoS One* 4, e4792.
9. Kaiser, J. (2008) Is the drought over for pharming? *Science* 320, 473–475.

10. McCormick, A. A., Reddy, S., Reinl, S. J., Cameron, T. I., Czerwinkski, D. K., Vojdani, F., Hanley, K. M., Garger, S. J., White, E. L., Novak, J., Barrett, J., Holtz, R. B., Tuse, D., and Levy, R. (2008) Plant-produced idiotype vaccines for the treatment of non-Hodgkin's lymphoma: safety and immunogenicity in a phase I clinical study, *Proc Natl Acad Sci USA* 105, 10131–10136.
11. Sparrow, P. A. and Twyman, R. M. (2009) Biosafety, risk assessment and regulation of plant-made pharmaceuticals, *Methods Mol Biol* 483, 341–353.
12. Vancanneyt, G., Dubald, M., Schroder, W., Peters, J., and Botterman, J. (2009) A case study for plant-made pharmaceuticals comparing different plant expression and production systems, *Methods Mol Biol* 483, 209–221.
13. Hellwig, S., Drossard, J., Twyman, R. M., and Fischer, R. (2004) Plant cell cultures for the production of recombinant proteins, *Nat Biotechnol* 22, 1415–1422.
14. Shaaltiel, Y., Bartfeld, D., Hashmueli, S., Baum, G., Brill-Almon, E., Galili, G., Dym, O., Boldin-Adamsky, S. A., Silman, I., Sussman, J. L., Futerman, A. H., and Aviezer, D. (2007) Production of glucocerebrosidase with terminal mannose glycans for enzyme replacement therapy of Gaucher's disease using a plant cell system, *Plant Biotechnol J* 5, 579–590.
15. Plasson, C., Michel, R., Lienard, D., Saint-Jore-Dupas, C., Sourrouille, C., de March, G. G., and Gomord, V. (2009) Production of recombinant proteins in suspension-cultured plant cells, *Methods Mol Biol* 483, 145–161.
16. Decker, E. L. and Reski, R. (2007) Moss bioreactors producing improved biopharmaceuticals, *Curr Opin Biotechnol* 18, 393–398.
17. Decker, E. L. and Reski, R. (2008) Current achievements in the production of complex biopharmaceuticals with moss bioreactors, *Bioprocess Biosyst Eng* 31, 3–9.
18. Leon-Banares, R., Gonzalez-Ballester, D., Galvan, A., and Fernandez, E. (2004) Transgenic microalgae as green cell-factories, *Trends Biotechnol* 22, 45–52.
19. Lico, C., Chen, Q., and Santi, L. (2008) Viral vectors for production of recombinant proteins in plants, *J Cell Physiol* 216, 366–377.
20. Gleba, Y., Klimyuk, V., and Marillonnet, S. (2007) Viral vectors for the expression of proteins in plants, *Curr Opin Biotechnol* 18, 134–141.
21. Ma, J. K., Drake, P. M., Chargelegue, D., Obregon, P., and Prada, A. (2005) Antibody processing and engineering in plants, and new strategies for vaccine production, *Vaccine* 23, 1814–1818.
22. Gomord, V., Chamberlain, P., Jefferis, R., and Faye, L. (2005) Biopharmaceutical production in plants: problems, solutions and opportunities, *Trends Biotechnol* 23, 559–565.
23. Daniell, H. (2006) Production of biopharmaceuticals and vaccines in plants via the chloroplast genome, *Biotechnol J* 1, 1071–1079.
24. Goodin, M. M., Zaitlin, D., Naidu, R. A., and Lommel, S. A. (2008) *Nicotiana benthamiana*: its history and future as a model for plant-pathogen interactions, *Mol Plant Microbe Interact* 21, 1015–1026.
25. Smith, M. L., Lindbo, J. A., Dillard-Telm, S., Brosio, P. M., Lasnik, A. B., McCormick, A. A., Nguyen, L. V., and Palmer, K. E. (2006) Modified tobacco mosaic virus particles as scaffolds for display of protein antigens for vaccine applications, *Virology* 348, 475–488.
26. Pogue, G. P., Lindbo, J. A., Garger, S. J., and Fitzmaurice, W. P. (2002) Making an ally from an enemy: plant virology and the new agriculture, *Annu Rev Phytopathol* 40, 45–74.
27. Gleba, Y., Klimyuk, V., and Marillonnet, S. (2005) Magnifection – a new platform for expressing recombinant vaccines in plants, *Vaccine* 23, 2042–2048.
28. Yusibov, V. and Rabindran, S. (2008) Recent progress in the development of plant derived vaccines, *Expert Rev Vaccines* 7, 1173–1183.
29. Ko, K., Brodzik, R., and Steplewski, Z. (2009) Production of antibodies in plants: approaches and perspectives, *Curr Top Microbiol Immunol* 332, 55–78.
30. O'Keefe, B. R., Vojdani, F., Buffa, V., Shattock, R. J., Montefiori, D. C., Bakke, J., Mirsalis, J., d'Andrea, A. L., Hume, S. D., Bratcher, B., Saucedo, C. J., McMahon, J. B., Pogue, G. P., and Palmer, K. E. (2009) Scaleable manufacture of HIV-1 entry inhibitor griffithsin and validation of its safety and efficacy as a topical microbicide component, *Proc Natl Acad Sci USA* 106(15), 6099–6104.
31. Clemente, T. (2006) Nicotiana (*Nicotiana tobaccum, Nicotiana benthamiana*), *Methods Mol Biol* 343, 143–154.
32. Matoba, N., Kajiura, H., Cherni, I., Doran, J. D., Bomsel, M., Fujiyama, K., and Mor, T. S. (2009) Biochemical and immunological characterization of the plant-derived candidate human immunodeficiency virus type 1 mucosal vaccine CTB-MPR(649–684), *Plant Biotechnol J* 7, 129–145.
33. Tackaberry, E. S., Prior, F., Bell, M., Tocchi, M., Porter, S., Mehic, J., Ganz, P. R., Sardana, R.,

Altosaar, I., and Dudani, A. (2003) Increased yield of heterologous viral glycoprotein in the seeds of homozygous transgenic tobacco plants cultivated underground, *Genome* 46, 521–526.

34. De Wilde, C., Van Houdt, H., De Buck, S., Angenon, G., De Jaeger, G., and Depicker, A. (2000) Plants as bioreactors for protein production: avoiding the problem of transgene silencing, *Plant Mol Biol* 43, 347–359.
35. Maliga, P. (2004) Plastid transformation in higher plants, *Annu Rev Plant Biol* 55, 289–313.
36. Verma, D., Samson, N. P., Koya, V., and Daniell, H. (2008) A protocol for expression of foreign genes in chloroplasts, *Nat Protoc* 3, 739–758.
37. Staub, J. M. and Maliga, P. (1995) Expression of a chimeric uidA gene indicates that polycistronic mRNAs are efficiently translated in tobacco plastids, *Plant J* 7, 845–848.
38. Kang, T. J., Han, S. C., Kim, M. Y., Kim, Y. S., and Yang, M. S. (2004) Expression of non-toxic mutant of *Escherichia coli* heat-labile enterotoxin in tobacco chloroplasts, *Protein Expr Purif* 38, 123–128.
39. De Cosa, B., Moar, W., Lee, S. B., Miller, M., and Daniell, H. (2001) Overexpression of the Bt cry2Aa2 operon in chloroplasts leads to formation of insecticidal crystals, *Nat Biotechnol* 19, 71–74.
40. Rich, J. R. and Withers, S. G. (2009) Emerging methods for the production of homogeneous human glycoproteins, *Nat Chem Biol* 5, 206–215.
41. Shih, S. M. and Doran, P. M. (2009) Foreign protein production using plant cell and organ cultures: advantages and limitations, *Biotechnol Adv* 27(6), 1036–1042.
42. Smetanska, I. (2008) Production of secondary metabolites using plant cell cultures, *Adv Biochem Eng Biotechnol* 111, 187–228.
43. Mishra, B. N. and Ranjan, R. (2008) Growth of hairy-root cultures in various bioreactors for the production of secondary metabolites, *Biotechnol Appl Biochem* 49, 1–10.
44. Gaume, A., Komarnytsky, S., Borisjuk, N., and Raskin, I. (2003) Rhizosecretion of recombinant proteins from plant hairy roots, *Plant Cell Rep* 21, 1188–1193.
45. Woods, R. R., Geyer, B. C., and Mor, T. S. (2008) Hairy-root organ cultures for the production of human acetylcholinesterase, *BMC Biotechnol* 8, 95.
46. Sourrouille, C., Marshall, B., Lienard, D., and Faye, L. (2009) From Neanderthal to nanobiotech: from plant potions to pharming with plant factories, *Methods Mol Biol* 483, 1–23.
47. Smith, M. L., Fitzmaurice, W. P., Turpen, T. H., and Palmer, K. E. (2009) Display of peptides on the surface of tobacco mosaic virus particles, *Curr Top Microbiol Immunol* 332, 13–31.
48. Wagner, B., Fuchs, H., Adhami, F., Ma, Y., Scheiner, O., and Breiteneder, H. (2004) Plant virus expression systems for transient production of recombinant allergens in *Nicotiana benthamiana*, *Methods* 32, 227–234.
49. Avesani, L., Marconi, G., Morandini, F., Albertini, E., Bruschetta, M., Bortesi, L., Pezzotti, M., and Porceddu, A. (2007) Stability of Potato virus X expression vectors is related to insert size: implications for replication models and risk assessment, *Transgenic Res* 16, 587–597.
50. Marillonnet, S., Giritch, A., Gils, M., Kandzia, R., Klimyuk, V., and Gleba, Y. (2004) In planta engineering of viral RNA replicons: efficient assembly by recombination of DNA modules delivered by Agrobacterium, *Proc Natl Acad Sci USA* 101, 6852–6857.
51. Marillonnet, S., Thoeringer, C., Kandzia, R., Klimyuk, V., and Gleba, Y. (2005) Systemic *Agrobacterium tumefaciens*-mediated transfection of viral replicons for efficient transient expression in plants, *Nat Biotechnol* 23, 718–723.
52. Giritch, A., Marillonnet, S., Engler, C., van Eldik, G., Botterman, J., Klimyuk, V., and Gleba, Y. (2006) Rapid high-yield expression of full-size IgG antibodies in plants coinfected with noncompeting viral vectors, *Proc Natl Acad Sci USA* 103, 14701–14706.
53. Hiatt, A. and Pauly, M. (2006) Monoclonal antibodies from plants: a new speed record, *Proc Natl Acad Sci USA* 103, 14645–14646.
54. Matoba, N., Magerus, A., Geyer, B. C., Zhang, Y., Muralidharan, M., Alfsen, A., Arntzen, C. J., Bomsel, M., and Mor, T. S. (2004) A mucosally targeted subunit vaccine candidate eliciting HIV-1 transcytosis-blocking Abs, *Proc Natl Acad Sci USA* 101, 13584–13589.
55. Vezina, L. P., Faye, L., Lerouge, P., D'Aoust, M. A., Marquet-Blouin, E., Burel, C., Lavoie, P. O., Bardor, M., and Gomord, V. (2009) Transient co-expression for fast and high-yield production of antibodies with human-like N-glycans in plants, *Plant Biotechnol J* 7, 442–455.
56. Villani, M. E., Morgun, B., Brunetti, P., Marusic, C., Lombardi, R., Pisoni, I., Bacci, C., Desiderio, A., Benvenuto, E., and Donini, M. (2009) Plant pharming of a full-sized, tumour-targeting antibody using different expression strategies, *Plant Biotechnol J* 7, 59–72.

57. Voinnet, O., Rivas, S., Mestre, P., and Baulcombe, D. (2003) An enhanced transient expression system in plants based on suppression of gene silencing by the p19 protein of tomato bushy stunt virus, *Plant J* 33, 949–956.

58. Ma, P., Liu, J., He, H., Yang, M., Li, M., Zhu, X., and Wang, X. (2009) A viral suppressor P1/HC-pro increases the GFP gene expression in agrobacterium-mediated transient assay, *Appl Biochem Biotechnol* 158, 243–252.

59. Wydro, M., Kozubek, E., and Lehmann, P. (2006) Optimization of transient Agrobacterium-mediated gene expression system in leaves of *Nicotiana benthamiana*, *Acta Biochim Pol* 53, 289–298.

60. Huang, Z., Chen, Q., Hjelm, B., Arntzen, C., and Mason, H. (2009) A DNA replicon system for rapid high-level production of virus-like particles in plants, *Biotechnol Bioeng* 103, 706–714.

61. Wroblewski, T., Tomczak, A., and Michelmore, R. (2005) Optimization of Agrobacterium-mediated transient assays of gene expression in lettuce, tomato and Arabidopsis, *Plant Biotechnol J* 3, 259–273.

62. Plesha, M. A., Huang, T. K., Dandekar, A. M., Falk, B. W., and McDonald, K. A. (2009) Optimization of the bioprocessing conditions for scale-up of transient production of a heterologous protein in plants using a chemically inducible viral amplicon expression system, *Biotechnol Prog* 25, 722–734.

63. Medrano, G., Reidy, M. J., Liu, J., Ayala, J., Dolan, M. C., and Cramer, C. L. (2009) Rapid system for evaluating bioproduction capacity of complex pharmaceutical proteins in plants, *Methods Mol Biol* 483, 51–67.

64. D'Aoust, M. A., Lavoie, P. O., Belles-Isles, J., Bechtold, N., Martel, M., and Vezina, L. P. (2009) Transient expression of antibodies in plants using syringe agroinfiltration, *Methods Mol Biol* 483, 41–50.

65. Sainsbury, F., Liu, L., and Lomonossoff, G. P. (2009) Cowpea mosaic virus-based systems for the expression of antigens and antibodies in plants, *Methods Mol Biol* 483, 25–39.

66. Sorensen, H. P. and Mortensen, K. K. (2005) Advanced genetic strategies for recombinant protein expression in *Escherichia coli*, *J Biotechnol* 115, 113–128.

67. Kost, T. A., Condreay, J. P., and Jarvis, D. L. (2005) Baculovirus as versatile vectors for protein expression in insect and mammalian cells, *Nat Biotechnol* 23, 567–575.

68. Chiba, Y. and Akeboshi, H. (2009) Glycan engineering and production of 'humanized' glycoprotein in yeast cells, *Biol Pharm Bull* 32, 786–795.

69. Mohan, C., Kim, Y. G., Koo, J., and Lee, G. M. (2008) Assessment of cell engineering strategies for improved therapeutic protein production in CHO cells, *Biotechnol J* 3, 624–630.

70. Fitzmaurice, W. P. (2002) Interspecific Nicotiana hybrids and their progeny, Large Scale Biology Corporation, United States.

71. Sheludko, Y. V., Sindarovska, Y. R., Gerasymenko, I. M., Bannikova, M. A., and Kuchuk, N. V. (2007) Comparison of several Nicotiana species as hosts for high-scale Agrobacterium-mediated transient expression, *Biotechnol Bioeng* 96, 608–614.

72. Silhavy, D., Molnar, A., Lucioli, A., Szittya, G., Hornyik, C., Tavazza, M., and Burgyan, J. (2002) A viral protein suppresses RNA silencing and binds silencing-generated, 21- to 25-nucleotide double-stranded RNAs, *EMBO J* 21, 3070–3080.

73. Dunoyer, P., Lecellier, C. H., Parizotto, E. A., Himber, C., and Voinnet, O. (2004) Probing the microRNA and small interfering RNA pathways with virus-encoded suppressors of RNA silencing, *Plant Cell* 16, 1235–1250.

74. Alvarez, M. L., Pinyerd, H. L., Topal, E., and Cardineau, G. A. (2008) P19-dependent and P19-independent reversion of F1-V gene silencing in tomato, *Plant Mol Biol* 68, 61–79.

75. Raju, T. S. (2008) Terminal sugars of Fc glycans influence antibody effector functions of IgGs, *Curr Opin Immunol* 20, 471–478.

76. Saint-Jore-Dupas, C., Faye, L., and Gomord, V. (2007) From planta to pharma with glycosylation in the toolbox, *Trends Biotechnol* 25, 317–323.

77. Jin, C., Altmann, F., Strasser, R., Mach, L., Schahs, M., Kunert, R., Rademacher, T., Glossl, J., and Steinkellner, H. (2008) A plant-derived human monoclonal antibody induces an anti-carbohydrate immune response in rabbits, *Glycobiology* 18, 235–241.

78. Bencurova, M., Hemmer, W., Focke-Tejkl, M., Wilson, I. B., and Altmann, F. (2004) Specificity of IgG and IgE antibodies against plant and insect glycoprotein glycans determined with artificial glycoforms of human transferrin, *Glycobiology* 14, 457–466.

79. Strasser, R., Stadlmann, J., Schahs, M., Stiegler, G., Quendler, H., Mach, L., Glossl, J., Weterings, K., Pabst, M., and Steinkellner, H. (2008) Generation of glyco-engineered *Nicotiana benthamiana* for the production of monoclonal antibodies with a homogeneous human-like N-glycan structure, *Plant Biotechnol J* 6, 392–402.

80. Strasser, R., Castilho, A., Stadlmann, J., Kunert, R., Quendler, H., Gattinger, P., Jez, J., Rademacher, T., Altmann, F., Mach, L., and Steinkellner, H. (2009) Improved virus neutralization by plant-produced anti-HIV antibodies with a homogeneous {beta}1,4-galactosylated N-glycan profile, *J Biol Chem* 284, 20479–20485.
81. Benchabane, M., Goulet, C., Rivard, D., Faye, L., Gomord, V., and Michaud, D. (2008) Preventing unintended proteolysis in plant protein biofactories, *Plant Biotechnol J* 6, 633–648.
82. Idiris, A., Bi, K., Tohda, H., Kumagai, H., and Giga-Hama, Y. (2006) Construction of a protease-deficient strain set for the fission yeast *Schizosaccharomyces pombe*, useful for effective production of protease-sensitive heterologous proteins, *Yeast* 23, 83–99.
83. Kuroda, K., Kitagawa, Y., Kobayashi, K., Tsumura, H., Komeda, T., Mori, E., Motoki, K., Kataoka, S., Chiba, Y., and Jigami, Y. (2007) Antibody expression in protease-deficient strains of the methylotrophic yeast *Ogataea minuta*, *FEMS Yeast Res* 7, 1307–1316.
84. Rivard, D., Anguenot, R., Brunelle, F., Le, V. Q., Vezina, L. P., Trepanier, S., and Michaud, D. (2006) An in-built proteinase inhibitor system for the protection of recombinant proteins recovered from transgenic plants, *Plant Biotechnol J* 4, 359–368.
85. Benchabane, M., Rivard, D., Girard, C., and Michaud, D. (2009) Companion protease inhibitors to protect recombinant proteins in transgenic plant extracts, *Methods Mol Biol* 483, 265–273.
86. Nuttall, J., Vine, N., Hadlington, J. L., Drake, P., Frigerio, L., and Ma, J. K. (2002) ER-resident chaperone interactions with recombinant antibodies in transgenic plants, *Eur J Biochem* 269, 6042–6051.
87. Sharma, S. K., Christen, P., and Goloubinoff, P. (2009) Disaggregating chaperones: an unfolding story, *Curr Protein Pept Sci* 10(5), 432–446.
88. Liu, D. (2009) Design of gene constructs for transgenic maize, *Methods Mol Biol* 526, 3–20.
89. Gustafsson, C., Govindarajan, S., and Minshull, J. (2004) Codon bias and heterologous protein expression, *Trends Biotechnol* 22, 346–353.
90. Sharp, P. M. and Li, W. H. (1987) The codon adaptation index – a measure of directional synonymous codon usage bias, and its potential applications, *Nucleic Acids Res* 15, 1281–1295.
91. Geyer, B. C., Fletcher, S. P., Griffin, T. A., Lopker, M. J., Soreq, H., and Mor, T. S. (2007) Translational control of recombinant human acetylcholinesterase accumulation in plants, *BMC Biotechnol* 7, 27.
92. Komar, A. A., Lesnik, T., and Reiss, C. (1999) Synonymous codon substitutions affect ribosome traffic and protein folding during in vitro translation, *FEBS Lett* 462, 387–391.
93. Thanaraj, T. A. and Argos, P. (1996) Ribosome-mediated translational pause and protein domain organization, *Protein Sci* 5, 1594–1612.
94. Joshi, C. P., Zhou, H., Huang, X., and Chiang, V. L. (1997) Context sequences of translation initiation codon in plants, *Plant Mol Biol* 35, 993–1001.
95. Kozak, M. (1984) Compilation and analysis of sequences upstream from the translational start site in eukaryotic mRNAs, *Nucleic Acids Res* 12, 857–872.
96. Nakagawa, S., Niimura, Y., Gojobori, T., Tanaka, H., and Miura, K. (2008) Diversity of preferred nucleotide sequences around the translation initiation codon in eukaryote genomes, *Nucleic Acids Res* 36, 861–871.
97. Kochetov, A. V., Palyanov, A., Titov, II, Grigorovich, D., Sarai, A., and Kolchanov, N. A. (2007) AUG_hairpin: prediction of a downstream secondary structure influencing the recognition of a translation start site, *BMC Bioinformatics* 8, 318.
98. Sawant, S. V., Kiran, K., Singh, P. K., and Tuli, R. (2001) Sequence architecture downstream of the initiator codon enhances gene expression and protein stability in plants, *Plant Physiol* 126, 1630–1636.
99. Burgyan, J. (2008) Role of silencing suppressor proteins, *Methods Mol Biol* 451, 69–79.
100. Streatfield, S. J. (2007) Approaches to achieve high-level heterologous protein production in plants, *Plant Biotechnol J* 5, 2–15.
101. Kermode, A. R. (2006) Plants as factories for production of biopharmaceutical and bioindustrial proteins: lessons from cell biology, *Can J Bot* 84, 679–694.
102. Conley, A. J., Joensuu, J. J., Menassa, R., and Brandle, J. E. (2009) Induction of protein body formation in plant leaves by elastin-like polypeptide fusions, *BMC Biol* 7, 48.
103. de Virgilio, M., De Marchis, F., Bellucci, M., Mainieri, D., Rossi, M., Benvenuto, E., Arcioni, S., and Vitale, A. (2008) The human immunodeficiency virus antigen Nef forms protein bodies in leaves of transgenic tobacco when fused to zeolin, *J Exp Bot* 59, 2815–2829.

104. Torrent, M., Llompart, B., Lasserre-Ramassamy, S., Llop-Tous, I., Bastida, M., Marzabal, P., Westerholm-Parvinen, A., Saloheimo, M., Heifetz, P. B., and Ludevid, M. D. (2009) Eukaryotic protein production in designed storage organelles, *BMC Biol* 7, 5.

105. Drake, P. M., Barbi, T., Sexton, A., McGowan, E., Stadlmann, J., Navarre, C., Paul, M. J., and Ma, J. K. (2009) Development of rhizosecretion as a production system for recombinant proteins from hydroponic cultivated tobacco, *FASEB J* 23(10), 3581–3589.

106. McCormick, A. A., Reinl, S. J., Cameron, T. I., Vojdani, F., Fronefield, M., Levy, R., and Tuse, D. (2003) Individualized human scFv vaccines produced in plants: humoral anti-idiotype responses in vaccinated mice confirm relevance to the tumor Ig, *J Immunol Methods* 278, 95–104.

107. McCormick, A. A., Kumagai, M. H., Hanley, K., Turpen, T. H., Hakim, I., Grill, L. K., Tuse, D., Levy, S., and Levy, R. (1999) Rapid production of specific vaccines for lymphoma by expression of the tumor-derived single-chain Fv epitopes in tobacco plants, *Proc Natl Acad Sci USA* 96, 703–708.

108. Du, H., Cameron, T. L., Garger, S. J., Pogue, G. P., Hamm, L. A., White, E., Hanley, K. M., and Grabowski, G. A. (2008) Wolman disease/cholesteryl ester storage disease: efficacy of plant-produced human lysosomal acid lipase in mice, *J Lipid Res* 49, 1646–1657.

109. Gegenheimer, P. (1990) Preparation of extracts from plants, *Methods Enzymol* 182, 174–193.

110. Platis, D. and Labrou, N. E. (2009) Application of a PEG/salt aqueous two-phase partition system for the recovery of monoclonal antibodies from unclarified transgenic tobacco extract, *Biotechnol J* 4(9), 1320–1327.

111. Platis, D. and Labrou, N. E. (2006) Development of an aqueous two-phase partitioning system for fractionating therapeutic proteins from tobacco extract, *J Chromatogr* 1128, 114–124.

112. Platis, D., Drossard, J., Fischer, R., Ma, J. K., and Labrou, N. E. (2008) New downstream processing strategy for the purification of monoclonal antibodies from transgenic tobacco plants, *J Chromatogr* 1211, 80–89.

113. Bordier, C. (1981) Phase separation of integral membrane proteins in Triton X-114 solution, *J Biol Chem* 256, 1604–1607.

114. Werner, S., Marillonnet, S., Hause, G., Klimyuk, V., and Gleba, Y. (2006) Immunoabsorbent nanoparticles based on a tobamovirus displaying protein A, *Proc Natl Acad Sci USA* 103, 17678–17683.

Chapter 12

Chromosome Analysis and Sorting Using Flow Cytometry

Jaroslav Doležel, Marie Kubaláková, Jarmila Číhalíková, Pavla Suchánková, and Hana Šimková

Abstract

Chromosome analysis and sorting using flow cytometry (flow cytogenetics) is an attractive tool for fractionating plant genomes to small parts. The reduction of complexity greatly simplifies genetics and genomics in plant species with large genomes. However, as flow cytometry requires liquid suspensions of particles, the lack of suitable protocols for preparation of solutions of intact chromosomes delayed the application of flow cytogenetics in plants. This chapter outlines a high-yielding procedure for preparation of solutions of intact mitotic chromosomes from root tips of young seedlings and for their analysis using flow cytometry and sorting. Root tips accumulated at metaphase are mildly fixed with formaldehyde, and solutions of intact chromosomes are prepared by mechanical homogenization. The advantages of the present approach include the use of seedlings, which are easy to handle, and the karyological stability of root meristems, which can be induced to high degree of metaphase synchrony. Chromosomes isolated according to this protocol have well-preserved morphology, withstand shearing forces during sorting, and their DNA is intact and suitable for a range of applications.

Key words: Accumulation of metaphases, Cell cycle synchrony, Cytogenetics stocks, Flow karyotype, Hydroponics, Fluorescence in situ hybridization, Mitotic synchrony, Plant chromosome isolation, Plant flow cytogenetics, Root-tip meristem

1. Introduction

The analysis of chromosomes using flow cytometry differs from microscopic techniques as the chromosomes are not attached to a firm surface, but are analyzed during their passage in a narrow stream of liquid. The chromosomes move in a single file, interact individually with a beam of excitation light, and can be classified according to light scatter and emitted fluorescence. As the liquid stream moves at high speed, large populations of chromosomes can be analyzed in a short time. The results are usually displayed

James A. Birchler (ed.), *Plant Chromosome Engineering: Methods and Protocols*, Methods in Molecular Biology, vol. 701, DOI 10.1007/978-1-61737-957-4_12, © Springer Science+Business Media, LLC 2011

as histograms of relative fluorescence intensity, called flow karyotypes. If the chromosomes in a species differ enough in size, each of them is represented by a single peak on a flow karyotype. Any change in relative frequency of a chromosome in the population will be reflected by a change in relative peak area while a change in chromosome size will result in altered peak position. Individual chromosomes may be purified by breaking the liquid stream into droplets and deflecting electrically charged droplets containing chromosomes of interest by a passage through an electrostatic field.

Flow cytometric chromosome analysis and sorting (flow cytogenetics) was originally developed for Chinese hamster and human, and subsequently for a variety of animal species (1–5). As the preparation of suspensions of meiotic chromosomes is not practical, only mitotic chromosomes have been used. While the early hopes for automatic detection of structural and numerical chromosome changes (6) were not met, the ability to dissect complex genomes of human and other animals to small parts by chromosome sorting greatly facilitated their genetics and genomics. Important uses included physical mapping (7), construction of chromosome-specific DNA libraries (8, 9), and development chromosome painting probes (10). More recently, next-generation sequencing methods were used with flow-sorted chromosomes to detect SNPs and map translocation breakpoints (11, 12).

In plants, the application of flow cytogenetics was delayed mainly due to difficulties in preparation of solutions of intact chromosomes (13, 14). In 1992, Doležel et al. (15) developed a procedure, which involves mechanical release of chromosomes from synchronized root tips of seedlings. To date, it has been developed for eight plant species, which include major legumes and cereals. The use of root tips for chromosome isolation offers several advantages over other systems: seedlings are easy to handle, root meristems are karyologically stable, and can be synchronized to obtain high proportion of metaphase cells (16–18). Other difficulty with the development of plant flow cytogenetics concerned discrimination of individual chromosomes. Most plant species have similarly sized chromosomes that cannot be resolved by analyzing relative DNA content. In human and some animals, the resolution of individual chromosomes is facilitated by bivariate analysis after staining with AT- and GC-binding fluorochromes. However, this approach does not improve the resolution of plant flow karyotypes (19). Thus, various cytogenetics stocks with altered chromosome size have been used (18, 20, 21). Flow-sorted plant chromosomes facilitate the analysis of complex plant genomes and their uses included physical mapping using PCR (22, 23) and DNA arrays (24), targeted development of molecular markers (25, 26), construction of chromosome-specific BAC libraries (27, 28), and next-generation sequencing (29).

2. Materials

2.1. Cell Cycle Synchronization and Metaphase Accumulation

1. Viable and healthy seeds. The quantity of seed needed depends on the number of chromosomes to be analyzed and sorted.
2. Hoagland's nutrient solution. Prepare solutions A, B, and C and stock solution as follows:
   (a) Solution A: 280 mg H_3BO_3 (45 mM), 340 mg $MnSO_4 \cdot H_2O$ (20 mM), 10 mg $CuSO_4 \cdot 5\,H_2O$ (0.4 mM), 22 mg $ZnSO_4 \cdot 7H_2O$ (0.8 mM), and 10 mg $(NH_4)_6Mo_7O_{24} \cdot 4H_2O$ (0.08 mM) in 100 ml deionized H_2O (final volume). Store at 4°C.
   (b) Solution B: 0.5 ml concentrated H_2SO_4 (0.05 mM) in 100 ml deionized H_2O (final volume). Store at 4°C.
   (c) Solution C: dissolve in 400 ml deionized H_2O 3.36 g Na_2EDTA (18 mM) and 2.79 g $FeSO_4$ (20 mM). Heat the solution to 70°C while stirring until the color turns yellow-brown. Cool down, adjust the volume to 500 ml, and store at 4°C.
   (d) Hoagland's stock solution (10×): 4.7 g $Ca(NO_3)_2 \cdot 4H_2O$ (40 mM), 2.6 g $MgSO_4 \cdot 7H_2O$ (20 mM), 3.3 g KNO_3 (65 mM), 0.6 g $NH_4H_2PO_4$ (10 mM), 5 ml solution A, and 0.5 ml solution B in 500 ml deionized H_2O (final volume). Store at 4°C.
   (e) Hoagland's nutrient solution (1×): 100 ml Hoagland's stock solution (10×) and 5 ml solution C in 1,000 ml deionized H_2O (final volume). Prepare just before use.
3. Hydroxyurea treatment solution: dissolve in 800 ml Hoagland's nutrient solution hydroxyurea (H8627, Sigma–Aldrich, Prague, Czech Republic) (Table 1). Prepare the solution just before use (see Note 1).
4. Amiprophos methyl treatment solution. Prepare the stock solution and treatment solutions as follows:
   (a) Amiprophos methyl stock solution (20 mM): dissolve in 10 ml ice-cold acetone 60.86 mg amiprophos methyl (A0185.1000, Duchefa Biochemie B.V., Haarlem, The Netherlands) and store at −20°C in 1 ml aliquots.
   (b) Amiprophos methyl treatment solution: add in 800 ml deionized H_2O amiprophos methyl stock solution (Table 2) with the pipette tip immersed in the solution and continuous stirring. Prepare the solution just before use (see Note 1).
5. Oryzaline solution. Prepare the stock solution and treatment solutions as follows:
   (a) Oryzaline stock solution (10 mM): dissolve in 25 ml acetone 86.59 mg oryzaline (O1318.0001, Duchefa

Table 1
Concentrations of hydroxyurea used to synchronize cell cycle

Species	Hydroxyurea	
	Amount dissolved (mg)	Final concentration (mM)
Cicer arietinum	76.0	1.25
Hordeum vulgare	121.6	2.0
Pisum sativum	76.0	1.25
Secale cereale	152.0	2.5
Triticum aestivum	121.6	2.0
Triticum durum	76	1.25
Vicia faba	76.0	1.25
Vicia sativa	152.0	2.5

Table 2
Concentration of amiprophos methyl used to accumulate dividing cells in metaphase

Species	Amiprophos methyl	
	Volume added (μl)	Final concentration (μM)
Vicia faba	101.3	2.5
Pisum sativum	405.2	10.0
Hordeum vulgare	101.3	2.5
Zea mays	101.3	2.5
Triticum aestivum	101.3	2.5
Triticum durum	101.3	2.5

Biochemie B.V., Haarlem, The Netherlands) and store at –20°C in 1 ml aliquots.

(b) Oryzaline treatment solution: add in 800 ml deionized H_2O oryzaline stock solution (Table 3) with the pipette tip immersed in the solution and continuous stirring. Prepare the solution just before use (see Note 1).

6. Ice water: add 1.6 kg ice cubes to 1,700 ml deionized water in a 3,000 ml glass vase and place in a refrigerator. Prepare just before use.

Table 3
Concentration of oryzaline used to accumulate dividing cells in metaphase

	Oryzaline	
Species	**Volume added (μl)**	**Final concentration (μM)**
Cicer arietinum	400	5.0
Vicia sativa	400	5.0

Table 4
Preparation of formaldehyde fixative

	Formaldehyde (see Note 2)	
Species	**Volume (ml)**	**Final concentration (%, v/v)**
Cicer arietinum	13.5	2.0
Hordeum vulgare	13.5	2.0
Pisum sativum	20.0	3.0
Secale cereale	13.5	2.0
Triticum aestivum	13.5	2.0
Triticum durum	13.5	2.0
Vicia faba	27.0	4.0
Vicia sativa	13.5	2.0

2.2. Preparation of Solutions of Mitotic Chromosomes

1. Tris buffer: dissolve in 500 ml deionized H_2O 0.606 g Tris (10 mM), 1.861 g Na_2EDTA (10 mM), 2.922 g NaCl (100 mM), and adjust pH to 7.5 using 1 N NaOH.
2. Formaldehyde fixative: dissolve in 200 ml deionized H_2O 0.242 g Tris (10 mM), 0.931 g Na_2EDTA (10 mM), 1.461 g NaCl (100 mM), 250 μl Triton X-100 (0.1%, v/v), and adjust pH to 7.5 using 1 N NaOH. Add formaldehyde (Table 4) and adjust volume to 250 ml with deionized H_2O. Prepare just before use.
3. LB01 lysis buffer: dissolve in 200 ml deionized H_2O 0.363 g Tris (15 mM), 0.149 g Na_2EDTA (2 mM), 0.0348 g spermine · 4HCl (0.5 mM), 1.193 g KCl (80 mM), 0.234 g NaCl (20 mM), 200 μl Triton X-100 (0.1%, v/v), and adjust pH to 9.0 using 1 M NaOH. Filter through a 0.22 μm filter to remove small particles. Add 220 μl β-mercaptoethanol and mix well. Store at –20°C in 10 ml aliquots.

4. LB01 HKS lysis buffer. Prepare the stock solution and treatment solutions as follows:
   (a) LB01 HKS lysis buffer stock solution (10×): dissolve in 50 ml deionized H_2O 0.606 g Tris (100 mM), 1.86 g Na_2EDTA (100 mM), 0.174 g spermine · 4HCl (10 mM) 0.127 g spermidine · 3HCl (10 mM), 4.84 g KCl (1.3 M), 0.585 g NaCl (200 mM), 500 μl Triton X-100 (1% v/v), and adjust pH to 9.4 using 1 M NaOH. Store at –20°C in 10 ml aliquots.
   (b) LB01 HKS lysis buffer working solution: Mix 1.5 ml LB01 HKS lysis buffer stock solution with 8.5 ml deionized H_2O and add 11 μl β-mercaptoethanol. Prepare just before use.

2.3. Flow Cytometric Analysis and Sorting

1. DAPI stock solution (0.1 mg/ml): dissolve 5 mg DAPI in 50 ml deionized H_2O by stirring for 60 min. Filter through a 0.22 μm filter to remove small particles. Store at –20°C in 0.5 ml aliquots.
2. Sheath fluid: dissolve 1.17 g NaCl in 2,000 ml deionized H_2O.

2.4. Identification of Sorted Chromosomes and Estimation of Purity in Sorted Chromosome Fractions by FISH

1. P5 buffer: dissolve 30.28 mg Tris (10 mM), 93.2 mg KCl (50 mM), 10.17 g $MgCl_2{\cdot}6H_2O$ (2 mM), and 1.25 g sucrose in 25 ml deionized H_2O, and adjust pH to 8 using 1 N HCl. Store at –20°C in 1 ml aliquots.
2. 20× SSC stock solution: dissolve 175.3 g NaCl (3 M) and 88.2 g $Na_3C_6H_5O_7 \cdot 2H_2O$ (300 mM) in 1,000 ml deionized H_2O (final volume), adjust pH to 7, and sterilize by autoclaving. Store at room temperature.
3. Hybridization mix: 10 μl 100% formamide (40% final), 1.25 μl 20× SSC (1×), 0.625 μl calf thymus (250 ng/μl), labeled probe(s) (1 ng/μl), add 50% dextrane sulfate to 25 μl. Prepare before use (see Note 3).
4. 4× SSC washing buffer: 200 ml 20× SSC and 2 ml Tween 20 (0.2% v/v) in 1,000 ml deionized H_2O (final volume).
5. 2× SSC washing buffer: mix 100 ml 20× SSC and 900 ml deionized H_2O. Prepare before use.
6. 0.1× SSC stringent washing buffer: 5 ml 20× SSC, 1 ml Tween 20 (0.1% v/v), and 406 mg $MgCl_2 \cdot 6H_2O$ (2 mM) in 1,000 ml deionized H_2O (final volume). Prepare before use.
7. Blocking buffer: dissolve 0.5 g (1%) blocking reagent (1,096,176, Roche, Mannheim, Germany) in 50 ml 4× SSC, mix 1 h at 70°C, and sterilize by autoclaving. Store at –20°C in 1 ml aliquots.
8. Vectashield antifade solution with DAPI (H-1200, Vector Laboratories, Burlingame, USA).

3. Methods

Chromosome analysis and sorting using flow cytometry requires the samples of chromosomes in solution. This protocol describes the preparation of solutions of intact mitotic chromosomes from root tips of young seedlings and for their analysis using flow cytometry and sorting. The procedure involves the accumulation of meristem root-tip cells at metaphase by a sequential treatment with a DNA synthesis inhibitor hydroxyurea and mitotic spindle poison amiprophos methyl or oryzaline. Root tips enriched for metaphase cells are fixed with formaldehyde and intact mitotic chromosomes are released into a lysis buffer by mechanical homogenization of root-tip meristems. Chromosomes in solution are stained by a DNA fluorochrome DAPI (4′,6-diamidino-2-phenylindole) and their fluorescence and light scatter properties are analyzed by flow cytometer. Chromosomes can be sorted onto a glass slide for microscopic observation and collected in appropriate vessels for other applications.

The quality of chromosome suspension is critical as it determines a probability of resolving individual chromosomes and their sorting without contamination by other particles. It is defined by the concentration of intact chromosomes and the presence of chromosome and cellular debris. It is important to minimize the occurrence of chromatids and doublets of chromosomes and chromosome fragments as they may have similar fluorescence and light scatter properties as chromosomes and contaminate sorted fractions. The proportion of intact chromosomes in solution depends on the proportion of metaphase cells in the root meristem and on the conditions under which chromosomes are released from root-tip cells. Thus, the procedure for cell cycle synchronization must be optimized to achieve high proportion of metaphase cells (above 50%) without the excessive occurrence of chromatids. The procedure for chromosome release must not damage chromosomes and avoid the presence of chromosome clumps. Chromosome solutions should contain at least 5×10^5 chromosomes/ml to obtain high-resolution flow karyotypes.

Apart from the quality of chromosome solutions, the ability to sort individual chromosomes depends on differences in size and hence DNA content among the chromosomes. The difference should be at least 10% to ensure reliable discrimination. However, in most plant species, some chromosomes do not differ enough in relative DNA content from other chromosome(s), leading to the appearance of composite peaks in flow karyotypes. This obstacle can be overcome by employing cytogenetic stocks such as chromosome translocation, deletion, and alien addition lines, which comprise chromosomes whose size differs enough to permit their discrimination and sorting.

Table 5
Seed germination time and optimal root length

Species	Germination (h)	Root length (mm)
Vicia faba	48	20
Pisum sativum	48	20

3.1. Seed Germination in Perlite (Vicia faba and Pisum sativum)

1. Imbibe the seeds for 24 h in deionized H_2O with aeration (approximately 30 seedlings are needed to prepare one sample) (see Note 4).
2. Wet perlite (800 ml for 110 seeds) with the Hoagland's nutrient solution and put it into a 4,000 ml plastic tray (e.g., 25 cm L, 25 cm W, 12 cm H) (see Note 5).
3. Wash the seeds in deionized H_2O, layer them onto the surface of perlite, and cover them with a layer of 1–2 cm of wet perlite.
4. Cover the tray with aluminum foil and leave the seeds to germinate at 25°C in the dark to achieve proper length of primary roots (Table 5).
5. Remove the seedlings from perlite and wash them in deionized H_2O.

3.2. Seed Germination on Filter Paper (Cicer arietinum, Hordeum vulgare, Secale cereale, Triticum aestivum, Triticum durum, and Vicia sativa)

1. Place several layers of paper towels into a glass petri dish (18 cm diameter), top them with a single sheet of filter paper.
2. Moisten the paper layers with deionized H_2O until runoff.
3. Spread the seeds on paper surface (approximately 30 seedlings are used to prepare one sample).
4. Cover the petri dish and leave the seeds to germinate at 25°C in the dark to achieve proper length of primary roots (Fig. 1a) (Table 6).

3.3. Cell Cycle Synchronization and Accumulation of Root-Tip Cells in Metaphases

1. Adjust the temperature of all solutions to 25 ± 0.5°C prior their use. Perform all incubations in the dark in a biological incubator at 25 ± 0.5°C. Aerate all solutions (see Note 4).
2. Select seedlings with proper length of their primary roots.
3. Thread seedling roots through the holes of the open-mesh basket positioned on a plastic tray filled with deionized H_2O adjusted to 25°C.
4. Transfer the basket with seedlings to a plastic tray containing the hydroxyurea treatment solution and incubate for periods according to Table 7.

Fig. 1. Preparation of solutions of intact mitotic chromosomes. (**a**) Seed germination in a petri dish. (**b**) Hydroponics system used to incubate roots of young seedlings in treatment solutions. (**c**) Fixation of detached roots in formaldehyde solution. (**d**) Detachment of meristem root tips prior to mechanical homogenization. (**e**) Release of chromosomes into a lysis buffer by mechanical homogenization.

5. Wash the roots in several changes of deionized H_2O.
6. Incubate in hydroxyurea-free Hoagland's nutrient solution for recovery periods specified in Table 7 (Fig. 1b).
7. Transfer the basket with seedlings to a tray filled with amiprophos methyl or oryzaline treatment solution and incubate for periods indicated in Table 7 (see Note 6).
8. If appropriate, transfer the basket with seedlings to a container filled with ice water (1–2°C). Place the container in a refrigerator and treat the roots overnight (Table 7) (see Note 7).

Table 6
Seed germination time and optimal root length

Species	Germination (h)	Root length (mm)
Cicer arietinum	30	20
Hordeum vulgare	60	30
Secale cereale	60	30
Triticum aestivum	60	30
Triticum durum	60	30
Vicia sativa	48	20

Table 7
Duration of hydroxyurea treatment, recovery time, treatment with antimicrotubular drugs and ice water

Species	Hydroxyurea treatment (h)	Recovery time (h)	Accumulation of metaphases (h)	Ice water (h)
Cicer arietinum	18	4.0	2.0	18.0
Hordeum vulgare	18	6.5	2.0	15.5
Pisum sativum	18	4.5	2.0	–
Secale cereale	18	6.5	2.0	15.5
Triticum aestivum	18	5.5	2.0	16.5
Triticum durum	18	5.0	2.0	17
Vicia faba	18	4.5	2.0	–
Vicia sativa	18.5	3.5	2.0	–

3.4. Preparation of Solutions of Intact Mitotic Chromosomes

1. Harvest root tips (1 cm) and transfer them into deionized H_2O.
2. Fix the roots immediately by transferring them into formaldehyde fixative and incubate them at 5°C for periods given in Table 8 (Fig. 1c) (see Note 8).
3. Wash the roots in Tris buffer three times for 5 min at 5°C (see Note 9).
4. Excise root meristems (1–2 mm, depending on a species) and transfer them in 5 ml polystyrene tube containing 1 ml LB01 or LB01 HKS lysis buffer (Fig. 1d) (see Note 10).
5. Isolate chromosomes by homogenizing using a mechanical homogenizer Polytron PT1200 fitted with a PT-DA 1205/5 probe as specified in Table 8 (Fig. 1e) (see Notes 11 and 12).

Table 8
The duration of fixation in formaldehyde and the extent of mechanical homogenization

Species	Formaldehyde fixation (min)	Homogenization	
		Speed (rpm)	Duration (s)
Cicer arietinum	20	13,000	18
Hordeum vulgare	20	15,000	13
Pisum sativum	30	13,000	18
Secale cereale	30	15,000	13
Triticum durum	20	20,000	13
Triticum aestivum	20	20,000	13
Vicia faba	30	15,000	18
Vicia sativa	25	13,000	18

6. Filter the crude solution through a 50 μm nylon mesh (4×4 cm^2) into a polystyrene tube.
7. Store the chromosome solution on ice for up to several hours (see Note 13).

3.5. Chromosome Analysis Using Flow Cytometry

1. This protocol describes the analysis using FACSVantage SE flow cytometer (BD Biosciences Immunocytometry Systems, San José, USA) equipped with argon laser operated at multi-UV mode (351.1–363.8 nm) with 200 mW output power.
2. Switch on the laser and the instrument.
3. Empty the waste container and fill the sheath container with sheath fluid.
4. Adjust sheath fluid pressure to 20 psi (pounds per square inch) and leave the fluid running to fill all plastic lines and filters of the instrument.
5. Install a nozzle (70 μm orifice), check for air bubbles, and position the liquid stream so that it enters the center of waste aspirator.
6. Install a band-pass filter 424/44 in front of the fluorescence 1 (FL1) detector.
7. Set up the trigger signal to fluorescence 1 pulse height (FL1-H), and select a proper threshold level (typically 20).
8. Run calibration beads (AlignFlow UV beads, Invitrogen, Cat. No. A7304) at a flow rate of 600–800 particles/s.

9. Display the data on a dot plot of FSC-H and FL1-H, and on histograms of FSC-H and FL1-H.
10. Align the instrument to achieve maximum signal intensity and minimum coefficient of variation of FSC-H and FL1-H signals (typically 2.5% on FSC-H and 2.2% on FL1-H) (see Note 14).
11. Run a dummy sample (1 ml LB01 buffer containing 2 μg/ml DAPI) to equilibrate the sample line for at least 15 min.
12. Stain the chromosome suspension by adding DAPI stock solution to final concentration of 2 μg/ml (see Note 15).
13. Filter the sample through a 20 μm nylon mesh (4×4 cm^2) and store the sample on ice until use (see Note 13).
14. Introduce the sample and let it stabilize at appropriate flow rate (e.g., 1,000 particles/s).
15. Set a gating region on a dot plot of FSC-H and FL1-H to exclude debris, nuclei, and large clumps.
16. Adjust photomultiplier voltage and amplification gains so that chromosome peaks are evenly distributed on a histogram of FL1 pulse area (FL1-A) (see Note 16).
17. Analyze 20–50 thousand chromosomes and save the result on a disk (Fig. 2).

3.6. Flow Cytometric Chromosome Sorting

1. Switch on the sorting module, adjust the drop drive frequency (typically 36,000 Hz), and drop drive amplitude to break the stream at suitable distance from the laser intercept point (check for satellite drops).
2. Switch on test mode and test sort and adjust the drop drive phase to obtain single-side streams.
3. Use the BD AccuDrop module to optimize drop delay. Verify the drop delay setting by sorting 20 UV beads onto a microscopic slide (use the counter mode for sorting).
4. Check the number of sorted beads with a fluorescence microscope. Typically, 19–20 beads should be sorted. Repeat drop delay optimization if lower number of beads is sorted.
5. Select the sort mode and sort envelope according to the required purity, number of chromosomes to be sorted, and desired volume for the sorted fraction (see Note 17).
6. Run the sample and display the signals on a dot plot of FL1 pulse width (FL1-W) versus FL1-A.
7. Adjust the FL1-W amplifier gain and width offset as needed to achieve optimal resolution of the width signal.
8. Define sorting regions on the FL1-W versus FL1-A dot plot.
9. Check for stability of the break-off point and of the side streams.
10. Sort the required number of chromosomes onto a microscopic slide or into a collection vessel as needed (see Note 18).

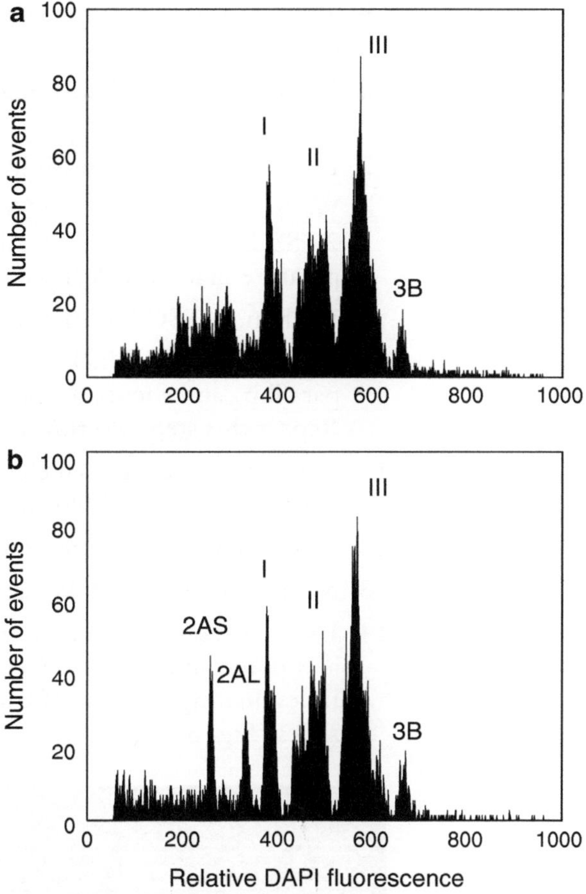

Fig. 2. Histograms of relative fluorescence intensities (flow karyotypes) obtained during the analysis of DAPI-stained mitotic chromosomes of bread wheat (*Triticum aestivum* L.). (**a**) Flow karyotype of cv. "Chinese Spring" ($2n=6x=42$) with a wild-type karyotype consists of three composite peaks (I–III) representing groups of chromosomes and a peak representing chromosome 3B. (**b**) Flow karyotype of a ditelosomic line of cv. "Chinese Spring" ($2n=40+2t2AS+2t2AL$) carrying a telocentric chromosomes 2AS and 2AL.

3.7. Identification of Sorted Chromosomes Using FISH

1. Pipette 10 μl P5 buffer onto a clean microscopic slide and immediately sort 1,000 chromosomes into the drop.
2. Air-dry the slide and store at room temperature (see Note 19).
3. Add 25 μl hybridization mix to the area containing flow-sorted chromosomes, cover with a glass coverslip, and seal up with rubber cement.
4. Denature for 45 s at 80°C by placing the slide on a temperature-controlled hot plate (see Note 20).

5. Transfer the slide to a wet chamber and incubate at 37°C overnight.
6. Remove carefully the coverslip and wash in 2× SSC for 10 min at 42°C.
7. Perform a stringent wash in 0.1× SSC for 5 min at 42°C.
8. Wash again 10 min in 2× SSC at 42°C and then for 10 min in 2× SSC (place the vial on laboratory table and leave the temperature gradually decrease to room temperature).
9. Wash in 4× SSC for 10 min at room temperature.
10. Apply 60 μl 1% blocking buffer onto the slide, cover with parafilm, and incubate for 10 min in at room temperature. Repeat this step two times.
11. Apply fluorescently labeled antibody and/or fluorescently labeled avidin in 1% blocking buffer for 1 h at 37°C.
12. Wash the slide three times in 4× SSC for 5 min at 40°C.
13. Mount the slide with Vectashield containing DAPI.
14. Use fluoresce microscope to identify sorted chromosomes and determine the presence of contaminating chromosomes by evaluating at least 100 chromosomes on three different slides (Fig. 3).

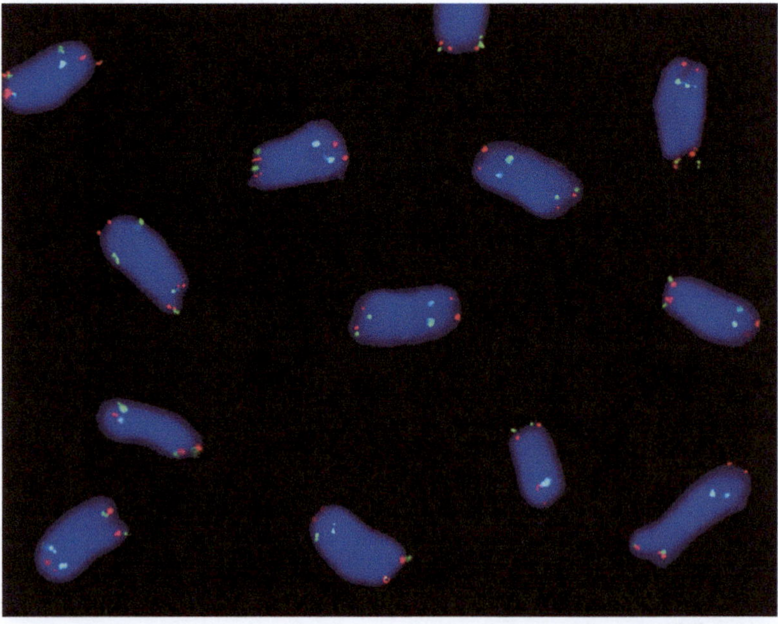

Fig. 3. Flow-sorted telocentric chromosomes 2AL of bread wheat (*Triticum aestivum* L.) after FISH with probes for GAA microsatellite (*yellow-green* signals) and telomeric repeat (*red* signals). While the distribution of GAA clusters serves to identify 2AL chromosome arm, the presence of telomeric DNA on both chromosome ends confirms that telocentric chromosomes and not chromosome arms separated from complete chromosomes were sorted.

4. Notes

1. This volume is needed for one plastic tray (e.g., 14 cm L, 8 cm W, 10 cm H) used for cell cycle synchronization treatments. The tray should include an open-mesh basket to hold germinated seeds and permit aeration (see Fig. 1b).
2. Use formaldehyde 37%, Catalog No. 1.04003, Merck, Darmstadt, Germany.
3. The choice of labeled probes for FISH depends on the chromosome, which needs to be identified. Telomeric probes are useful to distinguish telocentric chromosomes from broken chromosomes (telomeres must be present on both chromosome ends). The DNA probes can be labeled by biotin or digoxigenin.
4. Use aquarium bubbler with tubing and aeration stones. As the cell cycle kinetics depends on temperature, it is important that the temperature during all incubations is stable. Use a biological incubator (heating/cooling) with internal temperature adjusted to $25 \pm 0.5°C$. All solutions should be heated to 25°C prior to use.
5. Perlite is an inert substrate suitable for seed germination.
6. Although higher frequencies of metaphases may be obtained after longer treatments with mitotic spindle inhibitors, treatment for only 2 h is used to minimize the occurrence of chromatids and avoid extensive chromosome decondensation.
7. In some species, ice water treatment improves chromosome spreading and reduces the occurrence of chromosome clumps in chromosome solutions.
8. Place the vials with the fixative in cooled water bath to achieve temperature of the fixative $5 \pm 0.5°C$. The extent of formaldehyde fixation determines critically chromosome yield and morphology of isolated chromosomes. Weak fixation results in mechanically damaged chromosomes, and the solutions contain large amounts of chromosome debris. If the fixation is too strong, the samples obtained after tissue homogenization contain large numbers of chromosome clumps and intact cells.
9. Formaldehyde-fixed root tips cannot be stored for prolonged periods prior to chromosome isolation. They must be processed within a few hours of storage on ice.
10. The choice of the buffer depends on the use of flow-sorted chromosomes. LB01 HKS lysis buffer is used when intact high molecular weight DNA is needed, e.g., for BAC library construction.
11. The extent of homogenization determines critically chromosome yield and morphology of isolated chromosomes.
12. Chromosomes can be isolated also by chopping root tips with a sharp scalpel in a 6 cm glass petri dish containing 1 ml of

LB01 lysis buffer. In this case, the crude solution must be syringed once through a 22-gage needle to disperse chromosome clumps. This method is more laborious and inconvenient in species with small root tips.

13. In general, extended storage may compromise the quality of chromosomal DNA and the resolution of flow karyotypes.
14. The instrument must be aligned for the highest possible resolution of the FL1 and FSC signals to achieve good separation of chromosomes peaks on flow karyotypes.
15. DAPI-stained samples result in superior resolution of flow karyotypes as compared to those stained with DNA intercalators such as propidium iodide. This may be due to the formaldehyde fixation.
16. Data should be accumulated on a histogram of fluorescence pulse area, which gives better resolution of peaks representing large (long) chromosomes.
17. We have been using one-drop sort envelope and counter mode to sort small numbers of chromosomes (up to 10^3) e.g., for PCR mapping. Large-scale chromosome sorting (e.g., for BAC libraries) is done with one-drop sort envelope and normal-R mode.
18. Chromosomes are sorted onto a microscopic slide for microscopic observation. Smaller quantities of chromosomes to be used in physical mapping using PCR with specific primers and in DNA isolation and amplification are sorted into 0.5 ml (0.2 ml) PCR tubes. If large quantities of DNA are needed and large numbers of chromosomes have to be collected, 1.5 ml polystyrene cups (9000002, Deltalab, Rubí, Spain) are used.
19. Storage of slides at room temperature for a few days improves the morphology of chromosomes after FISH. However, longer storage may compromise the quality of FISH signals.
20. Chromosomes isolated according to our protocol are fixed only mildly and are easier to denature. Consequently, lower formamide concentration and shorter denaturation time is used as compared to traditional protocols for FISH.

Acknowledgments

We thank our colleagues Zdenka Dubská, Romana Šperková, and Petr Navrátil for excellent technical assistance. This work was supported by the Czech Science Foundation (award 521/07/1573), Czech Ministry of Education, Youth and Sports (award LC06004), and the European Community's Seventh Framework Programme (FP7/2007–2013) under grant agreement FP7-212019.

References

1. Gray, J.W., Carrano, A.V., Steinmetz, L.L., Van Dilla, M.A., Moore, H.H., Mayall, B.H., and Mendelsohn, M.L. (1975) Chromosome measurement and sorting by flow systems. *Proc. Natl. Acad. Sci. U. S. A.* **72**, 1231–1234.
2. Carrano, A.V., Gray, J.W., Langlois, R.G., Burkhart-Schultz, K.J., and Van Dilla, M.A. (1979) Measurement and purification of human chromosomes by flow cytometry and sorting. *Proc. Natl. Acad. Sci. U. S. A.* **76**, 1382–1384.
3. Dixon, S.C., Miller, N.G.A., Carter, N.P., and Tucker, E.M. (1992) Bivariate flow cytometry of farm animal chromosomes: a potential tool for gene mapping. *Anim. Genet.* **23**, 203–210.
4. Ferguson-Smith, M.A. (1997) Genetic analysis by chromosome sorting and painting: phylogenetic and diagnostic applications. *Eur. J. Hum. Genet.* **5**, 253–265.
5. Kulemzina, A.I., Trifonov, V.A., Perelman, P.L., Rubtsova, N.V., Volobuev, V., Ferguson-Smith, M.A., Stanyon, R., Yang, F.T., and Graphodatsky, A.S. (2009) Cross-species chromosome painting in Cetartiodactyla: reconstructing the karyotype evolution in key phylogenetic lineages. *Chromosome Res.* **17**, 419–436.
6. Boschman, G.A., Mandel, E.M.M., Rens, W., Slater, R., and Aten, J.A. (1992) Semi-automated detection of aberrant chromosomes in bivariate flow karyotypes. *Cytometry* **13**, 469–477.
7. Lebo, R.V. (1982) Chromosome sorting and DNA sequence localization: a review. *Cytometry* **3**, 145–154.
8. Van Dilla, M.A., and Deaven, L.L. (1990) Construction of gene libraries for each human chromosome. *Cytometry* **11**, 208–218.
9. Gingrich, J.C., Boehrer, D.M., Garnes, J.A., Johnson, W., Wong, B.S., Bergmann, A., Eveleth G.G., Langlois, R.G., and Carrano, A.V. (1996) Construction and characterization of human chromosome 2-specific cosmid, fosmid, and PAC clone libraries. *Genomics* **32**, 65–74.
10. Pinkel, D., Landegent, J., Collins, C., Fuscoe, J., Segraves, R., Lucas, J., and Gray J. (1988) Fluorescence in situ hybridization with human chromosome-specific libraries: detection of trisomy 21 and translocations of chromosome 4. *Proc. Natl. Acad. Sci. U. S. A.* **85**, 9138–9142.
11. Chen, W., Kalscheuer, V., Tzschach, A., Menzel, C., Ullmann, R., Schulz, M.H., Erdogan, F., Li, N., Kijas, Z., Arkesteijn, G., Pajares, I.L., Goetz-Sothmann, M., Heinrich, U., Rost, I., Dufke, A., Grasshoff, U., Glaeser, B., Vingron, M., and Ropers, H.H. (2008) Mapping translocation breakpoints by next-generation sequencing. *Genome Res.* **18**, 1143–1149.
12. Sudbery, I., Stalker, J., Simpson, J.T., Keane, T., Rust, A.G., Hurles, M.E., Walter, K., Lynch, D., Teboul, L., Brown, S.D., Li, H., Ning, Z., Nadeau, J.H., Croniger, C.M., Durbin, R., and Adams D.J. (2009) Deep short-read sequencing of chromosome 17 from the mouse strains A/J and CAST/Ei identifies significant germline variation and candidate genes that regulate liver triglyceride levels. *Genome Biol.* **10**, R112.1–R112.19.
13. Doležel, J., Lucretti, S., and Schubert, I. (1994) Plant chromosome analysis and sorting by flow cytometry. *Crit. Rev. Plant Sci.* **13**, 275–309.
14. Doležel, J., Kubaláková, M., Bartoš, J., and Macas, J. (2004) Flow cytogenetics and plant genome mapping. *Chromosome Res.* **12**, 77–91.
15. Doležel, J., Číhalíková, J., and Lucretti, S. (1992) A high-yield procedure for isolation of metaphase chromosomes from root tips of *Vicia faba* L. *Planta* **188**, 93–98.
16. Lysák, M.A., Číhalíková, J., Kubaláková, M., Šimková, H., Künzel, G., and Doležel, J. (1999). Flow karyotyping and sorting of mitotic chromosomes of barley (*Hordeum vulgare* L.). *Chromosome Res.* **7**, 431–444.
17. Vrána, J., Kubaláková, M., Šimková, H., Číhalíková, J., Lysák, M.A., and Doležel, J. (2000) Flow-sorting of mitotic chromosomes in common wheat (*Triticum aestivum* L.). *Genetics* **156**, 2033–2041.
18. Kubaláková, M., Valárik, M., Bartoš, J., Vrána, J., Číhalíková, J., Molnár-Láng, M., and Doležel, J. (2003) Analysis and sorting of rye (*Secale cereale* L.) chromosomes using flow cytometry. *Genome* **46**, 893–905.
19. Lucretti, S., and Doležel, J. (1997) Bivariate flow karyotyping in broad bean (*Vicia faba*). *Cytometry* **28**, 236–242.
20. Kubaláková, M., Vrána, J., Číhalíková, J., Šimková, H., and Doležel, J. (2002) Flow karyotyping and chromosome sorting in bread wheat (*Triticum aestivum* L.). *Theor. Appl. Genet.* **104**, 1362–1372.
21. Suchánková, P., Kubaláková, M., Kovářová, P., Bartoš, J., Číhalíková, J., Molnár-Láng, M., Endo, T.R., and Doležel, J. (2006) Dissection

of the nuclear genome of barley by chromosome flow sorting. *Theor. Appl. Genet.* **113**, 651–659.

22. Macas, J., Doležel, J., Lucretti, S., Pich, U., Meister, A., Fuchs, J., and Schubert, I. (1993) Localization of seed storage protein genes on flow-sorted field bean chromosomes. *Chromosome Res.* **1**, 107–115.
23. Kejnovský, E., Vrána, J., Matsunaga, S., Souček, P., Široký, J., Doležel, J., and Vyskot, B. (2001) Localization of male-specifically expressed MROS genes of *Silene latifolia* by PCR on flow-sorted sex chromosomes and autosomes. *Genetics* **158**, 1269–1277.
24. Šimková, H., Svensson, J.T., Condamine, P., Hřibová, E., Suchánková, P., Bhat, P.R., Bartoš, J., Šafář, J., Close, T.J., and Doležel, J. (2008) Coupling amplified DNA from flow-sorted chromosomes to high-density SNP mapping in barley. *BMC Genomics* **9**, 294.
25. Požárková, D., Koblížková, A., Román, B., Torres, A.M., Lucretti, S., Lysák, M., Doležel, J., and Macas, J. (2002) Development and characterization of microsatellite markers from chromosome 1-specific DNA libraries of *Vicia faba*. *Biol. Plant.* **45**, 337–345.
26. Kofler, R., Bartoš, J., Gong, L., Stift, G., Suchánková, P., Šimková, H., Berenyi, M., Burg, K., Doležel, J., and Lelley, T. (2008) Development of microsatellite markers specific for the short arm of rye (*Secale cereale* L.) chromosome 1. *Theor. Appl. Genet.* **117**, 915–926.
27. Šafář, J., Bartoš, J., Janda, J., Bellec, A., Kubaláková, M., Valárik, M., Pateyron, S., Weiserová, J., Tušková, R., Číhalíková, J., Vrána, J., Šimková, H., Faivre-Rampant, P., Sourdille, P., Caboche, M., Bernard, M., Doležel, J., and Chalhoub, B. (2004) Dissecting large and complex genomes: flow sorting and BAC cloning of individual chromosomes from bread wheat. *Plant J.* **39**, 960–968.
28. Šimková, H., Šafář, J., Suchánková, P., Kovářová, P., Bartoš, J., Kubaláková, M., Janda, J., Číhalíková, J., Mago, R., Lelley, T., and Doležel, J. (2008) A novel resource for genomics of Triticeae: BAC library specific for the short arm of rye (*Secale cereale* L.) chromosome 1R (1RS). *BMC Genomics* **9**, 237.
29. Mayer, K.F.X., Taudien, S., Martis, M., Šimková, H., Suchánková, P., Gundlach, H., Wicker, T., Petzold, A., Felder, M., Steuernagel, B., Scholz, U., Graner, A., Platzer, M., Doležel, J., and Stein N. (2009) Gene content and virtual gene order of barley chromosome 1H. *Plant Physiol.* **151**, 496–505.

Chapter 13

Super-Stretched Pachytene Chromosomes for Plant Molecular Cytogenetic Mapping

Dal-Hoe Koo and Jiming Jiang

Abstract

We developed a simple technique to mechanically stretch maize pachytene chromosomes more than 20 times longer than their original size. A modified Carnoy's II solution (6:3:1) ethanol:acetic acid:chloroform was used to fix the meiotic sample. The super-stretched pachytene chromosomes produced from this procedure can be directly used in conventional fluorescence in situ hybridization (FISH) experiments and also for the immunofluorescent in situ detection of DNA methylation. This technique adds a new dimension and higher resolving power to pachytene chromosome-based molecular cytogenetics research.

Key words: Pachytene chromosome, FISH, DNA methylation, Molecular cytogenetics

1. Introduction

Meiotic pachytene chromosomes are on average 10–50 times longer than somatic metaphase chromosomes and provide a superior resolution for cytogenetic mapping (1, 2). The euchromatin and heterochromatin are well differentiated on pachytene chromosomes in some plant species. Thus, pachytene chromosomes provide an excellent cytological target for molecular and biochemical characterization of these two distinct types of chromatin (3–6). In this chapter, we describe a technique that can be used to mechanically stretch plant pachytene chromosomes up to at least 20 times longer than their normal sizes (Fig. 1a). Such super-stretched pachytene chromosomes provide an unprecedented resolution for chromosome-based cytogenetic mapping. Regular fluorescence in situ hybridization (FISH) can be combined with the immunodetection of 5-methyl cytosine (5mC)

James A. Birchler (ed.), *Plant Chromosome Engineering: Methods and Protocols*, Methods in Molecular Biology, vol. 701, DOI 10.1007/978-1-61737-957-4_13,

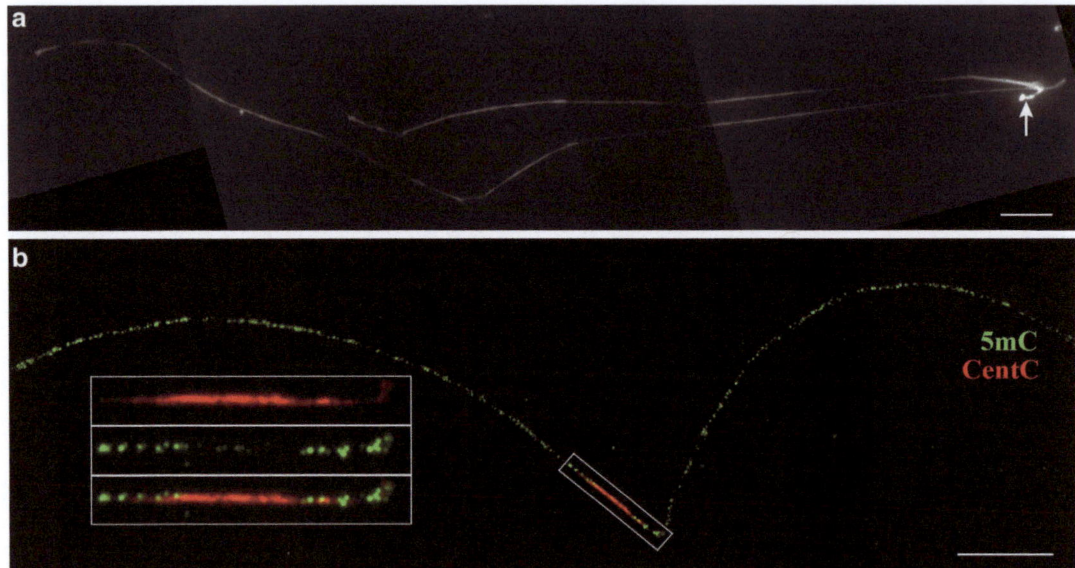

Fig. 1. Super-stretched maize pachytene chromosomes for cytogenetic mapping. (**a**) Two super-stretched pachytene chromosomes, including a chromosome that can be identified by its characteristic satellite. The satellite knob is resistant to stretching (*arrow*). (**b**) Mapping 5mC (*green*) and CentC (centromere satellite repeat – *red*) on a super-stretched pachytene chromosome. *Rectangle* delimits the centromere region of the chromosome. Inset: Signals from 5mC, CentC and a merged image are enlarged for clarity. Note the reduced methylation in the middle section of the centromere. Bars = 20 μm.

on super-stretched pachytene chromosomes (Fig. 1b), which provides a powerful tool to study DNA methylation associated with specific euchromatic and heterochromatic domains in plant genomes.

2. Materials

2.1. Fixatives

1. Modified Carnoy's II: 6:3:1 parts of absolute ethanol:glacial acetic acid:chloroform, respectively. For 100 ml: mix 60 ml of 100% ethanol with 30 ml of glacial acetic acid and 10 ml of chloroform. This fixative must be freshly prepared.
2. Carnoy's I: Three parts of ethanol and one part of glacial acetic acid. Prepare fresh.
3. 4% Formaldehyde: Heat 80 ml of ddH_2O to 50–55°C and add 4 g of paraformaldehyde (Sigma), add 0.5 ml of 1 N NaOH to solublize paraformaldehyde (formaldehyde), add 10 ml of 10× PBS, and adjust the final volume to 100 ml with ddH_2O. The solution can be stored at 4°C for several days.

4. 10× PBS: 1.4 M NaCl, 80 mM Na_2HPO_4, 18 mM KH_2PO_4, 27 mM KCl, pH 7.4. Dissolve the components in approximately 900 ml of H_2O. Adjust the pH to 7.2–7.4 with HCl and then adjust the final volume to 1 l with H_2O. Sterilize by autoclaving. Store at room temperature (see Note 1).

2.2. Denaturation

1. 70% Formamide in 2× SSC (10 ml): 7 ml deionized formamide, 1 ml 20× SSC, and 2 ml ddH_2O. Store at 4°C.

2.3. Detection

1. Mouse monoclonal antibody against 5mC (Eurogentec).
2. Alexa Fluor 488-conjugated rabbit anti-mouse IgG (Invitrogen).
3. Rhodamine anti-digoxigenin (Roche).
4. 0.5% BSA/PBST: 0.5% bovine serum albumin, 10× PBS, 0.5% Tween 20. Combine the following in 10 ml of 1× PBS: 50 μl of Tween 20 (Sigma) and 0.05 g BSA (Sigma). This solution should be freshly prepared.
5. TNB: 0.1 M Tris–HCl (pH 7.5), 0.15 M NaCl, 0.5% blocking reagent. For 100 ml: 10 ml of 1 M Tris–HCl (pH 7.5), 15 ml of 1 M NaCl, and 0.5 g of blocking reagent (Roche). Adjust the final volume to 100 ml of ddH_2O. Store at −20°C.

2.4. Posthybridization Washes

1. 2× SSC: Make the 10× dilution from a 20× SSC stock solution. 20× SSC stock: 3 M NaCl, 0.3 mM trisodium citrate, pH 7.0. Sterilize by autoclaving and store at room temperature.
2. 4 T: 4× SSC/0.05% Tween 20. Make the 5× dilution from a 20× SSC/0.25% Tween 20. Store at room temperature.
3. TNT: 0.1 M Tris–HCl (pH 7.5), 0.15 M NaCl, 0.05% Tween 20. For 1,000 ml: Mix 10 ml of 1 M Tris–HCl (pH 7.5), 30 ml of 5 M NaCl. Add distilled H_2O to 1 l. Sterilize by autoclaving. When cooled, add 0.05 ml of Tween 20. Store at room temperature.
4. 1× PBS: Make the 10× dilution from a 10× PBS stock solution. Store at room temperature.

3. Methods

3.1. Preparation of Super-Stretched Pachytene Chromosome

1. Choose panicles (or flower buds) at the pachytene stage.
2. Immerse the panicles in the modified Carnoy's fixative II solution for 3–4 h at room temperature, then re-fix with 3:1 ethanol:acetic acid and store at 4°C until use (see Note 2).
3. Extract one anther from a flower and extrude the pollen mother cells (PMCs) onto a glass slide containing 1% acetocarmine.

Make a slide preparation and check the stage of meiosis (see Note 3). Determine the proper floret size needed to recover meiotic chromosomes at pachytene stage, then continue to step 4.

4. In the same fashion as step 3, press out the PMCs from a size-selected anther onto a poly-L-lysine-coated slide (Sigma) containing 15 μl of 60% acetic acid/0.05% Triton X-100. Carefully dissect the anther using a needle. Do not let the cell suspension dry.
5. Remove the cellular detritus with a pair of fine forceps and place a coverslip (22 × 22 mm) over the cell suspension.
6. Examine the chromosomes using a phase contrast microscope to make sure the PMCs are well separated. If not, very lightly tapping on the coverslip with a needle should sufficiently separate the clumped cells.
7. To stretch the chromosomes: Place filter paper over the coverslip and while gently applying thumb pressure to the coverslip, roll the thumb one direction across the area of the coverslip (see Note 4).
8. Freeze the preparation immediately by placing the slide in a −80°C freezer (see Note 5).

3.2. Immunodetection of 5mC Combined with FISH

3.2.1. Pretreatment of Slides

1. Remove the coverslips from the frozen slides (stored in a −80°C freezer) with a razor blade.
2. Immediately immerse the slides in 3:1 ethanol:acetic acid for 2 min at room temperature and then bake for 10 min at 60°C.
3. Pipette 100 μl of a solution containing RNase A (Sigma, 100 μg/ml) in 2× SSC onto slides and cover with a coverslip. Place the slide in a humid chamber and incubate at 37°C for 1 h (see Note 6).
4. Remove the coverslip and immerse the slides in 1× PBS.
5. Wash slides in 1× PBS twice each with 5 min.
6. Dehydrate slides in 70 and 95% ethanol for 2 min each, and air dry.

3.2.2. DNA Denaturation

1. Pipette 100 μl of a 70% formamide/2× SSC solution directly on the slide and cover with a 22 × 44 mm coverslip.
2. Denature the chromosomal DNA on heat block at 80°C for 3 min.
3. Dehydrate slides in 70 and 95% cold ethanol for 3 min each, and air dry.

3.2.3. Immunodetection of 5mC on Chromosomes

1. Add 100 μl of 0.5% BSA/PBST, apply a clean coverslip and incubate for 30 min at room temperature (see Note 7).

2. Remove coverslip and apply 100 μl of a mouse monoclonal antibody against 5mC (1:250 dilution from an original stock solution, Eurogentec) in PBST and incubate at room temperature for less than 12 h, or alternatively, at 4°C for 24–48 h (see Note 8).
3. Remove coverslip and immediately place slides in 1× PBS.
4. Wash twice for 5 min each in 1× PBS at room temperature.
5. Protect slides from light during following detection procedures.
6. Add 100 μl of detection reagent mixture [Alexa Fluor 488-conjugated rabbit anti-mouse IgG (1:100 dilution from an original stock solution, Invitrogen) in TNB], and cover with a 22×44 mm coverslip.
7. Incubate at 37°C for 1 h in a humid chamber.
8. Wash slides twice, 5 min each in 1× PBS at room temperature.

3.2.4. Fixation of Immunostained Chromosomes

1. Protect slides from light during the following fixation procedure.
2. Add 600 μl of the 4% formaldehyde/1× PBS solution onto a slide and cover with plastic coverslip (see Note 9).
3. Incubate at room temperature for 20–30 min in a humid chamber.
4. Remove the coverslip and place the slide into a coplin jar in 1× PBS.
5. Wash slides twice, 5 min each in 1× PBS at room temperature, air dry.

3.2.5. FISH on Immunostained Chromosomes (see Note 10)

1. Protect the slides from light during the FISH procedure.
2. Pipette 100 μl of 70% formamide/2× SSC on the slide.
3. Cover with a 22×44 mm coverslip.
4. Denature the chromosomal DNA on a heat block at 70°C for 2 min.
5. Dehydrate slides in 70% ethanol (−20°C) for 5 min and 95% ethanol (room temperature) for 3 min, air dry.
6. Pipette 20 μl of denatured, digoxigenin-labeled FISH probe mix onto each slide and cover with a 22×44 mm coverslip. Seal the coverslip with rubber cement.
7. Place slides in a humid chamber at room temperature for 12–14 h.

3.2.6. Posthybridization Wash and Detection

1. Preheat a coplin jar containing 2× SSC to 42°C.
2. Remove slides from humid chamber and carefully remove the rubber cement.

3. Let coverslips fall in a coplin jar with 2× SSC. If necessary, the coplin jar can be placed on a rotary shaker on low speed to coax the coverslip into separating from the slide.
4. Wash the slide with the following steps. All washing steps are done with gentle agitation on a rotary shaker in dark condition (RT – room temperature).

2× SSC	RT	5 min
2× SSC	42°C	10 min
2× SSC	RT	5 min
4 T	RT	5 min

5. To detect the digoxigenin-labeled FISH probe, apply 100 μl of anti-digoxigenin rhodamine (1:100 dilution from an original stock solution, Roche) in TNB and cover with a 22 × 44 mm coverslip.
6. Incubate at 37°C for 1 h in moist chamber.
7. Wash twice for 5 min each in TNT at room temperature.

3.2.7. Visualization of Stretched Chromosome and Detection of Fluorescence Signals

1. Shake off excess wash buffer and apply a drop of Vectashield anti-fade solution containing DAPI (4',6-diamidino-2-phenylindole) (Vector Lab), and cover with a 22 × 44 coverslip. Squeeze out the excess Vectashield from the preparation by applying light pressure to the coverslip and slide with a piece of filter paper.
2. The 5mC and FISH signals are visualized using an epifluorescence microscope. Slides can be analyzed immediately or stored in the dark at 4C or –20°C for future analysis.

4. Notes

1. For in situ hybridization, some laboratories may use a different formulation of PBS, such as NaCl (7.59 g/l), Na_2HPO4 (1 g/l), and NaH_2PO_4 (0.42 g/l). The pH will be 7.4 based on the ratio of monobasic and dibasic sodium phosphate.
2. The process of fixing the chromosomes is critical for generating elasticity of the pachytene chromosomes. The impact of different fixatives on chromosome stretching was described previously (7).
3. Chromosomes at mid-pachytene stage are most suitable for super-stretching.
4. This preparation procedure is similar to the traditional squashing technique. However, the rolling motion causes displacement

of the coverslip, which moves approximately 1 mm after squashing. It is the actual movement of the coverslip over the preparation that results in the stretching of pachytene chromosomes. An illustration of the rolling motion is provided in Koo and Jiang (7).

5. Slides stored overnight generally retain a better morphology of the stretched chromosomes.
6. RNase A treatment removes endogenous RNA, which might otherwise contribute to background signal.
7. BSA treatment reduces nonspecific antibody binding to chromosome preparations. BSA solutions are filter-sterilized immediately before use.
8. A relatively long incubation time (≥12 h at 4°C) is necessary to ensure the proper and reproducible penetration of antibodies.
9. Chromosomes should be fixed to reduce 5mC signal loss during FISH procedures.
10. A conventional FISH procedure can be applied on immunostained chromosomes.

Acknowledgments

We thank Jason Walling for technical assistance and discussion on the data and manuscript. This research was supported by grant DBI-0421671 from the National Science Foundation.

References

1. de Jong, J.H., *et al.* (1999) High resolution FISH in plants – techniques and applications. *Trends Plant Sci.* 4, 258–263.
2. Cheng, Z.K., *et al.* (2002) Resolution of fluorescence in situ hybridization mapping on rice mitotic prometaphase chromosomes, meiotic pachytene chromosomes and extended DNA fibers. *Chromosome Res.* 10, 379–387.
3. Fransz, P.F., *et al.* (2000) Integrated cytogenetic map of chromosome arm 4 S of *A. thaliana*: Structural organization of heterochromatic knob and centromere region. *Cell* 100, 367–376.
4. McCombie, W.R., *et al.* (2000) The complete sequence of a heterochromatic island from a higher eukaryote. *Cell* 100, 377–386.
5. Cheng, Z., *et al.* (2001) A tandemly repeated DNA sequence is associated with both knob-like heterochromatin and a highly decondensed structure in the meiotic pachytene chromosomes of rice. *Chromosoma* 110, 24–31.
6. Shi, J. and Dawe, R.K. (2006) Partitioning of the maize epigenome by the number of methyl groups on histone H3 lysines 9 and 27. *Genetics* 173, 1571–1583.
7. Koo, D.-H. and Jiang, J.M. (2009) Super-stretched pachytene chromosomes for fluorescence *in situ* hybridization mapping and immunodetection of DNA methylation. *Plant J.* 59, 509–516.

Chapter 14

Cytological Dissection of the Triticeae Chromosomes by the Gametocidal System

Takashi R. Endo

Abstract

Triticeae species have a large and complex genome, which has made it difficult to obtain their sequence data. Some alien chromosomes called the gametocidal (Gc) chromosomes introduced into common wheat can induce chromosomal breakage resulting in the generation of deletions and translocations. The induced deletions have been established as deletion stocks in common wheat. This Gc system is also effective in inducing chromosomal breakages in Triticeae chromosomes added to common wheat. The induced aberrant chromosomes can be identified by chromosome banding and fluorescence in situ hybridization and can be established in common wheat as dissection lines. This Gs system will be useful to dissect the single chromosomes of Triticeae species.

Key words: Triticeae, Common wheat, Rye, Barley, C-banding, N-banding, FISH, GISH, Gametocidal chromosome, Alien addition, Deletion, Translocation, Dissection line

1. Introduction

Triticeae, a tribe in the grass family Poaceae, contains major crop genera including wheat, rye, and barley. These genera contain diploid, allotetraploid, and/or allohexaploid genomes with the basic chromosome number $x=7$. Most of the Triticeae species have a huge genome, e.g., bread wheat or common wheat (*Triticum aestivum*, $2n=6x=42$, 1C=16,979 Mb), rye (*Secale cereale*, $2n=2x=14$, 1C=8,110 Mb), and barley (*Hordeum vulgare*, $2n=2x=14$, 1C=5,439 Mb), when compared with those of rice (1C=490 Mb) and Arabidopsis (1C=157 Mb) (1). This large genome size has hindered the complete sequencing of the Triticeae genomes.

The polyploid plant species have tolerance to aneuploidy to some extent. Especially, the hexaploid nature has allowed a wide

James A. Birchler (ed.), *Plant Chromosome Engineering: Methods and Protocols*, Methods in Molecular Biology, vol. 701, DOI 10.1007/978-1-61737-957-4_14, © Springer Science+Business Media, LLC 2011

variety of aneuploid lines to be established in common wheat, i.e., monosomics, nullisomics, nullisomic–tetrasomics, and telosomics (2–4). In a sense, these aneuploids, which lack specific chromosomes or chromosome arms, dissected the wheat genomes and have been used to allocate genes and DNA markers for the specific missing chromosomes or chromosome arms (5).

Many Triticeae species have been successfully crossed with common wheat and the F_1 hybrids backcrossed to common wheat to produce wheat–alien chromosome addition and substitution lines (5). In these lines, the genomes of such Triticeae species have been dissected into single chromosomes or chromosome arms in the genomic background of common wheat. The presence of alien chromosome-specific genes and DNA markers would be indicated from the phenotypes and marker profiles of the alien addition and substitution lines, respectively.

Sub-arm dissection of the chromosome would undoubtedly be more useful in genome analysis. There are two genetic systems that induce chromosomal rearrangements lacking part of chromosome arms, namely, the homoeologous pairing (*Ph*) system and the gametocidal (Gc) system. The *Ph* system has been used to recombine the wheat and alien chromosomes by inducing homoeologous chromosome pairing (6). In the Gc system, the Gc chromosome is introduced into common wheat from various species of the genus *Aegilops*, and the Gc chromosome induces random chromosomal breakage resulting in the generation of deletions and translocations (7, 8). Figure 1 shows how the Gc chromosome induces chromosomal structural changes: When introduced

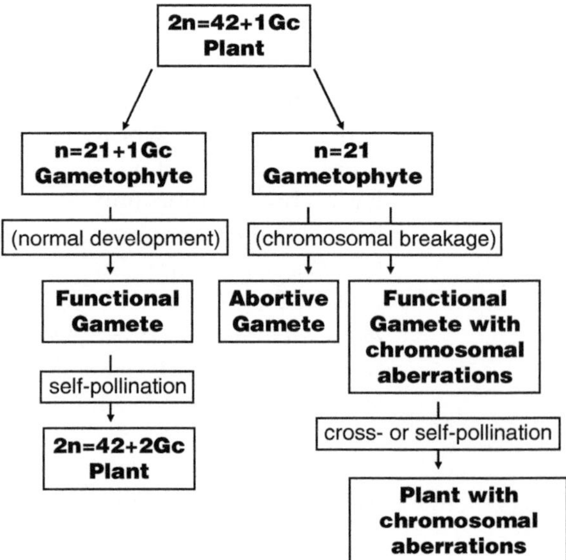

Fig. 1. Schematic illustration of the action of the Gc chromosome. See the text for details.

into common wheat, it induces chromosome breakage only in gametes without the Gc chromosome. The Gc system has been successfully used to produce deletion stocks of common wheat (9), and the deletion stocks have been used in the cytological chromosome mapping of DNA markers, such as RFLPs (10) and ESTs (11). The Gc system has also been proved to be useful for inducing chromosomal aberrations in rye and barley chromosomes added to common wheat (12, 13). The Gc system may also be useful for dissecting other alien chromosomes of Triticeae species, such as *Agropyron elongatum*. The following is described the procedure for using the Gc gametocidal system to dissect alien chromosomes added to common wheat, with examples of rye and barley chromosomes as targets of dissection. Important cytological protocols are also described because efficient and reliable chromosome identification is crucial to the successful application of the Gc system.

2. Materials

2.1. Chromosome Preparation

1. Fixative: one part glacial acetic acid, three parts ethanol (95–99%). Store at room temperature, need not be freshly prepared.
2. Acetocarmine stain solution: Dissolve 1 g carmine powder (Merck) in 100 mL 45% acetic acid and boil for 24 h, using a reflux condenser to prevent the solution from being boiled dry. Transfer to a bottle without filtration and store at room temperature. Use the clear layer on the top.

2.2. Chromosome Banding

1. Staining solution: Various stain solutions are available. Wright Stain Solution (Muto Pure Chemicals Co. LTD., Japan) is one of the recommended stain solutions (see Note 1). Store at room temperature.
2. Phosphate buffer: Prepare a stock solution with 0.1 M Na_2HPO_4 (14.2 g/1,000 mL) and 0.1 M KH_2PO_4 (13.6 g/1,000 mL).
3. 5% $Ba(OH)_2$ solution: Place 50 g $Ba(OH)_2$ in a glass container and pour hot tap water up to 1,000 mL. The container should be stopped tightly for storage at room temperature.
4. 2× SSC: Prepare a 20× SSC stock solution with 3 M NaCl and 0.3 M $C_6H_5Na_3O_7{\cdot}2H_2O$. Dilute tenfold with distilled water for use.

2.3. Fluorescence In Situ Hybridization

1. 0.15 N NaOH/ethanol solution: Dissolve 6 g NaOH in 1,000 mL 70% ethanol. Store at room temperature.
2. Fluorescence In Situ Hybridization (FISH) probe: Using DIG-High Prime (Roche Diagnostics) or Biotin-High Prime (Roche Diagnostics), label total genomic DNAs from the

alien species and PCR-amplified subtelomeric repeat sequences of alien chromosomes, such as pSc200 (14) and HvT 01 (15). Store the labeled probes at 4°C. FISH using total genomic DNA probes is called GISH (genomic in situ hybridization).

3. ISH solution: Prepare a solution with 50% (v/v) formamide and 5% (w/v) dextran sulfate in 2× SSC. Store at 4°C.
4. Hybridization mixture: Mix the labeled probe (1 part) and ISH solution (19 parts) (see Note 2). Heat the mixture at 90–100°C for 8–10 min and immediately freeze it until use.
5. Detection mixture: Anti-Digoxigenin-Fluorescein, Fab fragments (Roche Diagnostics) for Dig-labeled probe and Streptavidin-CY3 (Invitrogen) for Biotin-labeled probe diluted as recommended by the manufacturer (see Note 3). Store at 4°C or at −20°C for longer storage (do not repeatedly freeze after thawing).
6. Counter staining solution: DAPI (4, 6-diamidino-2-phenylindole) (Roche Diagnostics) diluted in an anti-fade solution as recommended by the manufacturer.

2.4. Wheat Stocks

1. Gametocidal lines: Addition lines of common wheat cv. "Chinese Spring" carrying a gametocidal chromosome, 2C or $3C^{SAT}$ (8) (see Note 4).
2. Nullisomic–tetrasomic lines: Aneuploids of "Chinese Spring" that lack one chromosome pair and have four doses of a homoeologous chromosome compensating for the missing chromosome pair.
3. Alien addition lines: Addition lines of common wheat cv. "Chinese Spring" carrying single alien chromosomes derived from rye, barley, and other Triticeae species.

The gametocidal and nullisomic–tetrasomic lines and many alien chromosome addition lines can be obtained from the National BioResource Project (NBRP) (http://www.shigen.nig.ac.jp/wheat/komugi/top/top.jsp).

3. Methods

3.1. Chromosome Preparation

1. Place root tips in small vials filled with distilled water and immerse them in ice water for 16–20 h to collect metaphase cells.
2. Fix the root tips in the fixative for 1 day, stain them in the acetocarmine solution for 12 h, return to the original fixative, and store for 2–3 days for chromosome banding and for 3–6 days for FISH. Perform all procedures at room temperature. The fixed and stained root tips can be stored at −20°C until use.

3. Stain the fixed and stained root tips again in the acetocarmine solution for 10–20 min.
4. Make chromosome preparations by the squash method from the stained root tips and immediately store them at −70°C or below until use.
5. Remove the cover slip quickly from the frozen slide, using a razor blade, immerse the slide in 45% acetic acid at 40–45°C for 2–3 min, and air-dry the slide at room temperature. The air-dried slide can be used immediately or stored in an airtight container at −20°C up to several months.

3.2. Chromosome Banding

3.2.1. N-Banding

1. Incubate the air-dried slide (see Subheading 3.1) in 1 M NaH_2PO_4 solution for 1.5 min at 92–95°C.
2. Wash the slide briefly with hard tap water (see Note 5).
3. Place the wet slide into the staining solution at room temperature until appropriate staining is achieved (usually about 2 h).

3.2.2. C-Banding

1. Place the air-dried slide (see Subheading 3.1) in a container with a lid.
2. Pour 5% $Ba(OH)_2$ solution into the container, put the lid on, and keep it for about 5 min at room temperature.
3. Take out the slide, quickly wash with hard tap water, and incubate in 2× SSC for 10 min at 42–45°C.
4. Wash the slide briefly with hard tap water.
5. Place the wet slide into the staining solution at room temperature until appropriate staining is achieved (usually about 50 min).

3.2.3. Microscopic Observation

1. Wash the stained slide briefly with hard tap water.
2. Air-dry the slide using a puffer (a camera tool).
3. Mount the slide in immersion oil with or without a cover slip (see Note 6).

Figure 2 shows a C-banded mitotic metaphase cell of rye (a) and N-banded mitotic metaphase cell of barley (b).

3.3. FISH/GISH

1. Immerse the air-dried slide (see Subheading 3.1) in the 0.15 N NaOH/ethanol solution for 5 min at room temperature.
2. Transfer the wet slide into a series of two 70% and one 99% ethanol each for 3 min at room temperature.
3. Dry the slide quickly with a puffer.
4. Apply the denatured hybridization mixture (10 μL per slide) onto the slide, and place a cover slip on it and incubate the slide in a moistened chamber for 6–24 h at 30°C (see Note 7).

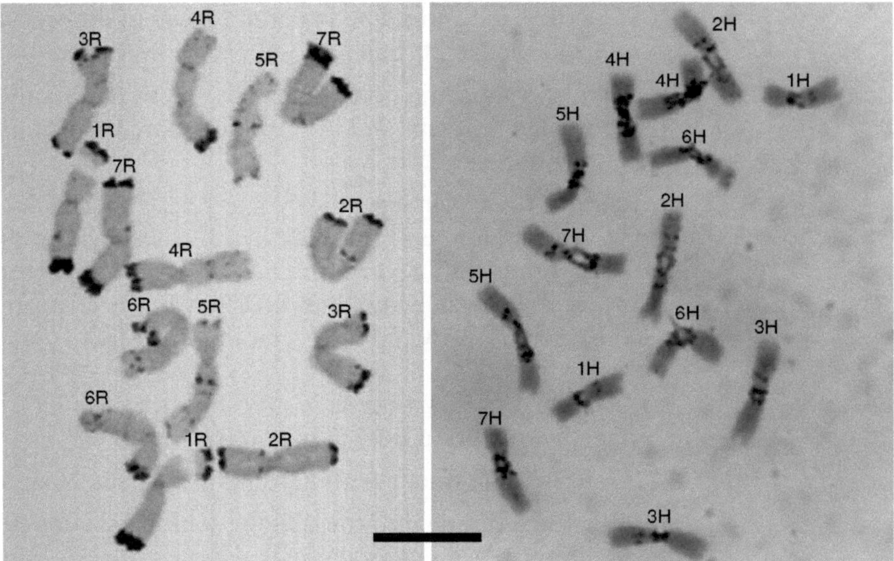

Fig. 2. A C-banded mitotic metaphase cell of "Prolific" rye (**a**) and an N-banded mitotic metaphase cell of "Shinebisu" barley (**b**). The chromosome designations of rye and barley are based on the karyotypes in previous studies (16, 17). The identification of chromosomes 2R, 3R and 7R is not clear. N- and C-banding patterns are basically the same in barley, but N-banding differentiates no terminal heterochromatic bands in rye as C-banding does.

5. Remove the cover slip with a pair of forceps (when the cover slip is firmly stuck on the slide, do not apply force to remove it; instead dip the slide in 2× SSC and let the cover slip fall off) and immerse the slide for 3 min in 2× SSC at room temperature.
6. Wash the slide briefly with distilled water and blow off water using a puffer.
7. Apply the detection mixture (10 µL per slide), place a cover slip on the slide and incubate in a wet chamber for about 1 h at 30°C.
8. Remove the cover slip with a pair of forceps (when the cover slip is firmly stuck on the slide, do not apply force to remove it, but dip the slide in 2× SSC and let the cover slip fall off) and wash the slide briefly with distilled water and blow off water using a puffer.
9. Apply the counter staining solution (5 µL per slide) and cover with a cover slip for fluorescence microscopic observation.

3.3.1. FISH/GISH After Chromosome Banding

1. After recording the images of chromosome banding, wash off immersion oil from the slide by dropping a mixture of xylene/99% ethanol (1:1) onto the slide three times and blowing off the mixture using a puffer.
2. Dip the slide in 70% ethanol for at least10 min and in 99% ethanol for at least 5 min at room temperature.
3. Air-dry the slide using a puffer.
4. Treat the slide as described in Subheading 3.3.

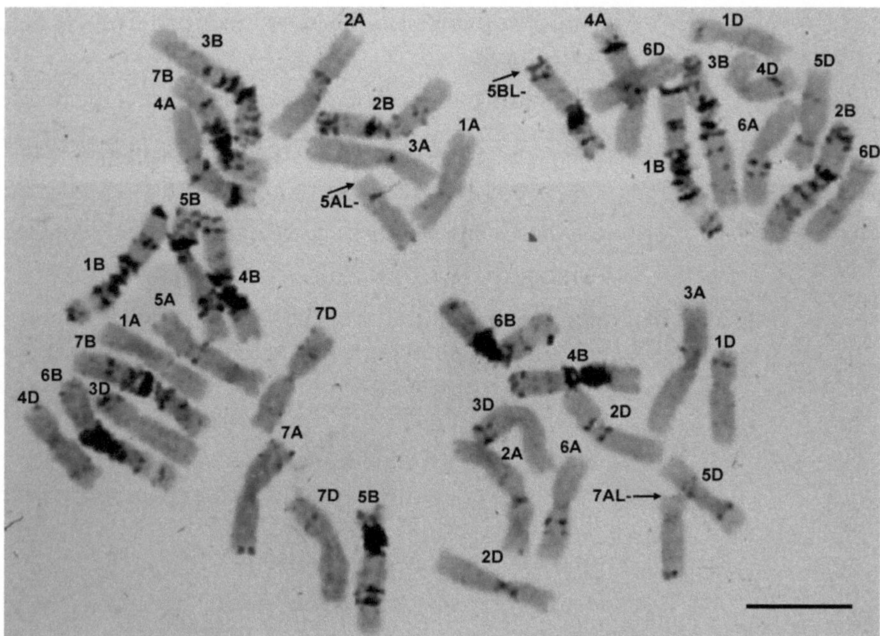

Fig. 3. A C-banded mitotic metaphase cell of a common wheat plant carrying deletions. Note this plant has three different deletions, 5A, 7A, and 5B (*pointed with arrows*) and is partially trisomic for 5B.

3.4. Dissection of Triticeae Chromosomes

3.4.1. Deletion Lines of Common Wheat

The deletion stocks of common wheat cultivar "Chinese Spring" have been produced from deletions induced mostly by the 2C chromosome and screened by C-banding (9). Figure 3 shows a C-banded mitotic metaphase cell of a "Chinese Spring" plant with three deletions. At present about 350 deletion-homozygous lines are available from NBRP, but the number of deletion stocks for each chromosome is at most 36 for 1B. Although an unlimited number of deletions can practically be produced in this way, the number becomes limited because the C-banding needs skill and is time consuming. However, the use of nullisomic–tetrasomic lines and PCR-based chromosome-specific markers would enable us to screen for chromosome-specific deletions on a large scale. The procedure is as follows.

1. Cross the disomic 2C line with euploid "Chinese Spring" to obtain monosomic 2C plants.
2. Cross the monosomic 2C plants as the female parent with one of the nullisomic–tetrasomic lines.
3. Grow the progeny plants from more or less shriveled seeds, which are liable to have chromosomal aberrations. Plump seeds tend to have the Gc chromosome and no chromosomal aberration.
4. Conduct PCR analysis of the plants using PCR markers specific to the chromosome that is missing in the nullisomic–tetrasomic line that was used to pollinate the monosomic 2C plants. The most distal markers on both chromosome arms

are most suitable for the detection of terminal deletions of the chromosome.

5. Select plants that lack any chromosome-specific marker. These plants can be regarded as lacking part of a critical chromosome in the hemizygous condition.

3.4.2. Developing Alien Addition Lines Carrying a Gc Chromosome

1. Cross the Gc lines with an alien addition line of common wheat.
2. Backcross the F_1s to the alien addition line.
3. Select plants disomic for the alien chromosome and monosomic for the Gc chromosome ($2n=45$) by C- or N-banding and GISH. An example of the 45-chromosome constitution is shown in Fig. 4.
4. Backcross the 45-chromosome plants to euploid "Chinese Spring" and grow the progeny plants from more or less shriveled seeds. Plump seeds tend to have the Gc chromosome and no chromosomal aberrations.
5. Harvest root tips and DNA from the progeny plants for screening.

3.4.3. Cytological Screening for and Characterization of Aberrant Alien Chromosomes

1. Check the chromosome constitutions of the progeny plants by FISH/GISH. The FISH probes of alien species-specific subtelomeric sequences, such as HvT 01 for barley chromosomes and pSc200 for rye chromosomes, are useful in detecting aberrations of the alien chromosomes. Figure 5 shows examples of aberrant alien chromosomes identified by FISH/GISH.

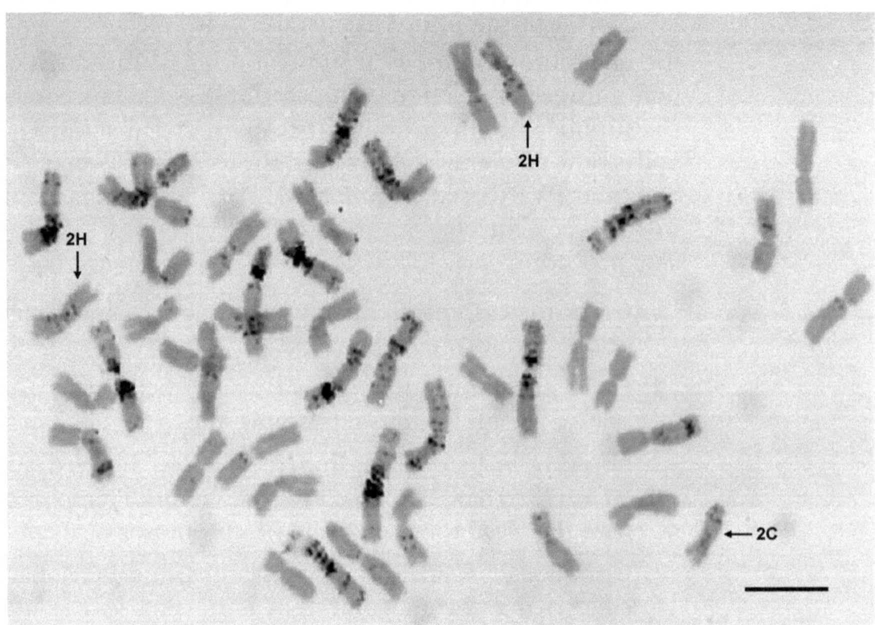

Fig. 4. A C-banded mitotic metaphase cell of a 45-chromosome plant of common wheat disomic for 2H and monosomic for 2C.

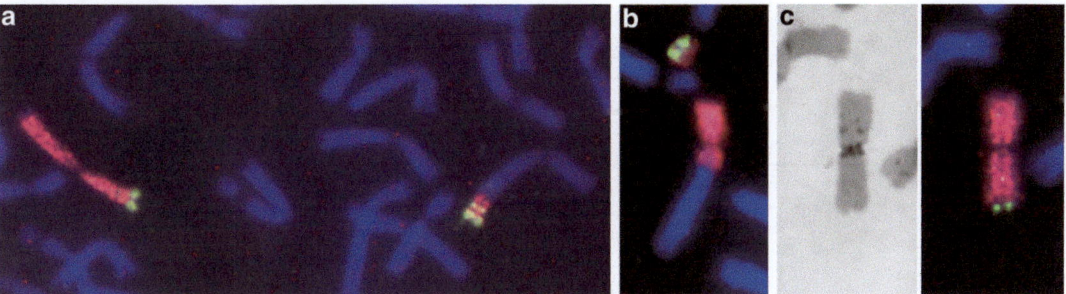

Fig. 5. The FISH/GISH images of reciprocal translocations involving 1R chromosome of "Imperial" rye (**a**) and of a translocation involving 5H chromosome of "Betzes" barley (**b**), and the sequential C-banding and FISH/GISH image of a deletion of 7H chromosome of "Betzes" barley (**c**). The brightest parts *(in green)* represent FISH signals of pSc200 repeats in (**a**) and HvT 01 repeats in (**b**) and (**c**), the less bright regions *(in pink)* show GISH signals of rye chromatin in (**a**) and of barley chromatin in (**b**) and (**c**), and the least bright regions (*in blue*) show wheat chromosomes.

2. When two alien chromosomal segments are present in a plant as shown in Fig. 5a, backcross the plant, and select plants with single alien chromosomal segments among the progeny.
3. The established lines with single alien chromosomal segments are called dissection lines of the alien chromosome.
4. In some aberrant alien chromosomes, the chromosome arm where the breakpoint is located can be identified from the FISH/GISH image (Fig. 5b). In some aberrant alien chromosomes, however, sequential C-(or N-) banding and FISH/GISH may be necessary to identify the chromosome arm where the breakpoint is located (Fig. 5c).

3.4.4. PCR Screening for Deletions of Alien Chromosomes

The terminal deletions of the alien chromosome can be detected by PCR using the most proximal marker among those mapped to an alien chromosome in the progeny of the 45-chromosome containing plants backcrossed to euploid "Chinese Spring". Although some deletions cannot be detected by PCR when the deleted segments are translocated to wheat chromosomes, PCR screening is the most efficient way to screen for the deletions of alien chromosomes added to common wheat on a large scale.

3.4.5. Deletion Mapping of DNA Markers Using Dissection Lines of Alien Chromosomes

1. Collect alien chromosome-specific markers such as RFLPs, SSRs and ESTs.
2. Check the polymorphism of the markers between the alien chromosome addition line and euploid "Chinese Spring".
3. Search for the polymorphic markers in an array of dissection lines.
4. Arrange the results into a matrix and construct a cytological map of the alien chromosome with the DNA markers.
5. The DNA markers can be used, instead of FISH/GISH, in screening for plants carrying aberrant alien chromosomes in the progeny of dissection lines (12, 13). The most proximal

DNA markers on both chromosome arms make universal markers to check for the presence of critical aberrant alien chromosomes in dissection lines.

4. Notes

1. Choice of stain solutions is crucial to successful chromosome banding. Therefore, different stain solutions from different manufacturers should be tried.
2. The proportion of the labeled probe to ISH solution should be kept minimal to attain the weakest FISH/GISH background signals, but FISH/GISH signals strong enough for observation and photography.
3. The detection mixture can be diluted much more than the manufacturer recommends.
4. Both Gc chromosomes induce deletions and translocations. However, the Gc action of $3C^{SAT}$ seems to be weaker than that of 2C and that $3C^{SAT}$ may induce more translocations than 2C does.
5. In Kyoto, Japan, 0.65 g $CaSO_4{\cdot}2H_2O$ is added to 1,000 mL tap water to make it hard water. The use of distilled water or soft tap water would deteriorate differential staining of chromosomes.
6. A special 100× objective that does not require a cover slip is available (OLYMPUS, MPLAPON 100XO), but ordinary 100× objectives can be used without a cover slip.
7. At the lower temperature (30°C) for hybridization, GISH without unlabeled blocking DNA generates as satisfactory signals as clear as those generated by other GISH protocols using 37°C for hybridization.

References

1. Bennett, M. D. and Leitch, I. J. (2005) Plant DNA C-values Database (release 4.0). http://data.kew.org/cvalues/.
2. Sears, E. R. (1954) The aneuploids of common wheat. *Missouri Agr. Expt. Sta. Res. Bull.* **572**, 1–59.
3. Sears, E. R. (1966) Nullisomic-tetrasomic combinations in hexaploid wheat. In *Chromosome Manipulations and Plant Genetics.* R. Riley and K. R. Lewis, Eds. Oliver and Boyd, Edinburgh, pp. 29–45.
4. Sears, E. R. and Sears L. M. S. (1978) The telocentric chromosomes of common wheat. *Proc. 5th Int. Wheat Genet. Sym.*, New Delhi, India, pp. 389–407.
5. Singh, R. J. (2003) Chromosomal aberrations – numerical chromosome changes. In *Plant Cytogenetics*, 2nd ed. CRC press, Boca Raton, FL, pp. 157–276.
6. Lukaszewski, A. J. (2000). Manipulation of the 1RS.1BL translocation in wheat by induced homoeologous recombination. *Crop Sci.* **40**, 216–225.
7. Endo, T. R. (1990) Gametocidal chromosomes and their induction of chromosome mutations in wheat. *Jpn. J. Genet.* **65**, 135–152.

8. Endo, T. R. (2007) The gametocidal chromosome as a tool for chromosome manipulation in wheat. *Chromosome Res.* **15**, 67–75.
9. Endo, T. R. and Gill, B. S. (1996) The deletion stocks of common wheat. *J. Hered.* **87**, 295–307.
10. Werner, J. E., Endo, T. R., and Gill, B. S. (1992) Toward a cytogenetically based physical map of the wheat genome. *Proc. Natl. Acad. Sci. USA* **89**, 11307–11311.
11. Qi, L. L., Eschalier, B., Chao, S., Lazo, G.R., Butler, G. E., Anderson, O. D., Akhunov, E. D., Dvorak, J., Linkiewicz, A. M., Ratnasiri, A., et al. (2004) A chromosome bin map of 16,000 expressed sequence tag loci and distribution of genes among the three genomes of polyploid wheat. *Genetics* **168**, 701–712.
12. Tsuchida, M., Fukushima, T., Nasuda, S., Masoudi-Nejad, A., Ishikawa, G., Nakamura, T., and Endo, T. R. (2008) Dissection of rye chromosome 1R in common wheat. *Genes Genet. Syst.* **83**, 43–53.
13. Sakai, K., Nasuda, S., Sato, K., and Endo, T. R. (2009) Dissection of barley chromosome 3H in common wheat and a comparison of 3H physical and genetic maps. *Genes Genet. Syst.* **84**, 25–34.
14. Vershinin, A. V., Schwarzacher, T., and Heslop-Harrison, J. S. (1995) The large-scale genomic organization of repetitive DNA families at the telomeres of rye chromosomes. *Plant Cell* 7, 1823–1833.
15. Belostotsky, D. A. and Ananiev, E. V. (1990) Characterization of relic DNA from barley genome. *Theor. Appl. Genet.* **80**, 374–380.
16. Mukai, Y., Friebe, B., and Gill, B. S. (1992) Comparison of C-banding patterns and *in situ* hybridization sites using highly repetitive and total genomic rye DNA probes of 'Imperial' rye chromosomes added to 'Chinese Spring' wheat. *Jpn. J. Genet.* **67**, 71–83.
17. Singh, R. J. and Tsuchiya, T. (1982) An improved Giemsa N-banding technique for the identification of barley chromosomes. *J. Hered.* **73**, 227–229.

Chapter 15

Development and Use of Oat–Maize Chromosome Additions and Radiation Hybrids

Ralf G. Kynast and Oscar Riera-Lizarazu

Abstract

Hybridization experiments of oat with maize require fastidious coordination of plant cultivation and flowering timing, meticulous crossing techniques, stimulation with plant growth substances, and in vitro rescue and culture of the hybrid embryos. The majority of hybrid offspring gradually lose all maize chromosomes consequently resulting in haploid oat plants. However, a minority of the offspring retain one or more maize chromosome(s) in addition to their haploid oat complements (partial hybrids). Oat haploids and partial hybrids with 1–3 maize chromosomes are partially fertile. Controlled self-fertilization of partial hybrids allows for the production of doubled haploid oat plants with an added single maize chromosome (monosomic addition) or an added pair of homologous maize chromosomes (disomic addition) among the inbred offspring. γ-Irradiation of monosomic oat–maize addition lines can be used to further dissect the maize chromosome in a given line. The lines with identified maize chromosome fragments (radiation hybrids) are the basis for establishing chromosome-specific panels. Although still in the experimental phase, the use of radiation hybrids has been useful and has widened the repertoire of maize genetics and genomics methodology.

Key words: Alien chromosome addition, *Avena sativa* L., γ-Irradiation, In situ DNA/DNA hybridization, In vitro embryo rescue culture, Interspecies crossing, Molecular marker, PCR assay, Physical chromosome mapping, *Zea mays* L

1. Introduction

Crossing plants of very remote relationship – such as common oat (*Avena sativa* L.; $2n=6x=42$; $1C\approx 12.961$ Gbp; fam. *Poaceae*, subfam. *Pooideae*) and cultivated maize (*Zea mays* L.; $2n=2x=20$; $1C\approx 2.671$ Gbp; fam. *Poaceae*, subfam. *Panicoideae*) – can combine unrelated genomes completely or partially into one hybrid genotype (1). Such interspecies hybrids can be used to expand the gene/allele pool beyond intraspecific variability (2) and open

James A. Birchler (ed.), *Plant Chromosome Engineering: Methods and Protocols*, Methods in Molecular Biology, vol. 701, DOI 10.1007/978-1-61737-957-4_15, © Springer Science+Business Media, LLC 2011

novel research opportunities on host–alien gene/allele interactions as well as genome dissection in an alien genetic background (3).

Crossing particular genotypes of oat by maize produces offspring of about 50–66% haploid oat plants ($2n_i - n_j = 3x = 21$) and 34–50% partial hybrids, haploid oat plants with one or more retained maize chromosome(s) ($2n_i - n_j = 3x + 1$ to $10 = 22$ to 31) owing to complete and partial loss, respectively, of the maize genome during embryogenesis (4, 5). Haploid oat and partial hybrids with 1–3 maize chromosomes can be partially fertile due to the formation of unreduced gametes. Partial hybrids with more than three maize chromosomes have low vigor and are sterile (1, 5). By employing the techniques described in this chapter, fertile and homozygous oat plants with an added pair of homologous maize chromosomes ($2n = 6x + 2 = 42 + 2 = 44$) can be identified and isolated among the inbred offspring after controlled self-fertilization of partial (oat × maize) hybrids (1). A series of these disomic oat–maize chromosome addition lines for each of the individual ten maize chromosomes has been established using various oat and maize genotype combinations (6).

Oat–maize addition (OMA) lines are unique and useful genetic stocks, for instance, where research requires a heterologous system to study, analyze, and manipulate a given maize chromosome separate from other chromosomes of its native genome. These materials can be used, e.g., to assign or physically localize molecular markers and genes to specific maize chromosomes (7, 8), to study chromosome pairing dynamics (9), and in the study of chromosome structure (10). Further dissection of the maize genome through the γ-irradiation of OMA lines can yield materials carrying maize subchromosomal segments (11). Like addition lines, subchromosomal stocks have been used in a variety of genomics and chromosome research applications (3, 12–14). A listing of addition lines made to date with sweet corn variety *Seneca60* and inbreds *B73* and *Mo17* as maize chromosome donors, phenotypes of OMA plants, and radiation hybrids produced is available at http://agronomy.cfans.umn.edu/Maize_Genomics.html.

2. Materials

2.1. Parent Plant Cultivation, (Oat × Maize) Crossing, In Vitro F_1-Embryo Rescue Culture, F_1-Hybrid Plant Cultivation, and I_1-Offspring Production

1. Plants: Maize inbred lines *B73* and *Mo17*, maize cultivar *Seneca60*; oat cultivars *Starter*, *Sun II*, and *GAF/Park*.
2. Soil/potting mix: Two parts of steam-sterilized soil and one part of Fison's LC1 mix (*Bellevue, WA*) supplemented with general commercial fertilizer, slow or quick release (e.g., one-half teaspoon soluble "20-20-20" per pot), or Hoagland's solution.

3. Plant growth substance cocktail: Dissolve 100 mg of 2,4-dichlorophenoxyacetic acid in ~2 ml of methanol and 50 mg of gibberellic acid in ~2 ml of methanol, then add both to 1 L ddH_2O and store at 4°C.
4. Diluted bleach: Mix 50 ml of any commercially available washing or household bleach with 50 ml sterile ddH_2O and use immediately.
5. In vitro culture medium (sterile "½MS" prepared under aseptic condition, autoclaved at 121°C for 15 min and cast into sterile Petri dishes and culture tubes): Dissolve 50 ml macronutrients, 5 ml Fe-EDTA, 29 ml of 102 mM $CaCl_2$, 1 ml micronutrients, 1 ml of 0.5 mM KI, 10 ml vitamins, 20 g sucrose, 2 g gelrite in ddH_2O up to 1 l.
6. Stock of macronutrients: 412 mM NH_4NO_3, 375 mM KNO_3, 30 mM $MgSO_4$, 25 mM KH_2PO_4 in ddH_2O.
7. Fe-EDTA solution: Dissolve 3.7 g $Na_2EDTA \cdot 2H_2O$ and 2.8 g $FeSO_4 \cdot 7H_2O$ in ddH_2O to 500 ml, boil for 5 min and stir while cooling down to RT for 1 h.
8. Stock of micronutrients: 100 mM $MnSO_4$, 100 mM H_3BO_3, 40 mM $ZnSO_4$, 1 mM Na_2MoO_4, 0.1 mM $CuSO_4$, 0.1 mM $CoCl_2$ in ddH_2O.
9. Stock of vitamins: 15 mg nicotinic acid, 30 mg thiamine-HCl, 15 mg pyridoxine-HCl, 60 mg glycine, 3 g myo-inositol in ddH_2O up to 300 ml.

2.2. F_1-Hybrid Plant Selection

1. Agarose: Standard melting and gelling temperature, multipurpose agarose.
2. ddH_2O: Deionized distilled water autoclaved and stored in aliquots at RT.
3. 70%$_{v/v}$ EtOH: Dilute 350 ml of absolute ethanol with 150 ml ddH_2O and store at RT.
4. 50× TAE (2 M Tris-acetate pH 8.0, 50 mM Na_2EDTA) buffer pH 8.0: Dissolve 242.24 g of Tris base and 18.6 g of Na_2EDTA in ~700 ml ddH_2O, titrate pH to 8.0 with glacial acetic acid (~57 ml), add ddH_2O up to 1 l, filter and store at RT.
5. 1× TAE (40 mM Tris-acetate pH 8.0, 1 mM Na_2EDTA) buffer pH 8.0: Dilute 100 ml of 50× TAE pH 8.0 with 4,900 ml ddH_2O and store at RT.
6. 6× SLB (Stop and Loading Buffer) pH 8.0: Dissolve 12 mg of bromophenol blue, 12 mg of xylene cyanol, 18 mg of orange G, and 2.23 g of Na_2EDTA in 1× TAE pH 8.0 up to 40 ml, add 60 ml of glycerol, mix and store at RT.
7. 10,000× SYBRSafe DNA Gel Stain (Invitrogen): Stain concentrate in DMSO stored in dark at RT.

8. 250 ng/µl DNA-ladder: Molecular weight marker (100 bp ladder) from 100 to 1,500 bp and stored in aliquots at −20°C.
9. REDExtract-N-Amp™ Plant PCR Kit with extraction tubes, extraction solution, dilution buffer, and red 2× PCR reaction mix (Sigma) stored in aliquots at −20°C.
10. 1 µM F/R primers mix (500 nM Forward primer + 500 nM Reverse primer): Dilute 20 µl of 100 µM forward primer and 20 µl of 100 µM reverse primer with 3,960 µl ddH_2O and store in aliquots at −20°C.

Marker	Primer	Sequence 5′ >>> 3′	*n*	GC%	T_m	ΔG
p-Grande1	Forward	AAAGA CCTCA CGAAA GGCCC AAGG	24	54	80.5	-49.2
	Reverse	AAATG GTTCA TGCCG ATTGC ACG	23	47	75.0	-47.0

2.3. Maize Chromosome Identification in Shoots and Roots of I_1 Plants

2.3.1. Germination and Cultivation of I_1 Offspring

1. Same soil/potting mix as described in Subheading 2.1.

2.3.2. DNA Extraction from I_1-Plant Leaf Tissues and PCR Assay Using Maize Chromosome Arm-Specific Microsatellite Markers

1. Same materials as described in Subheading 2.2.
2. Maize chromosome arm-specific microsatellite markers selected from the Maize Genetics and Genomics Database: http://www.maizegdb.org/.

2.3.3. DNA Extraction from Maize Leaf Tissues, Chemical Labeling, Somatic Chromosome Preparation, and Genomic In Situ Hybridization

1. CTAB (1% CTAB, 100 mM Tris–Cl pH 7.5, 10 mM Na_2EDTA pH 8.0, 700 mM NaCl, 140 mM β-mercaptoethanol) solution, e.g., 350 ml for 16 samples of ~20 ml for ~5 g leaf tissue: Dissolve 3.5 g of CetylTrimethylAmmonium Bromid (Sigma), 35 ml of 1 M Tris–Cl pH 7.5, 7 ml of 0.5 M Na_2EDTA pH 8.0, 49 ml of 5 M NaCl, and 3.5 ml of β-mercaptoethanol in ddH_2O up to 350 ml, warm up to 65°C and use immediately.
2. Chloroform-*n*-octanol mix: Mix 240 ml of chloroform with 10 ml of 1-octanol and store at RT.
3. $1\%_{w/v}$ RNase A in 10 mM Tris–Cl pH 7.5, 15 mM NaCl (DNase free): Dissolve 100 mg of RNase A, 100 µl of 1 M

Tris–Cl pH 7.5, and 30 µl of 5 M NaCl in ddH_2O to 10 ml, incubate in boiling water bath for 15 min, slowly cool down and store in 200-µl-aliquots (for DNA extraction) and 30-µl-aliquots (for in situ hybridization) at –20°C.

4. TE (10 mM Tris–Cl pH 8.0, 1 mM Na_2EDTA) buffer pH 8.0: Dilute 2 ml of 1 M Tris–Cl pH 8.0 and 400 µl of 0.5 M Na_2EDTA pH 8.0 with ddH_2O to 200 ml and store at 4°C.
5. 3 M Na-acetate pH 5.2: Dissolve 40.81 g of Na-acetate · $3H_2O$ in 60 ml ddH_2O, titrate pH to 5.2 with glacial acetic acid, add ddH_2O to 100 ml and store at 4°C.
6. $70\%_{v/v}$ EtOH: Dilute 350 ml of absolute ethanol with 150 ml ddH_2O and store at RT.
7. $80\%_{v/v}$ EtOH: Dilute 400 ml of absolute ethanol with 100 ml ddH_2O and store at RT.
8. $90\%_{v/v}$ EtOH: Dilute 450 ml of absolute ethanol with 50 ml ddH_2O and store at RT.
9. ULYSIS® Alexa Fluor® 488 Nucleic Acid Labeling Kit (Invitrogen), Alexa Fluor® 488: $Excitation_{max}$ = 490 nm and $Emission_{max}$ = 519 nm.
10. Farmer's Fixative Fluid: Mix 3× volume of absolute alcohol and 1× volume of glacial acetic acid and use immediately.
11. $45\%_{v/v}$ Acetic acid: Dilute 45 ml of glacial acetic acid in 55 ml ddH_2O and store in 20-ml aliquots at RT.
12. Diluted RNase A (for 15 slides): Thaw a 30-µl aliquot of 1% RNase A stock and dilute (1/100): Mix 30 µl of 1% RNase A stock, 300 µl of 20× SSC and 2,670 µl ddH_2O, and use immediately.
13. 500 mM PB (*p*hosphate *b*uffer, 350 mM Na_2HPO_4, 150 mM NaH_2PO_4) pH 7.4: Mix 500 mM Na_2HPO_4 and 500 mM NaH_2PO_4 together to pH 7.4 and store at RT.
14. 10× PBS (*p*hosphate-*b*uffered *s*aline, 70 mM Na_2HPO_4, 30 mM NaH_2PO_4, 1.3 M NaCl) pH 7.4: Mix 130 ml of 5 M NaCl, 100 ml of 500 mM PB pH 7.4 and 270 ml ddH_2O, filter and store at RT.
15. $4\%_{w/v}$ PFA: Warm 200 ml ddH_2O to 70–80°C, add 10 g of paraformaldehyde, mix, add 250 µl of 4 M NaOH, mix, add 25 ml of 10× PBS, mix, and cool to RT, add ddH_2O to 250 ml, mix, and use immediately.
16. 20× SSC (*s*aline *s*odium *c*itrate, 3 M NaCl, 0.3 M Na_3-citrate) pH 7.0: Dissolve 88.23 g of Na_3-citrate · $2H_2O$ and 175.32 g of NaCl in ~850 ml ddH_2O, titrate pH to 7.0 with NaOH/HCl, add ddH_2O to 1 l, filter and store at RT.
17. FA (formamide) for posthybridization wash: Use formamide of highest grade and store in 200-ml aliquots at RT.

18. FA (formamide) for hybridization solution: Deionize formamide of highest grade by stirring 5 g of in vacuo dried mixed bed exchange resin AG 501-X8(D), molecular biology grade (BIO-RAD) in 100 ml of formamide for ~3 h or until pH is neutral, check pH with pH paper (e.g., UNITEST), filter through Whatman No. 1 and store in 5-ml aliquots at –20°C.
19. 10 μg/μl Carrier DNA: Use sonicated (~150 bp), denatured DNA from salmon testes and store at –20°C.
20. 100 μg/ml PI (red counterstain; $Excitation_{max}$: 340 and 530 nm, $Emission_{max}$: 615 nm): Dissolve 1 mg of propidium iodide in 10 ml ddH_2O and store in 1-ml aliquots in amber tubes at –20°C.
21. Antifade Vectashield (Vector Laboratories).

2.4. Back-Crossing of Disomic Oat–Maize Chromosome Addition Plants to Oats

1. Soil/potting mix as described in Subheading 2.1.

2.5. Irradiation of Monosomic Oat–Maize Chromosome Addition Seeds by Gamma Rays, Cultivation and Self-Pollination of the M_1 Plants, and M_2-Offspring Production

1. ddH_2O/glycerol mix: Mix 60 ml ddH_2O and 40 ml glycerol thoroughly and use immediately.
2. Soil/potting mix as described in Subheading 2.1.
3. γ-Rays source: Use a ^{137}Cs or a ^{60}Co source and calibrate the irradiation with oat seeds.
4. Materials as described in Subheading 2.2.

2.6. Selection of Radiation Hybrids Among the M_2 Plants and Marker Allocation to Chromosome Segments

1. Same materials as described in Subheadings 2.1 and 2.2.

2.7. Panel Development for Maize Genetics/Genomics

1. Same materials as described in Subheadings 2.1 and 2.2.

3. Methods

3.1. Parent Plant Cultivation, (Oat × Maize) Crossing, In Vitro F_1-Embryo Rescue Culture, F_1-Hybrid Plant Cultivation and I_1-Offspring Production (Fig. 1)

3.1.1. Maize Plant Cultivation

1. Germinate and cultivate maize plants in soil/potting mix in 14-in. pots with ample supply of water and nutrients in a growth chamber with high (80%) relative air humidity for good vegetative growth in long days (16 | 8) for about 4 weeks (see Notes 1 and 5–7).
2. Shift plants to induce flowering into short days (10 | 14) with slightly lower (60%) relative air humidity and slightly increased temperature setting for the day phase to accelerate generative development (see Notes 1 and 4–7).

3.1.2 Oat Plant Cultivation

1. Germinate and cultivate oat plants in soil/potting mix in 7-in. pots with ample supply of water and nutrients in a growth chamber with high (80%) relative air humidity for good vegetative growth in short days (10 | 14) for about 6 weeks (see Notes 2, 5, 6, and 8).
2. Shift plants to induce flowering into long days (16 | 8) with slightly lower (60%) relative air humidity and shifted time setting (see Notes 2–6 and 8).

3.1.3. Oat × Maize Crossing

1. One day before anther dehiscence, remove secondary florets from the spikelets of the upper two-thirds of the oat panicle and emasculate primary florets of the same spikelets by pulling out all anthers by the use of a very fine forceps without touching the feathery stigmas and without wounding lemma, palea, and glumes.
2. Cut off all awns from emasculated spikelets.
3. Cut off all spikelets that are not emasculated (lower one-third of the panicle).
4. Isolate the panicle in a glassine bag and support it by fixing the bagged panicle to a sturdy stick; record date and time of emasculation.
5. One day after emasculation, collect freshly shed maize pollen onto a clean paper sheet by tapping the tassel branch with exposed anthers ready to dehisce (8:00 to 10:30 a.m.; see Notes 4 and 5).
6. Pour the pollen grains from the paper sheet into a glass Petri dish, and sprinkle pollen grains onto the feathery stigmas of the emasculated oat florets by using a fine soft camelhair brush.

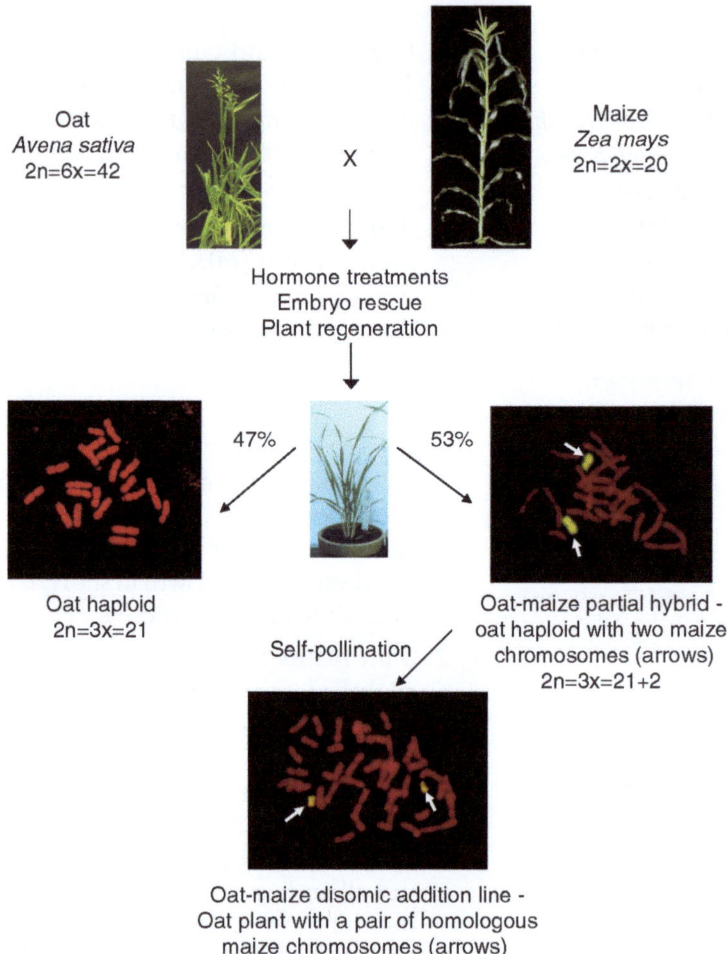

Fig. 1. Hybridization of oat and maize and production of oat–maize addition lines. Oat (female) is crossed by maize (male). One or 2 days after pollination, crossed florets are treated with hormones (auxin and gibberellic acid) to promote seed and embryo development. Fifteen to 18 days after pollination, seeds are collected, disinfected, and dissected under sterile conditions. Hybrid embryos formed are transferred to synthetic media for additional development and germination. In vitro-regenerated seedlings are transferred to pots and allowed to develop. Root tips from regenerated plants are collected for chromosome counting and GISH analysis. Simultaneously, leaf tissue is collected for DNA extraction and marker assays. About half of the plants will be oat haploids ($2n = 21$) and the rest will be oat haploids with one or more maize chromosomes (partial hybrids). Subsequently, partial hybrids are allowed to self-pollinate. Plants originating from seeds of partial hybrids are also analyzed cytologically and with DNA-based markers to determine their chromosome constitution. Oat–maize disomic addition lines are oat plants with a pair of homologous maize chromosomes ($2n = 6x + 2 = 42 + 2$).

7. Put the glassine bag back over the pollinated panicle and support it by fixing the bagged panicle to a sturdy stick; record genotype ID of pollen donor, date and time of pollination.
8. Two days after pollination, spray the upper half of the pollinated plant with the plant growth substance cocktail in order to support F_1-embryo development and to delay F_1-endosperm collapse; record date and time of treatment.

3.1.4. In Vitro Embryo Rescue Culture

1. Fifteen days after pollination, harvest thin green karyopses of pollinated florets at late afternoon and store them in sterile ddH_2O at 4°C overnight whereas discarding any plump starch-filled karyopses that may have resulted from oat self-fertilization.
2. Disinfect the karyopses by gentle shaking in diluted bleach at RT for 10 min.
3. Wash off bleach by 5× rinsing the karyopses in sterile ddH_2O at RT for 5 min, each.
4. Dismantle the immature embryo (proembryo) from the endosperm and put it with the scutellum onto the surface of culture medium in a Petri dish; label and record (oat × maize) ID, date and time of preparation.
5. In vitro dark culture: Cultivate the proembryos in permanent dark (24 h) consistently at 20 ± 1°C for ~14 days or till plantlets of ~2 cm shoot size with healthy roots of 1–2 cm length are formed.
6. Transfer plantlets into new culture medium in large culture tubes.
7. In vitro light culture: Cultivate the plantlets in short days (10 | 14) consistently at 20 ± 1°C for ~14 days or till plant height has passed ~6 cm (see Notes 5, 6, and 9).

3.1.5. F_1-Hybrid Plant Cultivation and I_1-Offspring Production

1. Transplant the F_1 plants into soil/potting mix in 7-in. pots and cultivate with ample supply of water and nutrients in a growth chamber with high (80%) relative air humidity for good vegetative growth in short days (10 | 14) for about 6 weeks (see Notes 5, 6, and 9).
2. Shift plants for inducing flowering in long days (16 | 8) with slightly lower (60%) relative air humidity (see Notes 5, 6, and 9).
3. Isolate four panicles (*a-, b-, c-, and d-panicle*) from the physiologically first four maize-positive tillers of the F_1 plant individually in glassine bags shortly before anthesis in order to ensure controlled self-pollination/self-fertilization for I_1-offspring production (see Note 10).
4. Support bagged panicles by fixing them to a sturdy stick; record date and time of isolation.

5. Continue cultivation and let plants completely mature before crop (see Notes 5, 6, and 9).
6. Harvest I_1 seeds from different tillers of the same F_1 plant separately and treat as distinctive I_1-offspring populations, and record data (see Note 11).

3.2. F_1-Hybrid Plant Selection (Fig. 1)

F_1 plants are selected on the basis of whether the shoot tissue cells carry in their complements one or more maize chromosome(s) in addition to the haploid oat genome (i.e., being maize-positive) or do not (i.e., being maize-negative) regardless of presence or absence of maize chromatin in their root tissues cells (see Note 10). Molecular markers for repetitive DNA segments that are dispersed in multiple clusters along all maize chromosome arms but absent from the total DNA of all oat genotypes used – such as, e.g., *p-Grande1* for the LTR-type retrotransposon Grande – are employed to verify presence vs. absence of maize chromatin.

The population of F_1 plants is tested in two steps (see Note 10):

At first, DNA is extracted from the completely unfolded blades of the juvenile plants' first leaves and assayed with the marker to discriminate maize-positive and maize-negative F_1 plants. Secondly, the maize-positive plants are continued to grow till panicles emerge from the physiologically first four tillers (*a-, b-, c-, and d-tiller*) and DNA is extracted from their flag leaves' blades and assayed likewise to discriminate maize-positive and maize-negative tillers.

DNA extractions and PCR assays using maize-specific repetitive DNA markers are accomplished by using the REDExtract-N-Amp™ Plant PCR Kit as follows:

1. Rinse paper punch in 70% EtOH, let dry and punch a 0.5 cm disc of leaf tissue into an extraction tube (kit component 1).
2. Add 100 µl of extraction solution (kit component 2) per tube, vortex and spin down droplets.
3. Incubate at 95°C for 10 min, let cool to RT for ~2 min and spin down droplets.
4. Add 100 µl of dilution buffer (kit component 3) per tube, vortex and spin down droplets (see Note 12).
5. Make PCR mixes (15 µl for one sample):

   7.5 µl of red 2× PCR reaction mix (kit component 4)

   4.5 µl of 1 µM F/R primers mix

   3.0 µl of leaf tissue disc extract mix
6. Incubate PCR mixes with the following thermal cycle program:

Lid	100°C
1×	94°C, 30 s
30×	92°C, 30 s; 62°C, 30 s; 72°C, 30 s
1×	72°C, 3 min 30 s
End	14°C

7. Add 3 μl of 6× SLB per mix and spin down droplets.
8. Prepare a 1.5% agarose gel in 1× TAE, add SYBRSafe, mix and cast it:

Mini gel (cm × cm)	Combs (8 teeth)	1× TAE (ml)	Agarose (g)	SYBRSafe (μl)
10.2 × 6.4	1	30	0.45	3

Maxi gel (cm × cm)	Combs (25 teeth)	1× TAE (ml)	Agarose (g)	SYBRSafe (μl)
11.1 × 23.0	1	100	1.50	10
21.0 × 23.0	2	190	2.85	19
30.9 × 23.0	3	280	4.20	28
40.8 × 23.0	4	370	5.55	37

9. Load stopped mixes (18 μl) into the gel wells and run electrophoresis at 80 V for ~30 min.
10. Check gel on a 470 nm blue light transilluminator for the DNA bands (PCR products).
11. Scan band pattern, select maize-positive vs. maize-negative plants/tillers, and record data.

3.3. Maize Chromosome Identification in Shoots and Roots of I_1 Plants and Oat–Maize Addition Production

3.3.1. Germination and Cultivation of I_1 Offspring, and Production of I_2 Offspring

1. Germinate I_1 seeds in permanent dark (24 h) on wet filter paper at 22 ± 2°C for 3–4 days or until seedlings have ~2 cm long shoots and roots, and harvest root tips for Genomic In Situ Hybridization (GISH)-assay (karyotype analysis).
2. Transplant the I_1 seedlings into soil/potting mix in 7-in. pots and cultivate them with ample supply of water and nutrients in a growth chamber with high (80%) relative air humidity for good vegetative growth in short days (10 | 14) for about 6 weeks (see Notes 5, 6, and 9).
3. Shift plants to induce flowering into long days (16 | 8) with slightly lower (60%) relative air humidity (see Notes 5, 6, and 9).
4. Isolate four panicles (*a-, b-, c-, and d-panicle*) developing from the physiologically first four maize chromatin-positive tillers with identified added maize chromosome(s) of the I_1 plant individually in glassine bags shortly before anthesis in order to ensure controlled self-pollination/self-fertilization for production of I_2 offspring with stable maize chromosome transmission (see Note 13).
5. Support bagged panicles by fixing them to a sturdy stick; record date and time of isolation.
6. Continue cultivation and let plants completely mature before crop (see Notes 5, 6, and 9).

7. Harvest I_2 seeds from different tillers of the same I_1 plant separately and treat as distinctive I_2-offspring populations (see Note 13).
8. Identify and select I_2 genotypes/karyotypes like for the I_1 plants, estimate the frequencies of transmission from I_1 to I_2 generation for the individual maize chromosomes, establish the disomic ($2n = 6x + 2 = 42 + 2 = 44$) and monosomic ($2n = 6x + 1 = 42 + 1 = 43$) oat–maize chromosome addition lines (OMAs), and record data.

3.3.2. DNA Extraction from I_1-Plant Leaf Tissues and PCR-Assay Using Maize Chromosome Arm-Specific Microsatellite Markers

The maize chromosomes sexually transmitted from the F_1 plant to the I_1 offspring are identified by a set of a minimum of 40 selected microsatellite markers representing at least two specific ones for each of the 20 maize chromosome arms. This is done to help determine along with later chromosome in situ hybridization if maize chromosomes are retained intact. Each individual I_1-plant tiller must be tested separately. Test the first four tillers by assaying DNA from their flag leaves and document the maize chromosome constitution of each plant tiller (see Notes 12 and 14).

3.3.3. DNA Extraction from Maize Leaf Tissues, Chemical Labeling, Somatic Chromosome Preparation, and Genomic In Situ Hybridization

3.3.3.1. Maize Plant Cultivation

1. Germinate and cultivate maize plants in soil/potting mix in 14-in. pots with ample supply of water and nutrients in a growth chamber with high (80%) relative air humidity for good vegetative growth in long days (16 | 8) for about 4 weeks (see Notes 1 and 5–7).
2. Put the plants into permanent dark (24 h) with the same temperature regime as in step 1 for about 4 days in order to etiolate the leaves.

3.3.3.2. Start Maize DNA Extraction, First Day

1. Freeze ~5 g of cut etiolated leaves (one sample) in liquid N_2 in a mortar.
2. Grind the frozen tissue in liquid N_2 with pestle to a very fine powder.
3. Transfer the powder into a pre-cooled 50-ml Falcon tube.
4. Add 20 ml of warm (65°C) CTAB solution and gently rock the sample at 65°C for 1 h.
5. Let sample cool down to RT.
6. Add 10 ml of chloroform-*n*-octanol mix and gently rock the sample at RT for 5 min.
7. Centrifuge the sample at RT for 20 min.
8. Transfer the top layer (~20 ml) into a new 50-ml Falcon tube (see Note 15).
9. Add 20 ml of chloroform-*n*-octanol mix and gently rock the sample at RT for 5 min.
10. Centrifuge the sample at RT for 20 min.

11. Transfer the top layer (~20 ml) into a new 50-ml Falcon tube (see Note 15).
12. Add 200 μl of 1% RNase A stock, mix and incubate the sample at 37°C for 30 min.
13. Add 20 ml of iso-propanol and *very gently mix* the sample at RT for ~5 min or until fine long DNA fibers are formed.
14. Spool the DNA fibers onto a sterile glass hook.
15. Repeatedly immerse the hook with DNA fibers in ~5 ml of 70% EtOH in a 15-ml Falcon tube until DNA fibers form a white fluffy cloud.
16. Spool the washed DNA fibers onto the same glass hook and air-dry.
17. Dissolve the DNA in 6 ml of TE buffer pH 8.0 at 4°C overnight in a 15-ml Falcon tube.

3.3.3.3. Continue Maize DNA Extraction, Second Day

1. Add 6 ml of chloroform-*n*-octanol mix and gently rock the sample at RT for 5 min.
2. Centrifuge the sample at RT for 20 min.
3. Transfer the top layer (~6 ml) into a new 15-ml Falcon tube (see Note 15).
4. Add 6 ml of chloroform-*n*-octanol mix and gently rock the sample at RT for 5 min.
5. Centrifuge the sample at RT for 20 min.
6. Transfer the top layer (~6 ml) into a new 15-ml Falcon tube (see Note 15).
7. Add 600 μl of 3 M Na-acetate pH 5.2 and mix.
8. Add 6 ml of iso-propanol and *very gently mix* the sample at RT for ~5 min or until fine long DNA fibers are formed.
9. Spool the DNA fibers onto a new sterile glass hook.
10. Repeatedly immerse the hook with DNA fibers in ~5 ml of 70% EtOH in a 15-ml Falcon tube until DNA fibers form a white fluffy cloud.
11. Spool the washed DNA fibers onto the same glass hook and air-dry.
12. Dissolve the DNA in 600 μl of TE buffer pH 8.0 at 4°C for ≥3 days in a 2-ml screw-cap tube (see Note 16).
13. Dilute DNA with TE buffer pH 8.0 to a final concentration of 1 μg/μl.

3.3.3.4. DNA Labeling Using the ULYSIS® Alexa Fluor® 488 Nucleic Acid Labeling Kit, First Day

1. Shear the extracted maize DNA (1 μg/μl) to a middle length of 400–500 bp by autoclaving in a small-size (3 L) pressure cooker at 121°C, 15 lb. for 4–12 min (Nominal pressure of 80 kPa = 11.6 psi).

2. Let cool down the autoclaved DNA to RT.
3. Dilute the DNA with ddH_2O to a final concentration of 100 ng/μl.
4. Add 10 μl of 100 ng/μl sheared DNA into an *amber* 0.5-ml tube on ice (see Note 17).
5. Add 1 μl of 3 M Na-acetate pH 5.2 and mix.
6. Add 27 μl of cold 100% EtOH, mix and incubate the sample at –20°C overnight.

3.3.3.5. DNA Labeling Using the ULYSIS® Alexa Fluor® 488 Nucleic Acid Labeling Kit, Second Day

1. Centrifuge the sample (38 μl) with 16,000×*g* at 4°C for 15 min.
2. Decant the supernatant and carefully wash the pellet in ~50 μl of cold 70% EtOH.
3. Centrifuge the sample with 16,000×*g* at 4°C for 3 min in order to refasten the pellet to the tube wall.
4. Decant the supernatant and air-dry the pellet in vacuo (e.g., Savant DNA 120 SpeedVac Concentrator) at RT for ~15 min.
5. Completely dissolve the pellet in 24 μl of 5 mM Tris–Cl pH 8.0, 1 mM Na_2EDTA (kit component C).
6. Denature sample at 95°C for 5 min, chill on ice for 5 min and spin down condensed water.
7. Add 1 μl of A/B mix to the denatured DNA sample, mix and incubate at 80°C for 15 min (see Note 18).
8. Stop incubation by setting the tube on ice.
9. Add 2.5 μl of 3 M Na-acetate pH 5.2 and mix.
10. Add 65 μl of cold 100% EtOH, mix and incubate at –80°C for 2 h.
11. Centrifuge the sample (92.5 μl) with 16,000×*g* at 4°C for 15 min.
12. Decant the supernatant and carefully wash the pellet in ~100 μl of cold 70% EtOH.
13. Centrifuge the sample with 16,000×*g* at 4°C for 3 min in order to refasten the pellet to the tube wall.
14. Decant the supernatant and carefully wash the pellet in ~100 μl of cold 80% EtOH.
15. Centrifuge the sample with 16,000×*g* at 4°C for 3 min in order to refasten the pellet to the tube wall.
16. Decant the supernatant and air-dry the pellet in vacuo (e.g., Savant DNA 120 SpeedVac Concentrator) at RT for ~15 min.
17. Completely dissolve the pellet in 20 μl of TE buffer pH 8.0 and store the F-probe (~50 ng/μl) at –20°C in the *amber* tube (see Notes 17 and 19).

3.3.3.6. Somatic Chromosome Preparation and Genomic In Situ Hybridization

1. Cut one or maximal two roots (about 1.5 cm long) per germinated I_1 seedling and incubate in 2 ml ddH_2O in 5-ml tubes in ample wet ice for 24 h (ice water treatment) (see Note 20).
2. Fix roots in Farmer's fixative fluid at RT for ~24 h (up to use store roots at about –10°C).
3. Wash roots in ddH_2O at RT for ~1 h in order to remove fixative from tissue.
4. Macerate roots in 45% acetic acid at RT for 3–5 min (see Note 21).
5. Gently strike out meristem cells from the root tip onto a slide.
6. Squash meristem cells in a drop of 45% acetic acid under a glass cover slip (18 mm × 18 mm).
7. Examine mitotic index and chromosome spreading with a phase-contrast microscope.
8. Freeze the slide upside-down on dry ice or in liquid N_2, flick off the cover slip and incubate the slide in 100% EtOH for ~10 min.
9. Air-dry the slide in an oven at 37°C for about ~20 min.
10. Add 200 µl of diluted RNase A onto a slide and cover with a plastic slip (20 mm × 20 mm).
11. Incubate slides in a wet chamber at 37°C for 30–45 min.
12. 3× gently shake slides in 2× SSC at RT for 5 min, each.
13. Gently shake slides in 4% PFA at RT for 10 min.
14. 3× gently shake slides in 2× SSC at RT for 5 min, each.
15. Dehydrate slides in 70%, then 90%, then 100% EtOH at RT for 3 min, each.
16. Air-dry slides in an oven at 37°C for about ~20 min.
17. Prepare 30 µl of hybridization solution (HS) per slide in an *amber* tube as follows (see Notes 17 and 22):

Volume (µl/slide)	Component	DNA mass (ng/slide)	Final concentration
9.10	ddH_2O	–	–
12.00	Formamide	–	40%
2.25	20× SSC pH 7.0	–	1.5× SSC = 0.2925 M Na^+
0.15	10 µg/µl Carrier (Salmon sperm DNA)	~1,500	~50.0 ng/µl
0.50	50 ng/µl F-probe (Maize DNA)	~25	~834 pg/µl
6.00	50% Dextran sulfate	–	10%

18. Denature HS at 95°C for 6 min, chill on ice for 6 min.
19. Add 30 µl of HS onto one slide, cover with plastic slip (20 mm × 20 mm).
20. Incubate slides in a wet chamber at 80°C for 6 min, cool down with 0.05°C/s to 37°C.
21. Incubate slides in a wet chamber at 37°C for 6–12 h.
22. 2× gently shake slides in 2× SSC at 42°C for 10 min, each.
23. 2× gently shake slides in 40% FA/1.5× SSC at 42°C for 10 min, each (see Note 22).
24. 2× gently shake slides in 2× SSC at 42°C for 10 min, each.
25. Gently shake slides in 2× SSC at RT for 10 min.
26. Add 2 ml of diluted PI onto one slide and incubate without cover slip at RT for 16 min.
27. Wash off excess of PI from slides with ~3 ml of 1× PBS from a pipette.
28. Mount slides with antifade (~20 µl/slide) and cover with thin glass slips (24 mm × 30 mm).
29. Analyze slides by using an epifluorescence microscope (see Note 23).

3.4. Back-Crossing of Disomic Oat–Maize Chromosome Addition Plants to Oat

3.4.1. Plant Cultivation

1. Germinate and cultivate the oat plants (to be emasculated mother plants) and identified disomic maize chromosome addition plants (to be pollen donors) in soil/potting mix in 7-in. pots with ample supply of water and nutrients in a growth chamber with high (80%) relative air humidity for good vegetative growth in short days (10 | 14) for about 6 weeks (see Notes 2, 5, 6, and 9).
2. Shift plants to induce flowering into long days (16 | 8) with slightly lower (60%) relative air humidity (see Notes 2, 5, 6, and 9).

3.4.2. (Oat × Disomic OMA) Crossing and BC_1-Offspring Production

1. One day before anther dehiscence, remove secondary florets from the spikelets of the upper two-thirds of the oat panicle and emasculate primary florets of the same spikelets by pulling out all anthers by the use of a very fine forceps without touching the feathery stigmas and without wounding the lemma, palea, and glumes.
2. Cut off all awns from emasculated spikelets.
3. Cut off all spikelets that are not emasculated (lower one-third of the panicle).
4. Isolate the panicle in a glassine bag and support it by fixing the bagged panicle to a sturdy stick; record date and time of emasculation.

5. One day after emasculation, collect freshly shed pollen from the disomic OMA plant into a glass Petri dish by tapping the exposed anthers ready to dehisce (11:30 a.m. to 12:30 p.m.; see Notes 4 and 5).
6. Sprinkle pollen grains onto the feathery stigmata of emasculated oat florets by using a fine and soft camelhair brush.
7. Put back the glassine bag over the pollinated panicle and support it by fixing the bagged panicle to a sturdy stick; record genotype ID of pollen donor, date and time of pollination.
8. Continue cultivation and let plants completely mature before crop (see Notes 5, 6, and 9).
9. Harvest BC_1-seeds and test a representative sample (about 20 seeds) from the population for monosomic condition of the added maize chromosome and the euploid condition for the oat complement ($2n=6x+1=42+1=43$) by GISH-assay, and record data.

3.5. Irradiation of Monosomic Oat–Maize Chromosome Addition Seeds by γ-Rays, Cultivation and Self-Pollination of the M_1 Plants, and M_2-Offspring Production (Fig. 2)

1. Store ~500 BC_1 seeds to be γ-irradiated – spread out in a single layer in a large Petri dish – together with a mixture of (60% ddH_2O + 40% glycerol) in a desiccator for 5 days in order to equilibrate seed moisture.
2. Irradiate the BC_1 seeds with γ-rays from a ^{137}Cs source in order to induce as much chromosome breakage as possible, while maintaining seed germination ability and plant viability; γ-rays from a different source (e.g., ^{60}Co) require recalibration (see Note 24).
3. Germinate the irradiated BC_1 seeds (=M_1 seeds) and cultivate the M_1 plants in soil/potting mix in 7-in. pots with ample supply of water and nutrients in a growth chamber with high (80%) relative air humidity for good vegetative growth in short days (10 | 14) for about 6 weeks (see Notes 5, 6, and 9).
4. Shift plants to induce flowering into long days (16 | 8) with slightly lower (60%) relative air humidity (see Notes 5, 6, and 9).
5. Test DNA of M_1 plants for presence of maize chromatin (chromosome segments) by the use of the PCR assay with a maize-specific dispersed marker (e.g., *p-Grande1*) in two steps as described in Subheading 3.2, and record data.
6. Isolate the four panicles (*a-, b-, c-, and d-panicle*) from the physiologically first four maize-positive tillers of the M_1 plants individually in glassine bags shortly before anthesis in order to ensure controlled self-pollination/self-fertilization for production of M_2 offspring with stable maize chromosome segment transmission (see Note 25).
7. Support bagged panicles by fixing them to a sturdy stick; record date and time of isolation.

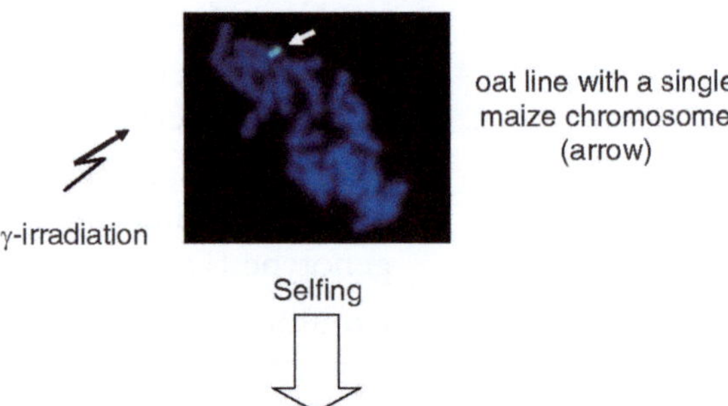

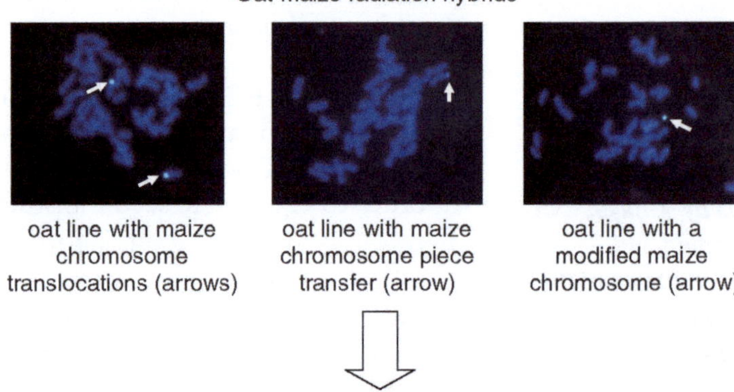

Fig. 2. Production of oat–maize radiation hybrids. Seed of a monosomic maize chromosome 9 addition line of oat was treated with gamma-rays (30–50 krad), subsequently planted, and surviving plants were self-pollinated. Progeny from self-pollination (the radiation hybrids) possessing maize chromosome 9 chromatin including plants with an apparently normal maize chromosome 9 as well as plants with various maize chromosome 9 rearrangements (intergenomic translocations, deletions, and a combination of both) are selected for analysis. This figure is based on Riera-Lizarazu et al. (15) and is reproduced with permission from the publisher (copyright © 2008 S. Karger AG, Basel).

8. Continue cultivation and let plants completely mature before crop (see Notes 5, 6, and 9).
9. Harvest M_2 seeds from different tillers of the same M_1 plant separately and treat as distinctive M_2-offspring populations, and record data (see Note 25).

3.6. Selection of Radiation Hybrids Among the M_2 Plants and Marker Allocation to Chromosome Segments (Fig. 3)

1. Plant a collection of independently derived M_2 seeds, and grow M_2 plants following the plant rearing conditions in Subheading 3.5.
2. Collect leaf tissue, isolate DNA, and test DNA for the presence of maize chromatin by the use of a PCR-based assay with a

Development of a radiation hybrid panel

Cytological (GISH) and molecular marker characterization

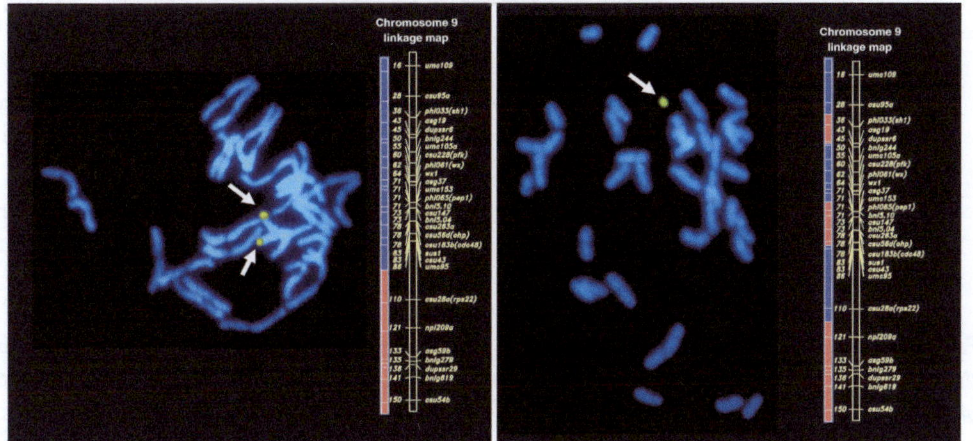

Line with a translocation involving the long arm of maize chromosome 9 (arrows)

Line with multiple interstitial deletions involving maize chromosome 9 (arrow)

Radiation hybrid panel assembly

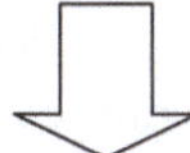

DNA-based markers

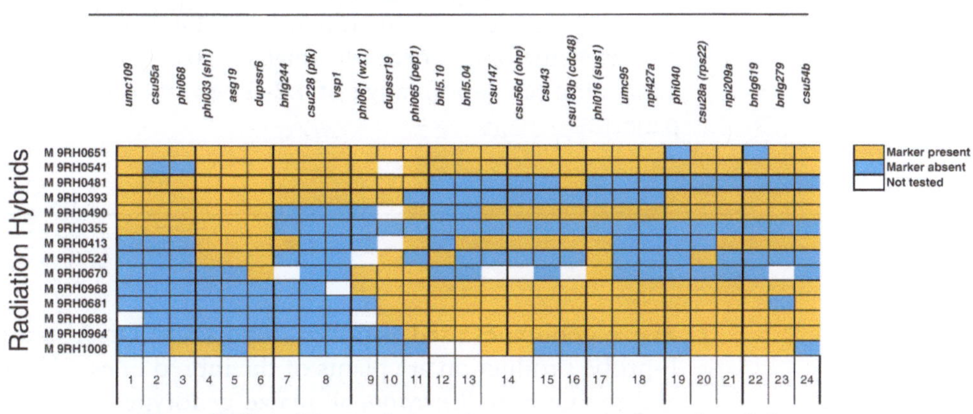

Bins defined by unique chromosome breakpoints

Fig. 3. Development of a radiation hybrid panel. Radiation hybrids are cytologically (GISH) characterized and analyzed with molecular markers to determine their maize chromosome constitution. These analyses provide a picture of the level and pattern of chromosome breakage that was induced in each line. Subsequently, radiation hybrids with defined constitutions are then assembled to form a panel. The collective pattern of chromosome breakpoints in a panel of radiation hybrids can be used to define chromosome segment bins for physical mapping and other applications.

maize-specific marker (maize-specific dispersed retrotransposon LTR) as described in Subheading 3.2.

3. Lines that tested positive for the maize-specific marker are tested with ~20-well-distributed markers (SSRs or ESTs) that span the pertinent maize chromosome.
4. Concurrently, lines that tested positive for the maize-specific marker should be analyzed cytologically (GISH) as described in Subheading 3.3.
5. These assays will reveal marker content and the pattern of maize chromosome breakage in each M_2 plant (or radiation hybrid).

3.7. Panel Development for Maize Genetics/ Genomics (Fig. 3)

A radiation hybrid panel is composed of individuals that have lost at least one marker due to radiation-induced chromosome breakage. Panels of varying levels of chromosome breakage or mapping resolution can be assembled depending on the intended use of the panel (see Note 26).

1. Select a small number of radiation hybrids (<100) where the maize chromosome in question was only involved in simple intergenomic translocations. These radiation hybrids will exhibit a low number of obligate chromosome breaks per individual (1–2). Typically, these panels have a mapping resolution in the range of 0.5–10 Mb (low resolution panel).
2. Collect larger number of individuals (>100) where the maize chromosome in question suffered multiple interstitial deletions. These radiation hybrids will exhibit a high number of chromosome breaks per individual (2–10). These panels have a mapping resolution range of 0.2–1 Mb (medium resolution panel).

4. Notes

1. Maize genotypes used as pollen donors and DNA source for F-probe production are plants of the inbred lines *B73*, *Mo17*, and the cultivar *Seneca60*; all maize genotypes are dark-germinating spring-type short-day plants.
2. Oat plants used to develop OMAs are offspring of single-plant selections from the cultivars *Starter*, *Sun II*, and *GAF/Park*; all oat genotypes are dark-germinating spring-type long-day plants.
3. An elevated temperature regime of about 3°C above the optimal growth conditions for oat during anthesis and early embryogenesis can foster (oat × maize) cross-fertilization and may reduce maize chromosome elimination.

4. In order to obtain hour-synchronized flowering time for crossing the two species, the parental oat and maize plants are grown with multiple planting dates ensuring properly overlapping the very diverse seasonal developments of oat and maize. Photoperiod and thermoperiod regimes are very critical for the oat and maize cultivation, especially when shifting from the vegetative to the generative phase of growth. Near the main flowering time, the relative air humidity must be reduced to support stamen exposure and anther dehiscence. Both, oat and maize have very limited time frames for pollen shedding and receptivity of their stigmas. When grown in the field, oat peaks stigma receptivity for pollen from 11:30 am to 12:30 pm, whereas maize peaks shedding of pollen grains in the field from 08:00 to 10:30 am. Maize pollen grains with high fertilization potency have a short life span – only 20–30 min. Taken these facts together, to synchronize the flowering time by the hour between maize and oat grown in chambers with conditions set as described in Notes 7 and 8, the entire oat life regime must be shifted about 3.5 h ahead. Hence, the "oat-day" starts 0:30 am and ends 4:30 pm provided that the "maize-day" is set from 7:00 am to 5:00 pm.
5. All time calculations are based on a standard time with the assumption, that 12:00 pm (noon) is when the sun is in the south (northern hemisphere) or north (southern hemisphere) regardless of whether the clock is changed to summer time/ daylight saving time. Thus, the local time settings for the chambers have to be adjusted accordingly.
6. Light of ~400 μE/m²/s at canopy height (=100%) is provided from a mixture of fluorescent and incandescent lamps to meet the spectral requirements for the two species. Introducing transition phases of 30 min of 50% light intensity on each end of the light cycle seems to generate more vigorous plants with higher seed sets.
7. Growth conditions for maize plants:

	↗Ramp (sunup)	Light (day)	↘Ramp (sundown)	Dark (night)
16 | 8	30 min, 20 ± 1 24 ± 1°C 50% Light	15 h, 24 ± 1°C 100% Light	30 min, 24 ± 1 ↘ 20 ± 1°C 50% Light	8 h, 20 ± 1°C 0% Light
Start	4:00 am	4:30 am	7:30 pm	8:00 pm
10 | 14	30 min, 20 ± 1 ↗ 26 ± 1°C 50% Light	9 h, 26 ± 1°C 100% Light	30 min, 26 ± 1 ↘ 20 ± 1°C 50% Light	14 h, 20 ± 1°C 0% Light
Start	7:00 am	7:30 am	4:30 pm	5:00 pm

8. Growth conditions for oat plants (shifted for cross-pollination by maize):

	↗Ramp (sunup)	Light (day)	↘Ramp (sundown)	Dark (night)
10 |14	30 min, 14 ± 1 ↗ 18 ± 1°C 50% Light	9 h, 18 ± 1°C 100% Light	30 min, 18 ± 1 ↘ 14 ± 1°C 50% Light	14 h, 14 ± 1°C 0% Light
Start	7:00 am	7:30 am	4:30 pm	5:00 pm
16 | 8	30 min, 20 ± 1 ↗ 24 ± 1°C 50% Light	15 h, 24 ± 1°C 100% Light	30 min, 24 ± 1 ↘ 20 ± 1°C 50% Light	8 h, 20 ± 1°C 0% Light
Start	0:30 am	1:00 am	4:00 pm	4:30 pm

9. Growth conditions for oat plants (regular) and OMAs:

	↗Ramp (sunup)	Light (day)	↘Ramp (sundown)	Dark (night)
10 |14	30 min, 14 ± 1 ↗ 18 ± 1°C 50% Light	9 h, 18 ± 1°C 100% Light	30 min, 18 ± 1 ↘ 14 ± 1°C 50% Light	14 h, 14 ± 1°C 0% Light
Start	7:00 am	7:30 am	4:30 pm	5:00 pm
16 | 8	30 min, 17 ± 1 ↗ 21 ± 1°C 50% Light	15 h, 21 ± 1°C 100% Light	30 min, 21 ± 1 ↘ 17 ± 1°C 50% Light	8 h, 17 ± 1°C 0% Light
Start	4:00 am	4:30 am	7:30 pm	8:00 pm

10. Because uniparental genome loss (UGL) of the maize genome in F_1 (oat × maize) hybrids can be incomplete and last through a number of cell cycles during early embryogenesis, different individual cells of one and the same F_1 embryo can have different individual karyotypes, including those cells being the origin for the tillers to build. As a consequence, the individual tillers of one F_1 plant can carry different maize chromosomes retained together with the oat chromosomes, and F_1 plants frequently become mosaics. Thus, avoid cross-pollination among different panicles of the same F_1 plant! For practicability, the flag leaves of the physiologically first four tillers (*a-, b-, c-, and d-tiller*) of an F_1 plant are considered for DNA analysis, and the corresponding panicles for I_1 offspring production. Mosaicism occasionally occurs also between shoot and root tissues in F_1 plants.
11. Do not pool I_1 seeds from different panicles (tillers) of the same F_1 plant; each of the seeds must be treated as individual genotype with potentially different karyotype!
12. Store leaf tissue disc extract mix (200 μl) at 2–8°C no longer than 4 days. For long-term storage, remove leaf tissue disc and store extract mix at –20°C.

13. It is crucial to avoid cross-pollination among different panicles of the same I_1 plant. Pollen grains from panicles of different tillers with different chromosome constitution (e.g., euploid oat vs. monosomic maize addition vs. disomic maize addition of the same maize chromosome) may compete with one another and as a result diminish or even prevent paternal transmission of the added maize chromosome to I_2 offspring. Hence, I_2 seeds from different panicles of the same I_1 plant must not be pooled. Consecutive offspring generations need to be treated likewise until steady transmission of the added maize chromosome without any residual mosaicism is reached. Mixing seed batches from monosomic and disomic maize addition tillers with each other and/or with oat seeds without maize addition disguise frequencies of maternal and paternal maize chromosome transmission to offspring.
14. DNA extraction, PCR, and electrophoresis as in Subheading 3.2, except that PCR with primers for maize chromosome arm-specific microsatellite markers is accomplished with 35 cycles (92°C, 30 s, 62°C, 30 s, 72°C, 40 s) instead in 30 cycles used for PCR with *p-Grande1* primers.
15. Use a large orifice plastic 5-ml transfer pipette in order to avoid shearing the DNA.
16. Measure DNA with spectrometer as follows:

    Dilute (1/20) a 5 μl DNA sample with 95 μl ddH_2O

    Measure the diluted sample at 260 nm UV light (DNA)

    Measure the diluted sample at 280 nm UV light (protein)

    Measure the diluted sample at 320 nm UV light (background)

    Calculate:

    X = (absorption at 260 nm) – (absorption at 320 nm)

    Y = (absorption at 280 nm) – (absorption at 320 nm)

    X = amount (OD)

    $X \times 20 \times 50$ (μg/ml) ÷ 1,000 = concentration (μg/μl)

    $Z = X \div Y$ = purity (≥1.8)
17. Use amber tubes for all steps of DNA labeling in order to prevent ULS agent (Alexa Fluor® 488) from bleaching by extended exposures to artificial and/or daylight.
18. Immediately prior to use, add 5 μl of DMSO (kit component B) to one vial of the ULS labeling reagent (kit component A) and protect from light. Vortex A/B mix until all of the ULS labeling reagent has dissolved and no particulate matter remains. The A/B mix is stable, when stored at 4°C and protected from light, for 3–4 weeks.

19. Spot 0.5 μl of the labeled DNA (F-probe) onto a sheet (2 cm × 2 cm) of nylon membrane. Air-dry sheet for ~10 min and examine fluorescence intensity using blue excitation light with appropriate wavelength (Alexa Fluor® 488 excitation$_{max}$: 490 nm/emission$_{max}$: 519 nm).
20. Root meristems can also be treated with similar effect of arresting cells at metaphase with good chromosome morphology in a mixture of 1.25 mM colchicine and 2 mM quinolinol in 0.1%$_{v/v}$ DMSO at RT for 3 h.
21. Very tough roots can be macerated by washing roots in 50 mM citrate buffer pH 4.5 at RT for ~20 min, then incubating roots in an enzyme solution (450 U/ml cellulase + 150 U/ml pectinase in 50 mM citrate buffer pH 4.5) at 37°C for 40–180 min, and again washing roots in 50 mM citrate buffer pH 4.5 at RT for ~20 min to stop maceration and remove enzymes from tissue; very efficient enzymes are CELLULYSIN® (*Calbiochem*) and MACERASE® (*Calbiochem*).
22. Stringency [%] = 100 [%] − $f_m(T_m$ [°C] − T_h [°C])

    T_m = melting temperature [°C] = 81.5 [°C] + 16.6 × $\log_{10} M$ + 0.41 × (% G + C) − 500/n − 0.61 × (% formamide)

    On the assumption that: f_m = mismatch factor = 1, M = 1.5× SSC = 0.2925 M Na^+, (% G + C) = 45.5%, n = 450, (% formamide) = 40%

    GISH stringency = 100 − 66 + 37 = 71%

    Wash stringency = 100 − 66 + 42 = 76%
23. Epifluorescence microscope equipped with filter sets suitable for:

	F-probe (Alexa Fluor® 488)		PI counterstaining	
Excitation$_{max}$	490 nm	Blue	340 nm + 530 nm	UV + green
Emission$_{max}$	519 nm	Green	615 nm	Red

    Maize chromatin will emit green fluorescence under blue light, whereas oat chromosomes will emit red fluorescence under UV or green light.
24. The appropriate dosage of γ-rays to be applied to a batch of seeds needs to be determined empirically. The dosage that is applied is dependent on the source of γ-rays (^{60}Co or ^{137}Cs) and seed quality. Thus, it is advisable that the kill or survival rates as a function of radiation dose (krad) is established beforehand. First, apply dosages ranging from 0 to 50 krad and choose a dose that results in 20–50% plant survival. Based on our experience, treatments that result in 20–50% plant survival produces a population of plants where 90% of plants

will be self-fertile and will yield progeny carrying chromosomes with 0–10 radiation-induced chromosome breaks.

25. γ-Irradiation is intended to generate random breaks in the added monosomic maize chromosomes of an OMA embryo (seed). There are three groups of chromosome mutations to consider (in the context of RH production): (1) "Background", M_1 seeds with lesions on oat chromosomes without involvement of the added maize chromosome. Provided a physiological neutrality of the rearranged oat genome, the unchanged maize chromosome is expected to transmit to M_2 offspring maternally at a usual frequency for alien plant chromosomes of about 10%, whereas paternal transmission is inhibited. (2) "Deficiency," M_1 seeds with truncations in the added maize chromosome and rearrangements among oat chromosomes only. Provided the rearranged oat genome stays physiologically neutral, a truncated maize chromosome with an active maize centromere is expected to transmit to M_2 offspring at similar maternal and paternal frequencies as those of the added unchanged maize chromosome. (3) "Translocation," M_1 seeds with reciprocal translocations between the added maize chromosome and one oat chromosome, and independent oat background rearrangements. Provided the oat background acts physiologically neutral, the frequency of maternal transmission to M_2 offspring is expected to reflect the segregation pattern of the heteromorphic trivalent (centric maize/noncentric oat translocation – oat homolog – reciprocal centric oat/noncentric maize translocation) during meiosis I. Paternal transmission is expected to be merely inhibited due to the competitive superiority of euploid gametes without translocations. Thus, it is crucial to prevent any cross-pollination among different panicles of the same M_1 plant in order to avoid further increase of segregation complexity for transmitted maize chromosome segments. M_2 seeds from the a-, b-, c-, and d-panicle of the same M_1 plant must not be pooled! Mixing the seed batches disguise frequencies of maternal and paternal transmission of maize chromosome segments. Consecutive offspring generations need to be back-crossed by/to oat until oat–oat chromosome translocations and any residual karyotype segregation are eliminated.

26. The production of oat–maize radiation hybrids can be viewed as a way to produce partial maize chromosome hypo-aneuploids where a maize chromosome has been dissected and distributed into various lines. In this context, a collection of lines containing overlapping pieces of a maize chromosome can be used to define chromosome segments or bins. The average physical size of these chromosome bins will depend on the number of lines and unique chromosome breaks

present in each line. Radiation hybrids can also be viewed as substrates for high resolution mapping if a sufficiently large number of lines with high levels of chromosome breakage is available.

References

1. Riera-Lizarazu, O., Rines, H. W., and Phillips, R. L. (1996) Cytological and molecular characterization of oat x maize partial hybrids. *Theor Appl Genet* 93, 123–135.
2. Kynast, R. G., Okagaki, R. J., Rines, H. W., and Phillips, R. L. (2002) Maize individualized chromosome and derived radiation hybrid lines and their use in functional genomics. *Funct Integr Genomics* 2, 60–69.
3. Kynast, R. G., Okagaki, R. J., Galatowitsch, M. W., Granath, S. R., Jacobs, M. S., Stec, A. O., Rines, H. W., and Phillips, R. L. (2004) Dissecting the maize genome by using chromosome addition and radiation hybrid lines. *Proc Natl Acad Sci USA* 101, 9921–9926.
4. Rines, H. W., and Dahleen, L. S. (1990) Haploid oat plants produced by application of maize pollen to emasculated oat florets. *Crop Sci* 30, 1073–1078.
5. Kynast, R. G., Riera-Lizarazu, O., Vales, M. I., Okagaki, R. J., Maquieira, S. B., Chen, G., Ananiev, E. V., Odland, W. E., Russell, C. D., Stec, A. O., Livingston, S. M., Zaia, H. A., Rines, H. W., and Phillips, R. L. (2001) A complete set of maize individual chromosome additions to the oat genome. *Plant Physiol* 125, 1216–1227.
6. Rines, H. W., Phillips, R. L., Kynast, R. G., Okagaki, R. J., Galatowitsch, M. W., Huettl, P. A., Stec, A. O., Jacobs, M. S., Suresh, J., Porter, H. L., Walch, M. D., and Cabral, C. B. (2009) Addition of individual chromosomes of maize inbreds B73 and Mo17 to oat cultivars Starter and Sun II: maize chromosome retention, transmission, and plant phenotype. *Theor Appl Genet* 119, 1255–1264.
7. Amarillo, F. I., and Bass, H. W. (2007) A transgenomic cytogenetic sorghum (*Sorghum propinquum*) bacterial artificial chromosome fluorescence in situ hybridization map of maize (*Zea mays* L.) pachytene chromosome 9, evidence for regions of genome hyperexpansion. *Genetics* 177, 1509–1526.
8. Okagaki, R. J., Kynast, R. G., Livingston, S. M., Russell, C. D., Rines, H. W., and Phillips, R. L. (2001) Mapping maize sequences to chromosomes using oat-maize chromosome addition materials. *Plant Physiol* 125, 1228–1235.
9. Bass, H. W., Riera-Lizarazu, O., Ananiev, E. V., Bordoli, S. J., Rines, H. W., Phillips, R. L., Sedat, J. W., Agard, D. A., and Cande, W. Z. (2000) Evidence for the coincident initiation of homolog pairing and synapsis during the telomere-clustering (bouquet) stage of meiotic prophase. *J Cell Sci* 113, 1033–1042.
10. Jin, W., Melo, J. R., Nagaki, K., Talbert, P. B., Henikoff, S., Dawe, R. K., and Jiang, J. (2004) Maize centromeres: organization and functional adaptation in the genetic background of oat. *Plant Cell* 16, 571–581.
11. Riera-Lizarazu, O., Vales, M. I., Ananiev, E. V., Rines, H. W., and Phillips, R. L. (2000) Production and characterization of maize chromosome 9 radiation hybrids derived from an oat-maize addition line. *Genetics* 156, 327–339.
12. Phillips, R. L., and Rines, H. W. (2009) Genetic analyses with oat-maize addition and radiation hybrid lines, in *Maize Handbook – Volume II: Genetics and Genomics,* (Bennetzen, J. L., and Hake, S., Eds.), pp 523–538, Springer Science, New York.
13. Topp, C. N., Okagaki, R. J., Melo, J. R., Kynast, R. G., Phillips, R. L., and Dawe, R. K. (2009) Identification of a maize neocentromere in an oat-maize addition line. *Cytogenet Genome Res* 124, 228–238.
14. Okagaki, R. J., Jacobs, M. S., Stec, A. O., Kynast, R. G., Buescher, E., Rines, H. W., Vales, M. I., Riera-Lizarazu, O., Schneerman, M., Doyle, G., Friedman, K. L., Staub, R. W., Weber, D. F., Kamps, T. L., Amarillo, I. F., Chase, C. D., Bass, H. W., and Phillips, R. L. (2008) Maize centromere mapping: a comparison of physical and genetic strategies. *J Hered* 99, 85–93.
15. Riera-Lizarazu, O., Vales, M. I., and Kianian, S. F. (2008) Radiation hybrid (RH) and HAPPY mapping in plants. *Cytogenet Genome Res* 120, 233–240.

Chapter 16

Enhancer Trapping in Plants

Sivanandan Chudalayandi

Abstract

Advances in sequencing technology have led to the availability of complete genome sequences of many different plant species. In order to make sense of this deluge of information, functional genomics efforts have been intensified on many fronts. With improvements in plant transformation technologies, T-DNA and/or transposon-based gene and enhancer-tagged populations in various crop species are being developed to augment functional annotation of genes and also to help clone important genes. State-of-the-art cloning and sequencing technologies, which would help identify T-DNA or transposon junction sequences in large genomes, have also been initiated. This chapter gives a brief history of enhancer trapping and then proceeds to describe gene and enhancer tagging in plants. The significance of reporter gene fusion populations in plant genomics, especially in important cereal crops, is discussed.

Key words: Enhancer trap, T-DNA mutagenesis, Transposon tagging, Gene trap, Functional genomics

1. Introduction

Enhancers are stretches of DNA sequences that regulate genes lying far downstream or upstream. They are capable of activating genes from long distances. There is no doubt that these sequences that mostly occur in the vast noncoding part and sometimes in the coding regions of the genome have a great deal of influence on gene expression. Although the exact mechanism of enhancer–promoter interaction remains elusive, several models have been proposed (1). Among these models, the most plausible is the looping model. The intervening sequence between the enhancer and the downstream gene is looped out so that the enhancer is in close proximity to the target gene. However, to date, no technical advance has been made to unequivocally test the model, though some really promising advances have been made toward fulfilling this objective. Carter et al. (2) devised RNA TRAP (RNA tagging

James A. Birchler (ed.), *Plant Chromosome Engineering: Methods and Protocols*, Methods in Molecular Biology, vol. 701, DOI 10.1007/978-1-61737-957-4_16, © Springer Science+Business Media, LLC 2011

and recovery of associated proteins) strategy to examine promoter–enhancer interaction of the mouse *Hbb* locus. A high throughput technology called capturing chromosome conformation, also called 3C (3), was developed to detect the frequency of interaction between any two genomic loci. Derivatives of 3C, viz. circular chromosome capture conformation, 4C (4), and chromosome conformation capture carbon copy, 5C (5), have made it possible to study interchromosomal interactions on a large scale in mammalian systems. These state-of-the-art technologies have been thoroughly described in a recent review (6).

While amazing strides have been made in studying inter and intrachromosomal interactions, a three-decade-old technology, enhancer trapping, is still relevant especially in this postgenomic era. Enhancer trapping is a method of gene regulatory sequences detection. A directly or indirectly detectable "reporter" gene under the control of a minimal promoter is distributed throughout the genome of the organism. When the minimal promoter comes under the influence of endogenous enhancers, the reporter gene is activated and faithfully reports the presence of the enhancers. Although the term enhancer trap is common in the scientific literature, it essentially amounts to "enhancer detection" as the system does not trap sequences as such unlike a promoter or gene trap but helps detect sequences that amplify the expression of the reporter gene that it carries (7). In plants, enhancer detection using reporter genes has been available for the last two decades. In the postgenomic era, these populations of enhancer-trapped plants could be more relevant in determining the role and functions of enhancer elements in orchestrating the expression of the whole genome. Over the course of the last two decades, various gene detection/tagging systems have been developed, which can be broadly divided into three categories, and are summarized in Fig. 1. In this chapter, I cover the technique of enhancer trapping in plants and how it has evolved over the years and discuss its significance to plant genomics.

2. A Brief History of Enhancer Trapping

Enhancer/gene trapping was first carried out in bacteria more than 30 years ago (8). Promoterless *beta galactosidase* (*lacZ*) reporter gene was randomly inserted into the *Escherichia coli* genome. By looking for expression of *lacZ* gene in the bacterial cell, gene fusions were identified. Because the bacterial genome is small and gene-rich, promoterless constructs worked very well in acting as promoter detectors. This allowed for screening classes of constitutive as well as various stress-induced genes. Apart from being adapted to other bacterial systems (9), this system of gene

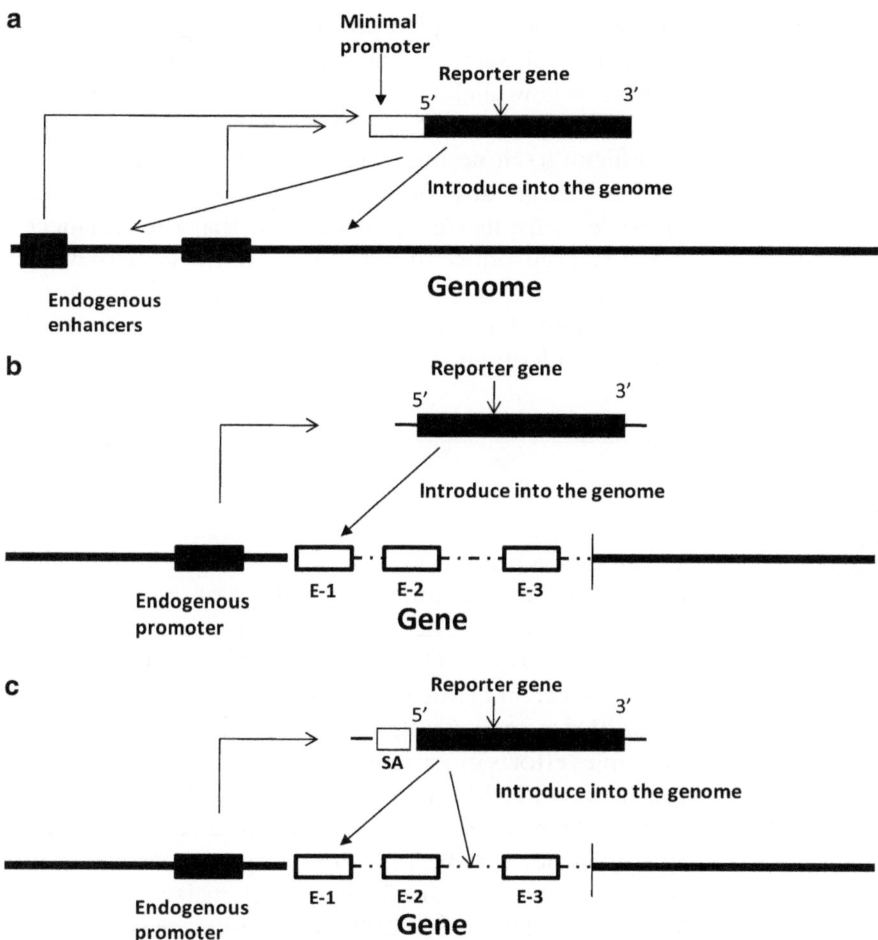

Fig. 1. Gene and regulatory sequence detection and trapping systems have been used in prokaryotes and eukaryotes to augment functional genomic studies. This figure summarizes the three broad types of traps. (**a**) An enhancer detector: a reporter gene under the control of a minimal promoter (TATA) is introduced randomly into the genome. When the minimal promoter comes under the influence of endogenous enhancers, it is activated thus leading to expression of the reporter gene in the target tissues. Thus, the reporter gene faithfully reports the presence of enhancers. (**b**) A promoter trap: a promoterless reporter gene is introduced randomly into a genome. When the reporter integrates into the exon of an endogenous gene, a transcriptional fusion is formed that faithfully reports the expression of the endogenous promoter. E1, E2, and E3 represent three exons of the hypothetical gene. The *dotted lines* represent the introns. Note if a promoter trap is inserted into an intron of a gene, it gets spliced out, and there will be no reporter gene expression. (**c**) A gene trap: a promoterless reporter gene equipped with a splice acceptor at the 5' end is introduced randomly into a genome. When the reporter integrates into the exon or intron of an endogenous gene, the reporter is properly spliced, resulting in a transcriptional fusion. Thus, the reporter faithfully reports the expression of the endogenous gene. E1, E2, and E3 represent three exons of the hypothetical gene; the *dotted lines* represent the introns. SA refers to the 5' splice acceptor.

tagging was also extended to higher eukaryotes, namely, fruit flies (*Drosophila*), using the weak P element (7). The construct used in flies had a weak P element promoter controlling the *lacZ* gene to detect enhancers in the fly genome (10). Eukaryotic genomes are much bigger and not as gene-rich, so it is not surprising that minimal promoters have to be used to detect enhancers. These constructs had an added advantage in that the P element could excise and reinsert elsewhere in the genome, leading to the formation of new alleles. It may be easy to detect enhancer expression in eukaryotes using suitable reporter genes; however, it might be difficult to clone and characterize an enhancer because enhancers can act in any orientation and at considerable distances. However, various studies confirmed that the enhancer detectors did in fact reproduce the expression of the gene they regulate ((7) and references therein). The technique of enhancer detection using minimal promoter-driven reporter genes has been very fruitful in identifying several molecular markers in *Drosophila*.

3. Enhancer Trapping and Insertional Mutagenesis in Plants

In the postgenomic era, the plant research community has been besieged by an ever-increasing deluge of sequence data from different organisms. The draft sequence of *Arabidopsis* was published in 2000 (11). This was followed by an increase of sequencing efforts in cereal plants such as rice and corn. The first announcement of rice sequencing (12, 13) shifted the focus to functional genomics efforts in *Arabidopsis*, rice, and other crop plants including maize (14, 15) in many labs around the world. Unlike rice and *Arabidopsis*, *Zea mays* (maize) has a very large, mostly repetitive genome, yet the first draft sequence of maize inbred B73 was announced at the 50th annual maize genetics meeting at Washington DC in Feb 2008. The sequence is now being further refined for gaps (for latest information see http://www.maizegdb.org/sequencing_project.php).

It may be impossible to assign a biological function to every gene of higher plants considering the fact that most crop genomes are full of duplications leading to the presence of many genes with redundant functions (see e.g., refs. 16, 17) and conserved noncoding sequences even among disparate species (18). It is important, however, to comprehend the biological significance of gene sequences. Understanding the biological function of important genes in a pathway is one major step toward metabolic engineering. Insertional mutagenic screens have been developed for many plants since the late 1980s (19). Over the years, several improvements to plant transformation techniques and in cloning techniques have resulted in improved ability to make larger and beneficial genetic screens. Additionally, many technical improvements to sequencing, especially from the high-throughput nongel-based massively parallel signature sequencing (20) to the latest direct RNA sequencing (reviewed in ref. 21), have revolutionized the way scientists look at eukaryotic transcriptomes and hence significantly improved our understanding of gene regulation.

3.1. T-DNA Genetic Screens

The growth of the science and art of *Agrobacterium*-mediated plant transformation with foreign DNA grew hand in hand with the evolution of gene traps and comprehensive genetic screens. Early experiments were conducted on tobacco protoplasts with *Agrobacterium* T-DNA containing promoterless antibiotic resistance gene at one of the T-DNA borders (summarized in ref. 22 and references therein). In these cases, only successful plant-bacterial gene fusions expressing antibiotic resistance that were able to regenerate in tissue culture were produced. Further development of tissue-culture-mediated plant transformation saw the addition of a second antibiotic resistance marker within the T-DNA (e.g., ref. 23); later, *beta glucuronidase* (*Gus*) reporter genes (24, 25) were included. The *Gus* reporter from *E. coli* has been a common reporter for gene fusions in plants since it was first reported (26). Plants lack endogenous *Gus* activity; hence it became the popular reporter for gene fusions.

The earliest experiments in nontissue-culture-based transformation, *in planta*, of *Arabidopsis* were performed by Feldmann and Marks (27). *Agrobacterium* solution was applied to *Arabidopsis* seeds and planted. The seeds from these plants were screened and determined to have a low percentage of transformants. Nevertheless, since this transformation circumvented the need for plant tissue culture, it significantly accelerated the development of *Arabidopsis* gene tagging (28). Eventually, the vacuum infiltration technique of *Arabidopsis* (29) was developed, where *Arabidopsis* plants were uprooted and placed in a bell jar containing a solution of *Agrobacterium*. Vacuum was then applied and released and the plants were transferred back to soil and allowed to set seed. Many transformed seeds could be selected on a medium containing an antibiotic selection marker. This technique was rapid and successful, so it became possible to generate and screen a large number of transformed *Arabidopsis* plants, leading to the development of many gene trap systems (reviewed in ref. 28). Subsequently, *Arabidopsis in planta* transformation became even easier when transformation by simple floral dip in *Agrobacterium* solution containing a surfactant (Silwet) was reported (30). Thousands of promoter trap lines of *Arabidopsis* using *Gus* have been very helpful in embryology in plants (31–33), by identifying several molecular markers of embryogenesis.

T-DNA tagging is still common in plants, but the technique suffers from some disadvantages. T-DNA does not always insert a single copy. Multiple loci integrations have often been reported, while at other times complex integrations (tandem insertions) are found at the same locus (34). Barbara McClintock first identified DNA elements that she termed controlling elements, i.e., the autonomous activator (Ac) and the nonautonomous dissociator (Ds) in maize, and showed that in the presence of Ac, Ds can excise from its original place in the genome (35). Ac and Ds in maize were identified and cloned more than three decades later (36). Ac encodes an enzyme called the transposase that catalyzes the transposition of the nonautonomous Ds element, which is a variant of Ac that cannot transpose by itself. In addition to an explosion of genetic and mutant screens, the mid-1980s also saw the genetic characterization of many plant transposons, notably the Ds, dSpm, and Mu element from maize (36–38). Subsequently, Ds transposition has also been characterized in heterologous systems like tobacco (39) and tomato (40). Most eukaryotic genomes are full of inactive and active transposons (reviewed thoroughly in ref. 41). The Ac/Ds system and En/Spm-i/dSpm system from maize have been adapted in T-DNA vector delivery systems to generate enhancer and gene traps in *Arabidopsis* (42) and rice (43, 44) efficiently. There are many advantages of using transposons within T-DNA. Starting with a relatively small T-DNA population, it is possible to generate several lines containing unique transposons insertions by simply crossing these lines to plants containing a stable source of transposase.

3.2. Enhancer Detection in Arabidopsis and Rice

In *Arabidopsis*, the first comprehensive enhancer trap collection was the Cold Spring Harbor Lab (CSHL) enhancer trap lines (42). The Ds-E (enhancer trap) construct carried the *Gus* gene downstream of a minimal 35S promoter. In addition to enhancer traps, a gene trap (Ds-G) was also constructed where a promoterless *Gus* gene was cloned immediately downstream of an intron sequence and two slice acceptors. By selecting for kanamycin resistance and against another closely linked marker gene (iaah; indole acetic acid hydrolase gene), the population was enriched for unlinked Ds transpositions. This collection of enhancer trap lines was the first that used Ds transposons effectively in conjunction with T-DNA-mediated transformation. Use of transposons ensures relatively large number of new insertions from simple genetic cross with a line containing the stable transposase source. The transposed Ds insertions (reinsertion) are single copy, and isolation of sequences flanking a Ds insertion is relatively easy, since the Ds sequence is known. It was demonstrated using this collection that it is possible to identify patterns of gene expression in almost all developmental processes.

In the mid-1990s, the green fluorescent protein (GFP) from Jellyfish was codon-optimized and adapted to plants and was shown to be successful as a fusion reporter (45). When exposed to UV light, GFP fluoresces in the visual range. Screening for GFP is thus remarkably easier and nondestructive compared with the destructive histochemical staining for detecting *beta glucuronidase*. Developments in *Drosophila* research have immensely benefited the *Arabidopsis* research community. In *Drosophila*, the P element transposon-based gene detection approach has been improved upon by using the yeast transcriptional factor, Gal4, mediated tagging and by using enhancer suppressors to increase gene knockouts. Also, more recently, different recombinase systems have improved transgenesis in *Drosophila* (46). An *Arabidopsis* enhancer trap population was generated using minimal 35S promoter controlling modified yeast *Gal4* with Herpes Simplex virus transcriptional activator (*Gal4–VP16*) in conjunction with upstream activation sequence (UAS) cloned upstream of 35S TATA: GFP (*35S TATA refers to the 90 to +8 sequence of the CaMV35S promoter, please refer to ref.* 47 *for further details*). Note that UAS from yeast can only activate the downstream gene if GAL4 protein binds to it. Thus, GFP expression effectively and faithfully reports the enhancers that activate *Gal4*. This collection was employed to screen 250 lines with root-specific GFP expression patterns (48). Subsequently, the enhancer trap vector was used to generate the University of Pennsylvania enhancer trap collection in *Arabidopsis* (http://www.enhancertraps.bio.upenn.edu/default.html). The University of Pennsylvania and the Cambridge collections were used to analyze stomatal guard cell development in *Arabidopsis* (49). One of these lines tracked stomatal development, since GFP expression was detected in the early stages of guard cell development but not in the mature guard cells. This Gal4:VP16-GFP line could be used to devise strategies involving mistargeting genes involved in early stomatal development or tissue-specific silencing by controlling the expression of Gal4. A comprehensive listing of collections of enhancer trap lines in *Arabidopsis* and rice has been reported in an earlier review (50). In addition to GFP, other reporters, namely, luciferase and GUS, have also been used to generate *Gal4*-mediated enhancer trap systems in *Arabidopsis* (51). Recently, codon-optimized luciferase has been used to generate a promoter trap collection of banana (52) which was used to characterize a promoter that was activated during low-temperature stress.

Rice has been a model system for cereal genomics (53) due to its small genome, the availability of finished quality sequence (54), and the relative ease of *Agrobacterium*-mediated transformation (see ref. 55 for indica rice *Agrobacterium* transformation protocols). Several groups around the world have generated many rice T-DNA insertion lines (56) including some high-throughput targeted insertion population. For example, enhancer trap lines have been generated using *Gus* with minimal 35S promoter (57). More than 13,000 lines of rice have also been generated using a combination of minimal 35S promoter: GAL4 vp16 and UAS:GFP (58). A good percentage of these two enhancer-tagged rice populations have been organized into a phenotypic mutant database (59). About 30% of the GAL4:GFP enhancer trap lines showed GFP expression. Much of the GFP expression was found in specific cell types. Some of these lines could be useful tools in studying developmental progression of specific organs and in studies related to identification of gene targets. Separately, Gal4 and GUS Plus (from *Staphylococcus*) systems have been used to generate 30,000 rice enhancer trap lines (60). In addition to T-DNA-mediated enhancer tagging (43), maize En/Spm–i/dSpm in conjunction with DsRed and s-GFP has been produced to develop a library of insertional mutations in rice (43). Separately, the same group (44) also used the Ac–Ds system to generate a large library of insertional mutants in rice. The Ac–Ds system was enriched for unlinked Ds transposition by selecting for bialaphos (Basta) resistance and selecting against closely linked expression of a closely linked GFP expression. In comparing the two populations, they reported that the Spm–dSpm system is as good as the Ac–Ds system. Further, dSpm tended to transpose more often to unlinked locations. Thus, the Spm/dSpm system would give a wider genomic coverage. Several thousands of lines of rice insertional mutants and enhancer trap population are already available to the plant scientific community to augment and complement functional genomics efforts in rice. Because rice chromosomes are syntenic to maize and other grass genomes, interesting results from the study of the rice genome could be applied to or at least form the basis of similar experiments on other cereal crops.

3.3. Insertional Mutagenesis and Functional Genomics in Maize

Functional genomics efforts in maize as in other crops predate the genomic era. Maize genetics is robust and has a very rich history. The availability of several polymorphic inbreds, mapping populations, and chromosomal stocks make designs of many elegant genetic screens possible (61 and references therein). The ease of crossing and the ability to score several visible traits on the maize kernel are just some of the many advantages of using corn. Barbara McClintock followed the progenies of spotted and purple kernels (due to the expression of C1 gene in the aleurone) over generations to arrive at her prediction about the maize dissociator and activator controlling elements.

Many active transposable elements have been cloned and characterized in maize. The active DNA transposons consisting of the two component autonomous–nonautonomous element pair, such as the Ac–Ds originally described by McClintock (35), the Spm–dSpm independently reported by McClintock(62) and Peterson (63), or the Robertson's Mutator (64), among others, have been thoroughly characterized. The Mu transposon has been the most widely used to create insertion tagged lines of maize because it accumulates to very high copy numbers and transposes to unlinked loci. Sequences from many such populations e.g., the uniform Mu population where Mu transposition is maintained at a steady state by repeated backcrossing to an inbred (65) and rescue-Mu (for details see http://www.maizegdb.org/

rescuemu-phenotype.php), a transgenic Mu1 that allows for plasmid rescue. Maize genome sequences flanking the site of insertion/transposition are publicly available. Mu transposes at a very high rate and is thus very much suitable for saturating the maize genome. Though there is a problem with high levels of somatic insertion, the availability of complete maize genome sequence and improvements in sequencing will make these populations really important resources for maize functional genomics.

The maize Ac–Ds two component tagging system has been used as a heterologous tool to create populations of knockout/insertion mutants that have been routinely used to clone and identify many genes in *Arabidopsis* and rice, as discussed in the previous subsections. Since Ds elements transpose mostly to nearby sites in maize, they have been useful for regional mutagenesis (see ref. 66). An ambitious Ac–Ds tagging system that uses the Ac-immobilized (Ac-im) source has been developed (67). The sequences flanking the Ds insertion and the details about the project are posted on (http://www.plantgdb.org/prj/AcDsTagging/#Contents).

The collections of En/Sp–I/dSpm tagged maize plants have facilitated the identification and cloning of many genes in maize such as *ramosa* (68) and *opaque2* gene (69). Ac–Ds and Spm–dSpm have been the most widely used transposons systems to tag genes in cereals. Table 1 summarizes the various collections using Ac–Ds and En/spm transposon-tagged populations in various cereal crops constructed for the purposes of gene tagging, enhancer trapping, and insertional mutagenesis.

3.4. Enhancer Trapping in Maize

Compared with *Arabidopsis* and rice, transformation of maize inbreds is still in its infancy. However, *Agrobacterium*-mediated transformation of Maize Hi II hybrids using immature embryos is well established (70) and has been used for introduction of a Cre–*lox*-site-specific recombinase system in maize (71). There have been no reports of enhancer trap screens in maize developed using reporter gene fusions. Our laboratory has used the *Agrobacterium* transformation of Hi II hybrids of maize to develop a screen of maize enhancer traps using a Ds-based minimal 35S: mGFP5. The scorable marker is the *C1* gene of maize responsible for anthocyanin pigmentation in the aleurone under the control of the CaMV35S promoter. To determine the location of transgene insertion, we employed chromosomal FISH (72). Figure 2 shows a chromosome spread of a T_1 transgenic line (Hi II A×B hybrid) with a transgene (Texas red) insertion in chromosome 7. The immobile Ac transposase (73) is used to launch the Ds elements from their original place of insertion. Somatic excision of Ds occurs, and germinal Ds reinsertions are being characterized.

Table 1
Ac–Ds and Spm/dSpm transposon-tagged insertional mutant/regulatory sequence trap population developed in cereal crops

Cereal crop	Type of population and reporter gene if applicable	Details of construct and crosses	Reference/Web site/database
Ac–Ds-tagged population			
Barley	Gene tag/test population	Ds-Ubi Bar Ac Gene Gun	Koprek et al. 2000 (77) University of California, Berkeley, CA
Barley	Gene trap	Ds-Ubi Bar Ac from maize (Agrobacterial transformation)	Zhao et al. 2006 (78) MPI for Plant Breeding Research, Cologne, Germany
Barley	Gene trap GUS	Ds-Gus cw-AC Gene Gun	Lazarow et al. 2009 (79) University of Hamburg, Hamburg, Germany
Maize	Insertional mutants/no reporter	Natural Ds Spontaneous mutant Ac-im	Ahern et al. (66) http://www.plantgdb.org/prj/AcDsTagging/ Boyce Thompson Institute, Ithaca, NY Iowa State University, IA, USA
Rice	Gene trap	Ds-Ubi BAR GFP for counterselection 35S-Ac from maize	Kolesnik et al. (44) UC Davis, CA
Rice	Gene trap Gus	Derived from Kolenik et al. 2004	Jiang et al. 2007 (82) Temasek Life Sciences laboratory, Singapore and China
Rice Japonica Indica	Enhancer trap	Ds GFP Ac	Van Encevort et al. 2005 (80) OS-TID database Plant Research International, Wageningen, The Netherlands
Rice	Gene trap	Ubi:Ds-Bar-Ds Gus reporter Ubi-Ac	Luan et al. 2008 (81) Rice Research Institute, Zheijiang, China
Rice	Enhancer trap	Ds-Gus Ubi Bar 35S-Ac	Ito et al. 2004 (84) National Institute of Genetics, Shizuoka-ken, Japan
Rice Japonica and indica	Gene trap; GUS	Callus cultures of Ac and Ds containing lines	Park et al. 2003 (83) Korea
En/Spm-tagged population			
Rice	Gene trap GFP and Ds red	dSpm-Ds-Red En/Spm-GFP	Kumar et al. (43) Sundaresan Lab, UC Davis, Davis, California
Rice	Gene trap Enhancer trap	dSpm-Bar-min 35 Gus 35S-En/Spm	Greco et al. 2003 (85) Plant Research International, Wageningen, The Netherlands

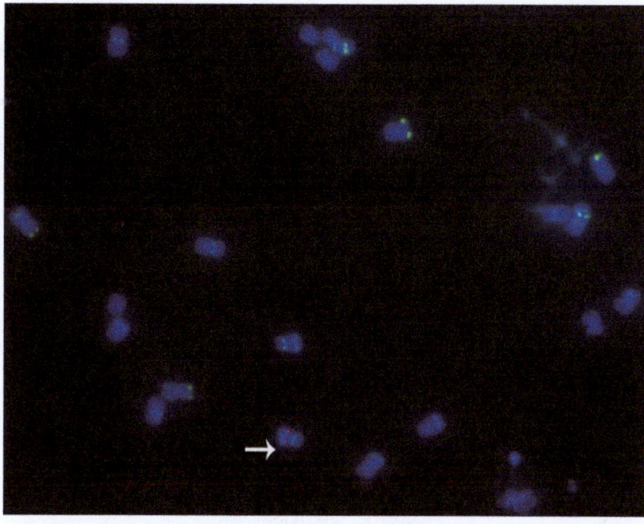

Fig. 2. FISH on transgenic Hi II maize plant: a chromosome spread of Transgenic Hi II showing transgene insertion in the seventh chromosome. Chromosomes are stained with 4' 6-diamidino-2-phenylindole (DAPI). The centromeres and TAG microsatellite repeats are labeled *green* and the "transgene" is labeled *red*. The location of transgene is marked with an *arrow*.

4. Postgenomic Perspectives and Applications

The postgenomic era has seen the rise of high-throughput strategies for studying the biology of organisms, leading to the advent of systems biology. The vast amounts of data have led to the development of numerous databases including those integrating semantics and biology, for example, BioGateway (http://www.semantic-systems-biology.org/biogateway). In agriculture, the comparative genomics of cereals could take us closer to manipulate genetic pathways that might eventually help improve cereal yield and quality. Modernized high-throughput ventures support and immensely benefit from the painstaking work that needs to be carried out in the lab or field by scientists and breeders. Development of newer tools has opened new vistas for traditional analytical methods. Since the 1980s, the discoveries of abundant polymorphic markers in natural populations of plants and the vast improvement in computational powers have greatly improved the science of QTL detection (74). Vast populations of insertion mutants in model plants such as *Arabidopsis* will continue to complement the efforts of functional genomics and advance studies on epigenetics. For example, a whole genome transposon tagging in *Arabidopsis* revealed location-dependent effects on transcription and chromatin organization (75). Great advances in sequencing have provided a wealth of sequence data and also made it possible to sequence difficult targets. As an example, amplifying unknown sequences flanking a known target in large genomes such as maize has been difficult. However, recently, a digestion–ligation–amplification-based strategy was used to clone the *gl4* gene from a maize mutator-tagged population (76). A transposon-based insertional mutant population can always be genetically (by crossing to the line containing transposase) made to produce newer alleles and thus tag newer genes. The Gal4:GFP-based enhancer trap populations developed in *Arabidopsis* and rice could be extremely important tools in learning about the nuances of gene action at the tissue and cellular levels in developmental pathways. Newer and more optimized reporter genes would enable researchers to track gene expression changes in specific cell types throughout the life of a plant. The potential role of enhancers in speciation and evolution has been the subject of a few studies (74). Reporter-gene fusions have a great potential to provide information regarding subtle gene expression changes in intra and interspecific hybrid backgrounds.

Acknowledgments

Research supported by National Science Foundation grant DBI 0733857 to James Birchler.

References

1. Bulger, M. and Groudine, M. (2002) TRAPping enhancer function. *Nat Genet.* **32**, 555–6.
2. Carter, D., Chakalova, L., Osborne, C.S., Dai, Y.F. and Fraser, P. (2002) Long-range chromatin regulatory interactions in vivo. *Nat Genet.* **32**, 623–6.
3. Dekker, J., Rippe, K., Dekker, M. and Kleckner, N. (2002) Capturing chromosome conformation. *Science.* **295**, 1306–11.
4. Zhao, Z., Tavoosidana, G., Sjölinder, M., Göndör, A., Mariano, P., Wang, S., Kanduri, C., Lezcano, M., Sandhu, K.S., Singh, U., Pant, V., Tiwari, V., Kurukuti, S. and Ohlsson, R. (2006) Circular chromosome conformation capture (4C) uncovers extensive networks of epigenetically regulated intra- and interchromosomal interactions. *Nat Genet.* **38**, 1341–7.
5. Dostie, J., Richmond, T.A., Arnaout, R.A., Selzer, R.R., Lee, W.L., Honan, T.A., Rubio, E.D., Krumm, A., Lamb, J., Nusbaum, C., Green, R.D. and Dekker, J. (2006) Chromosome conformation capture carbon copy (5C): a massively parallel solution for mapping interactions between genomic elements. *Genome Res.* **16**, 1299–309.
6. Vassetzky, Y., Gavrilov, A., Eivazova, E., Priozhkova, I., Lipinski, M. and Razin, S. (2009) Chromosome conformation capture (from 3C to 5C) and its ChIP-based modification. *Methods Mol Biol.* **567**, 171–88.
7. Bellen, H.J. (1999) Ten years of enhancer detection: lessons from the fly. *Plant Cell.* **11**, 2271–81.
8. Casadaban, M.J. and Cohen, S.N. (1979) Lactose genes fused to exogenous promoters in one step using a Mu-lac bacteriophage: in vivo probe for transcriptional control sequences. *Proc Natl Acad Sci U S A.* **76**, 4530–3.
9. O'Kane, C., Stephens, M.A. and McConnell, D. (1986) Integrable alpha-amylase plasmid for generating random transcriptional fusions in *Bacillus subtilis. J Bacteriol.* **168**, 973–81.
10. O'Kane, C.J. and Gehring, W.J. (1987) Detection in situ of genomic regulatory elements in *Drosophila. Proc Natl Acad Sci U S A.* **84**, 9123–7.

11. Arabidopsis Genome Initiative. (2000) Analysis of the genome sequence of the flowering plant *Arabidopsis thaliana*. *Nature*. **408**, 796–815.
12. Yu, J. Hu, S., Wang, J., et al. (2002) A draft sequence of the rice genome (*Oryza sativa* L. ssp. *indica*). *Science*. **296**, 79–92.
13. Goff, S.A., Ricke, D., Lan, T.H., et al. (2002) A draft sequence of the rice genome (*Oryza sativa* L. ssp. *japonica*). *Science*. **296**, 92–100.
14. Lister, R., O'Malley, R.C., Tonti-Filippini, J., Gregory, B.D., Berry, C.C., Millar, A.H. and Ecker, J.R. (2008) Highly integrated single-base resolution maps of the epigenome in *Arabidopsis*. *Cell*. **133**, 523–36.
15. Yamamoto, T., Yonemaru, J. and Yano, M. (2009) Towards the understanding of complex traits in rice: substantially or superficially? *DNA Res*. **16**, 141–54.
16. Adams, K.L. (2007) Evolution of duplicate gene expression in polyploid and hybrid plants. *J Hered*. **98**, 136–41.
17. Blanc, G. and Wolfe, K.H. (2004) Widespread paleopolyploidy in model plant species inferred from age distributions of duplicate genes. *Plant Cell*. **16**, 1667–78.
18. Freeling, M. and Subramaniam, S. (2009) Conserved noncoding sequences (CNSs) in higher plants. *Curr Opin Plant Biol*. **12**, 126–32.
19. Koncz, C., Mayerhofer, R., Koncz-Kalman, Z., Nawrath, C., Reiss, B., Redei, G.P. and Schell, J. (1990) Isolation of a gene encoding a novel chloroplast protein by T-DNA tagging in *Arabidopsis thaliana*. *EMBO J*. **9**, 1337–46.
20. Brenner, S., Johnson, M., Bridgham. J., Golda, G., Lloyd, D.H., Johnson, D., Luo, S., McCurdy, S., Foy, M., Ewan, M., Roth, R., George, D., Eletr, S., Albrecht, G., Vermaas, E., Williams, S.R., Moon, K., Burcham, T., Pallas, M., DuBridge, R.B., Kirchner, J., Fearon, K., Mao, J. and Corcoran, K. (2000) Gene expression analysis by massively parallel signature sequencing (MPSS) on microbead arrays. *Nat Biotechnol*. **18**, 630–4.
21. Wang, Z., Gerstein, M. and Snyder, M. (2009) RNA-Seq: a revolutionary tool for transcriptomics. *Nat Rev Genet*. **10**, 57–63.
22. Springer, P.S. (2000) Gene traps: tools for plant development and genomics. *Plant Cell*. **12**, 1007–20.
23. Koncz, C., Martini, N., Mayerhofer, R., Koncz-Kalman, Z., Körber, H., Redei, G.P. and Schell J. (1989) High-frequency T-DNA-mediated gene tagging in plants. *Proc Natl Acad Sci U S A*. **86**, 8467–71.
24. Kertbundit, S., De Greve, H., Deboeck, F., Van Montagu, M. and Hernalsteens, J.P. (1991) In vivo random beta-glucuronidase gene fusions in *Arabidopsis thaliana*. *Proc Natl AcadSci U S A*. **88**, 5212–6.
25. Fobert, P.R., Miki, B.L. and Iyer, V.N. (1991) Detection of gene regulatory signals in plants revealed by T-DNA-mediated fusions. *Plant Mol Biol*. **17**, 837–51.
26. Jefferson, R.A., Kavanagh, T.A. and Bevan, M.W. (1987) GUS fusions: beta-glucuronidase as a sensitive and versatile gene fusion marker in higher plants. *EMBO J*. **6**, 3901–7.
27. Feldmann, K.A. and Marks, M.D. (1987) *Agrobacterium*-mediated transformation of germinating seeds of *Arabidopsis thaliana*: a non-tissue culture approach. *Mol Gen Genet*. **208**, 1–9.
28. Azpiroz-Leehan, R. and Feldmann, K.A. (1997) T-DNA insertion mutagenesis in *Arabidopsis*: going back and forth. *Trends Genet*. **13**, 152–6.
29. Bechtold, N., Ellis, J. and Pelletier, G. (1993) *In planta Agrobacterium* mediated gene transfer by infiltration of adult *Arabidopsis thaliana* plants. *C R Acad Sci Paris Life Sci*. **316**, 1194–9.
30. Clough, S.J. and Bent, A.F. (1998) Floral dip: a simplified method for *Agrobacterium*-mediated transformation of *Arabidopsis thaliana*. *Plant J*. **16**, 735–43.
31. Lindsey, K., Wei, W., Clarke, M.C., McArdle, H.F., Rooke, L.M. and Topping J.F. (1993) Tagging genomic sequences that direct transgene expression by activation of a promoter trap in plants. *Transgenic Res*. **2**, 33–47.
32. Topping, J.F., Agyeman, F., Henricot, B. and Lindsey, K. (1994) Identification of molecular markers of embryogenesis in *Arabidopsis thaliana* by promoter trapping. *Plant J*. **5**, 895–903.
33. Topping, J.F. and Lindsey, K. (1997) Promoter trap markers differentiate structural and positional components of polar development in *Arabidopsis*. *Plant Cell*. **9**, 1713–25.
34. De Buck, S., Jacobs, A., Van Montagu, M. and Depicker, A. (1999) The DNA sequences of T-DNA junctions suggest that complex T-DNA loci are formed by a recombination process resembling T-DNA integration. *Plant J*. **20**, 295–304.
35. McClintock, B. (1948) Mutable loci in maize. *Carnegie Inst Wash Yearb*. **47**, 155–69.
36. Fedoroff, N., Wessler, S. and Shure, M. (1983) Isolation of the transposable maize controlling elements Ac and Ds. *Cell*. **35**, 235–42.

37. Schwarz-Sommer, Z., Gierl, A., Klösgen, R.B., Wienand, U., Peterson, P.A. and Saedler, H. (1984) The Spm (En) transposable element controls the excision of a 2-kb DNA insert at the wx allele of *Zea mays*. *EMBO J*. **3**, 1021–8.
38. Strommer, J.N., Hake, S., Bennetzen, J., Taylor, W.C. and Freeling, M. (1982) Regulatory mutants of the maize Adhl gene caused by DNA insertions. *Nature* **300**, 542–4.
39. Hehl, R. and Baker, B. (1990) Properties of the maize transposable element activator in transgenic tobacco plants: a versatile inter-species genetic tool. *Plant Cell*. **2**, 709–21.
40. Osborne, B.I., Corr, C.A., Prince, J.P., Hehl, R., Tanksley, S.D., McCormick, S. and Baker, B. (1991) Ac transposition from a T-DNA can generate linked and unlinked clusters of insertions in the tomato genome. *Genetics*. **129**, 833–44.
41. Wessler, S.R. (2006). Eukaryotic transposable elements: teaching old genomes new tricks. In *The Implicit Genome* (Caporale L, ed.). Oxford University Press, USA.
42. Sundaresan, V., Springer, P., Volpe, T., Haward, S., Jones, J.D., Dean, C., Ma, H. and Martienssen, R. (1995) Patterns of gene action in plant development revealed by enhancer trap and gene trap transposable elements. *Genes Dev*. **9**, 1797–810.
43. Kumar, C.S., Wing, R.A. and Sundaresan, V. (2005) Efficient insertional mutagenesis in rice using the maize En/Spm elements. *Plant J*. **44**, 879–92.
44. Kolesnik, T., Szeverenyi, I., Bachmann, D., Kumar, C.S., Jiang, S., Ramamoorthy, R., Cai, M., Ma, Z.G., Sundaresan, V. and Ramachandran, S. (2004) Establishing an efficient Ac/Ds tagging system in rice: large-scale analysis of Ds flanking sequences. *Plant J*. **37**, 301–14.
45. Haseloff, J., Siemering, K.R., Prasher, D.C. and Hodge, S. (1997) Removal of a cryptic intron and subcellular localization of green fluorescent protein are required to mark transgenic *Arabidopsis* plants brightly. *Proc Natl Acad Sci U S A*. **94**, 2122–7.
46. Venken, K.J. and Bellen, H.J. (2007) Transgenesis upgrades for *Drosophila melanogaster*. *Development*. **134**, 3571–84.
47. Benfey, P.N., Ren, L. and Chua, N.H. (1990) Tissue-specific expression from CaMV 35S enhancer subdomains in early stages of plant development. *EMBO J*. **9**, 1677–84.
48. Laplaze, L., Parizot, B., Baker, A., Ricaud, L., Martinière, A., Auguy, F., Franche, C., Nussaume, L., Bogusz, D. and Haseloff J. (2005) GAL4-GFP enhancer trap lines for genetic manipulation of lateral root development in *Arabidopsis thaliana*. *J Exp Bot*. **56**, 2433–42.
49. Gardner, M.J., Baker, A.J., Assie, J.M., Poethig, R.S., Haseloff, J.P. and Webb, A.A. (2009) GAL4 GFP enhancer trap lines for analysis of stomatal guard cell development and gene expression. *J Exp Bot*. **60**, 213–26.
50. Acosta-García, G., Autran, D. and Vielle-Calzada, J.P. (2004) Enhancer detection and gene trapping as tools for functional genomics in plants. *Methods Mol Biol*. **267**, 397–414.
51. Engineer, C.B., Fitzsimmons, K.C., Schmuke, J.J., Dotson, S.B. and Kranz, R.G. (2005) Development and evaluation of a Gal4-mediated LUC/GFP/GUS enhancer trap system in *Arabidopsis*. *BMC Plant Biol*. **5**, 9.
52. Santos, E., Remy, S., Thiry, E., Windelinckx, S., Swennen, R. and Sági L. (2009) Characterization and isolation of a T-DNA tagged banana promoter active during in vitro culture and low temperature stress. *BMC Plant Biol*. **9**, 77.
53. Goff, S.A. (1999) Rice as a model for cereal genomics. *Curr Opin Plant Biol*. **2**, 86–9.
54. International Rice Genome Sequencing Project. (2005) The map-based sequence of the rice genome. *Nature*. **436**, 793–800.
55. Datta, K. and Datta S.K. (2006) Indica rice (*Oryza sativa*, BR29 and IR64). *Methods Mol Biol*. **343**, 201–12.
56. An, G., Lee, S., Kim, S.H. and Kim, S.R. (2005) Molecular genetics using T-DNA in rice. *Plant Cell Physiol*. **46**(1), 14–22.
57. Sallaud, C., Gay, C., Larmande, P., Bès, M., Piffanelli, P., Piégu, B., Droc, G., Regad, F., Bourgeois, E., Meynard, D., Périn, C., Sabau, X., Ghesquière, A., Glaszmann, J.C., Delseny, M. and Guiderdoni, E. (2004) High throughput T-DNA insertion mutagenesis in rice: a first step towards in silico reverse genetics. *Plant J*. **39**, 450–64.
58. Johnson, A.A., Hibberd, J.M., Gay, C., Essah, P.A., Haseloff, J., Tester, M. and Guiderdoni, E. (2005) Spatial control of transgene expression in rice (*Oryza sativa* L.) using the GAL4 enhancer trapping system. *Plant J*. **41**, 779–89.
59. Larmande, P., Gay, C., Lorieux, M., Périn, C., Bouniol, M., Droc, G., Sallaud, C., Perez, P., Barnola, I., Biderre-Petit, C., Martin, J., Morel, J.B., Johnson, A.A., Bourgis, F., Ghesquière, A., Ruiz, M., Courtois, B. and Guiderdoni, E. (2008) *Oryza* Tag Line, a phenotypic mutant database for the Genoplante rice insertion line library. *Nucleic Acids Res*. **36**(Database issue), D1022–7.

60. Wu, C., Li, X., Yuan, W., Chen, G., Kilian, A., Li, J., Xu, C., Li, X., Zhou, D.X., Wang, S. and Zhang, Q. (2003) Development of enhancer trap lines for functional analysis of the rice genome. *Plant J.* **35**, 418–27.
61. Candela, H. and Hake, S. (2008) The art and design of genetic screens: maize. *Nat Rev Genet.* **9**, 192–203.
62. McClintock, B. (1954) Mutations in maize and chromosomal observations in *Neurospora*. *Carnegie Inst Wash Yearb.* **53**, 254–60.
63. Peterson, P.A. (1953) A mutable pale green locus in maize. *Genetics.* **38**, 682–3.
64. Robertson, D.S. (1978) Characterization of a mutator system in maize. Mutat Res. **51**, 21–8.
65. McCarty, D.R., Settles, A.M., Suzuki, M., Tan, B.C., Latshaw, S., Porch, T., Robin, K., Baier, J., Avigne, W., Lai, J., Messing, J., Koch, K.E. and Hannah, L.C. (2005) Steady-state transposons mutagenesis in inbred maize. *Plant J.* **44**, 52–61.
66. Ahern, K.R., Deewatthanawong, P., Schares, J., Muszynski, M., Weeks, R., Vollbrecht, E., Duvick, J., Brendel, V.P. and Brutnell, T.P. (2009) Regional mutagenesis using dissociation in maize. *Methods.* **49**(3), 248–54.
67. Conrad, L.J. and Brutnell, T.P. (2005) Ac-immobilized, a stable source of activator transposase that mediates sporophytic and gametophytic excision of dissociation elements in maize. *Genetics.* **171**, 1999–2012.
68. Vollbrecht, E., Springer, P.S., Goh, L., Buckler, E.S., 4th and Martienssen, R. (2005) Architecture of floral branch systems in maize and related grasses. *Nature.* **436**, 1119–26.
69. Schmidt, R.J., Burr, F.A. and Burr, B. (1987) Transposon tagging and molecular analysis of the maize regulatory locus opaque-2. *Science.* **238**, 960–3.
70. Frame, B.R., Shou, H., Chikwamba, R.K., Zhang, Z., Xiang, C., Fonger, T.M., Pegg, S.E., Li, B., Nettleton, D.S., Pei, D. and Wang K. (2002) *Agrobacterium tumefaciens*-mediated transformation of maize embryos using a standard binary vector system. *Plant Physiol.* **129**, 13–22.
71. Veg, J.M., Yu, W., Han, F., Kato, A., Peters, E.M., Zhang, Z.J. and Birchler, J.A. (2008) *Agrobacterium*-mediated transformation of maize (*Zea mays*) with Cre-lox site specific recombination cassettes in BIBAC vectors. *Plant Mol Biol.* **66**, 587–98.
72. Kato, A., Lamb, J.C. and Birchler, J.A. (2004) Chromosome painting using repetitive DNA sequences as probes for somatic chromosome identification in maize. *Proc Natl Acad Sci U S A.* **101**, 13554–9.
73. Xiao, Y.L. and Peterson, T. (2002). Ac transposition is impaired by a small terminal deletion. *Mol Genet Genomics.* **266**, 720–31.
74. Mackay, T.F., Stone, E.A. and Ayroles, J.F. (2009) The genetics of quantitative traits: challenges and prospects. *Nat Rev Genet.* **10**, 565–77.
75. Rosin, F.M., Watanabe, N., Cacas, J.L., Kato, N., Arroyo, J.M., Fang, Y., May, B., Vaughn, M., Simorowski, J., Ramu, U., McCombie, R.W., Spector, D.L., Martienssen, R.A. and Lam, E. (2008). Genome-wide transposon tagging reveals location-dependent effects on transcription and chromatin organization in *Arabidopsis*. *Plant J.* **55**, 514–25.
76. Liu, S., Dietrich, C.R. and Schnable, P.S. (2009) DLA-based strategies for cloning insertion mutants: cloning the gl4 locus of maize using mu transposon tagged alleles. *Genetics.* **183**, 1215–25.
77. Koprek, T., McElroy, D., Louwerse, J., Williams-Carrier, R. and Lemaux, P.G. (2000) An efficient method for dispersing Ds elements in the barley genome as a tool for determining gene function. *Plant J.* **24**, 253–63.
78. Zhao, T., Palotta, M., Langridge, P., Prasad, M., Graner, A., Schulze-Lefert, P. and Koprek, T. (2006) Mapped Ds/T-DNA launch pads for functional genomics in barley. *Plant J.* **47**, 811–26.
79. Lazarow, K. and Lütticke, S. (2009) An Ac/Ds-mediated gene trap system for functional genomics in barley. *BMC Genomics.* **10**, 55.
80. van Enckevort, L.J., Droc, G., Piffanelli, P., Greco, R., Gagneur, C., Weber, C., González, V.M., Cabot, P., Fornara, F., Berri, S., Miro, B., Lan, P., Rafel, M., Capell, T., Puigdomènech, P., Ouwerkerk, P.B., Meijer, A.H., Pe', E., Colombo, L., Christou, P., Guiderdoni, E., Pereira, A. (2005) EU-OSTID: a collection of transposon insertional mutants for functional genomics in rice. *Plant Mol Biol.* **59**, 99–110.
81. Luan, W.J., He, C.K., Hu, G.C., Dey, M., Fu, Y.P., Si, H.M., Zhu, L., Liu, W.Z., Duan, F., Zhang, H., Liu, W.Y., Zhuo, R.Y., Garg, A., Wu, R. and Sun, Z.X. (2008) An efficient field screening procedure for identifying transposants for constructing an Ac/Ds-based insertional-mutant library of rice. *Genome.* **51**, 41–9.
82. Jiang, S.Y., Bachmann, D., La, H., Ma, Z., Venkatesh, P.N., Ramamoorthy, R. and Ramachandran, S. (2007) Ds insertion mutagenesis as an efficient tool to produce diverse variations for rice breeding. *Plant Mol Biol.* **65**, 385–402.
83. Park, S.H., Jun, N.S., Kim, C.M., Oh, T.Y., Huang, J., Xuan, Y.H., Park, S.J., Je, B.I.,

Piao, H.L., Park, S.H., Cha, Y.S., Ahn, B.O., Ji, H.S., Lee, M.C., Suh, S.C., Nam, M.H., Eun, M.Y., Yi, G., Yun, D.W. and Han, C.D. (2007) Analysis of gene-trap Ds rice populations in Korea. *Plant Mol Biol.* **65**, 373–84.

84. Ito, Y., Eiguchi, M. and Kurata, N. (2004) Establishment of an enhancer trap system with Ds and GUS for functional genomics in rice. *Mol Genet Genomics.* **271**, 639–50.

85. Greco, R., Ouwerkerk, P.B., Taal, A.J., Sallaud, C., Guiderdoni, E., Meijer, A.H., Hoge, J.H. and Pereira, A. (2004) Transcription and somatic transposition of the maize En/Spm transposons system in rice. *Mol Genet Genomics.* **270**, 514–23.

Chapter 17

Chromatin Beacons: Global Sampling of Chromatin Physical Properties Using Chromatin Charting Lines

Aniça Amini, Chongyuan Luo, and Eric Lam

Abstract

The extent to which physical properties and intranuclear locations of chromatin can influence transcription output remains unclear and poorly quantified. Because the scale and resolution at which structural parameters can be queried are usually so different from the scale that transcription outputs are measured, the integration of these data is often indirect. To overcome this limitation in quantifying chromatin structural parameters at different locations in the genome, a Chromatin Charting collection with 277 transposon-tagged *Arabidopsis* lines has been established in order to discover correlations between gene expression and the physical properties of chromatin loci within the nuclei.

In these lines, dispersed loci in the *Arabidopsis* genome are tagged with an identical transgene cassette containing a luciferase gene reporter, which permits the quantification of gene expressions in real time, and an ~2 kb *LacO* repeat that acts as a "chromatin beacon" to facilitate the visual tracking of a tagged locus in living plants via the expression of LacI–GFP fusion proteins in *trans*. In this chapter, we describe the methods for visualizing and tracking these insertion loci in vivo and illustrate the potential of using this approach to correlate chromatin mobility with gene expression in living plants.

Key words: *Arabidopsis*, Epigenetics, Luciferase, Fluorescence microscopy, 3D nuclear space

1. Introductions

Eukaryotic genomes are organized into morphologically distinct subnuclear structures (e.g., nucleoli) and functional domains (e.g., heterochromatin vs. euchromatin). Even in the interphase cells, where chromatin is known to be decondensed and relatively accessible as compared to those in the mitotic nuclei, the existence of these structural and functional organization units raises the possibility that large-scale physical organization of the genome could be a mechanism through which processes such as transcription can be regulated (1). However, direct causal relationships between the physical and functional properties of chromatin have been difficult

James A. Birchler (ed.), *Plant Chromosome Engineering: Methods and Protocols*, Methods in Molecular Biology, vol. 701, DOI 10.1007/978-1-61737-957-4_17, © Springer Science+Business Media, LLC 2011

to resolve. For example, the structural integrity of the nucleolus apparently depends on an intact RNA Polymerase I complex (2), while expressed genes can be found within pericentromeric regions that are heterochromatic (3), albeit at much lower density. These observations serve as examples that illustrate the dilemma of distinguishing whether an apparent subnuclear structure is a consequence of the organized functionality of macromolecular machines or the spatial organization of chromatin is in fact a distinct type of global control mechanism. The existence of possible microheterogeneities within large heterochromatic domains further complicates the analyses (4). Nevertheless, recent technological advances have provided additional supporting evidence, although not definitive proof, for the existence of some level of three-dimensional (3D) order of genome organization in interphase cells and its importance for global gene regulations (1, 5).

Deployment of multicolored fluorescence in situ hybridization (FISH) technology in the past decade has revealed that spatial localization of genomic loci can be highly correlated with gene expression in eukaryotic cells (6). Domains occupied by chromosomes in interphase nuclei are called chromosome territories (CTs), with little evidence for intermixing between chromosomes. Each CT is separated from its neighbors by interchromatin domains (7) that are thought to represent the "accessible" regions within the nucleus where most of the active genes and decondensed facultative heterochromatin reside. CTs are organized in 50–200 kbp loops called small loops (SLs), and these SLs form a chromatin rosette-like structure where the conformation of each SL may change according to the functional status of the genes residing within (8). In contrast, constitutive heterochromatic regions are organized as chromocenters (CCs) that contain high levels of cytosine methylation as well as repetitive sequences, whereas euchromatin typically emanates from CCs as loops at the periphery of the CT (8). In addition to heightened DNA methylation, heterochromatin and repressive domains of the genome are also known to correlate with specific types of posttranslational modifications on the N-terminal tails of core histones in nucleosomes (9). Diverse types of covalent modifications such as phosphorylation, ubiquitinylation, acetylation, and methylation on these proteins have been characterized. For example, histone H3 can be methylated at different lysines or arginines and to different extents on the free amino groups on each of these lysine and arginine residues (e.g., mono-, di-, and trimethylation). Suppression of transcription in heterochromatin is generally correlated with the trimethylation of histone H3 at lysine 9 (H3K9me3) in animal cells but with the dimethylation of H3K9 in plants (10). In contrast, transcription repression induced by Polycomb-group protein (PcG) typically correlates with trimethylation of lysine 27 (H3K27me3) and are also accompanied by a depletion of dimethylation of lysine 4 (H3K4me2) and lysine 36 (H3K36me2) at the promoter region and the 3′ portion of the transcription unit respectively (11). In *Arabidopsis*, H3K27me3 is specifically associated with the protein TERMINAL FLOWER2 (TFL2)/LIKE HETEROCHROMATIN PROTEIN1 (LHP1). LHP1 together with AtRING1a and AtRING1B forms the polycomb repressive complex 1 (PRC1)-like complex, which can bind H3K27me3 via the chromodomain of LHP1 (12). The PRC1-like complex is not directly involved in the deposition of H3K27me3 mark, but it represses the transcriptional activities of genes targeted by the polycomb repressive complex 2 (PRC2) such as FLOWERING LOCUS T (FT) or FLOWERING LOCUS C (FLC) genes that are involved in flowering time determination (13). The exact mechanism of gene repression by the PRC complex remains to be worked out in plants, with the formation of more compacted chromatin domains a probable scenario.

Using genome-wide methods that examine associations of genomic sequences to various subnuclear compartments in yeasts, it has been found that suppressed genes are predominantly localized to the nuclear periphery, similar to other cases that have been reported for individual loci in animal models (14). However, it has also been observed that gene activation in yeasts can result in the directed association of the activated locus to the nuclear pore (15), thus indicating that the nuclear periphery can also be heterogeneous with some locations containing actively expressing genes, while others contain inhibitory functions. In addition to the possible role of nuclear periphery tethering in regulating gene expression, nucleolar associations via linkage to tRNA genes have also been reported to result in silencing of RNA Pol II-dependent genes in yeast (16). These and other studies indicate that higher order genome organizations can be dynamic and may correlate with changes in gene expression patterns as well as cell type specifications that associate with reversible chromatin modifications (17). Elucidating the mechanisms of how different regions of the genome are localized to distinct compartments and regions within the nucleus will thus be important for delineating the causes and effects for these correlative observations. A number of recent studies using nuclear lamina tethering to remodel the locations of target loci indicate that nuclear periphery localization can result in silencing of gene expression (14, 18). In addition, other approaches aimed at disrupting key proteins involved in creating and/or maintaining specific types of large-scale chromatin structures will also be useful for determining the importance of spatial organization for global gene regulation (19).

Aside from methods aimed at perturbing the "normal" organization of the genome in the interphase nucleus, methods that can allow the quantification of physical properties of chromatin at multiple loci can also reveal evidence for global controls through genome-wide association analyses. Recently, adaptation of NextGen ultrahigh-throughput sequencing technology (Illumina GA2) to map nearest neighbor relationships by the so-called Chromosome

Conformation Capture strategy (20) has successfully generated the first glimpse of the 3D packing of the human genome at a resolution of about 1 Mbp (21). In addition to confirming the existence of CTs, this Hi-C method revealed clear evidence for two genome-wide compartments representing active and repressed portions of the genome, which map respectively to the open and closed chromatin domains as assayed by sensitivity to DNaseI digestion in two different cell types. Further refinements and applications of this technology should reveal much about the spatial organization of chromatin and how it may be controlled dynamically.

Implicit in the description of transcriptionally repressed chromatin domains and heterochromatin as "closed" or "condensed" is the prediction that these silenced loci are spatially more compacted and less accessible to proteins. In addition, one may also expect these loci to have slower diffusion rates as well as more constrained mobility relative to open chromatin within the nucleoplasm. The direct measurement of local chromatin mobility at the scale of a single gene (i.e., 10 kb or less) has only been possible since 1997, using a visual tracking system (also referred to as a chromatin beacon) consisting of a tandem array of the *LacO* operator sequence and the regulated expression of a LacI–GFP fusion protein (22). Using this approach, the movement of chromatin has been studied in intact cells of yeast, animal, and plant model systems. In the cases that have been reported, the data is best fitted to a constrained random diffusion model and diffusion coefficients observed typically in the range of 10^{-4} $\mu m^2 s^{-1}$, with rates as fast as 1×10^{-3} $\mu m^2 s^{-1}$ (23) and as slow as 0.3×10^{-4} $\mu m^2 s^{-1}$ having been reported (24). The confinement radius in the *Arabidopsis* nucleus also ranges from 0.1 to 0.4 μm, depending on the ploidy and activity state of the locus (24, 25). This chromatin beacon technique thus provides the means by which the physical location and dynamic behavior of specific insertions in a genome can be measured nondestructively. Coupling this tool with a convenient reporter gene for the quantification of transcriptional activity should allow one then to correlate some of the physical parameters of the insert's location to its associated gene expression potential. In our lab, we have recently characterized a collection of 277 transgenic *Arabidopsis* lines, called Chromatin Charting (CC) lines, in which a common, single-copy insertion element containing a 35S-Luc reporter gene and a *LacO* array of 2–2.2 kb are present (24). This construct thus permits one to visually track the insertion loci via the chromatin beacon and to quantify the linked Luc gene's expression at the corresponding locations in the genome of living plants. To construct the CC lines, wild-type *Arabidopsis thaliana* (ecotype Columbia) plants were first transformed with the CCP4 construct. CCP4 construct contains a Luciferase gene (Luc) driven by the constitutive CaMV 35S promoter for quantification of gene expression, a *LacO* tandem array to tag the locus of insertion, a Nos:NptII:nos expression cassette

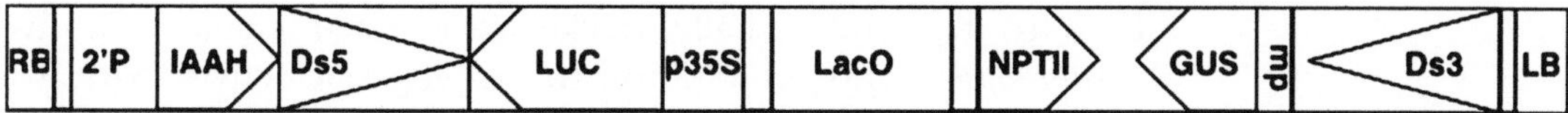

Fig. 1. Chromatin charting vector and its uses. *RB and LB* right and left border of T-DNA; *2′P* promoter for the IAAH selection gene; *IAAH* indole acetamide hydrolase gene; *DS5 and DS3 5′* and 3′ border sequence for maize Ds element; *LUC* firefly luciferase gene; *35S* CaMV 35S promoter; *LacO* Lac operator array; *NPTII* neomycin phosphotransferase gene; *GUS* β-glucuronidase gene; *Mini* minimal CaMV 35S promoter (−46 to +8). Sizes of elements are not drawn to scale.

for positive selection, and the GUS gene under the control of a minimal 35S promoter to detect possible enhancer elements near the insert sites (Fig. 1). This gene unit is flanked by two maize Ds sequences to create a nonautonomous transposable element. At the proximal end of the Ds5 sequence is placed a 2P′:IAAH marker gene expressing an indole acetamide hydrolase gene that confers 1-naphthaleneacetamide (NAM) sensitivity to plants. Homozygous transgenic plants containing single copies of the CCP4 construct were then crossed with homozygous CCP5 transgenic plants, which express constitutively the maize Ac transposase (24), and F1 progeny were obtained and screened by double selection for novel transposants (CCT lines). This transposon-aided tagging system has previously been described for the creation of gene-trap and enhancer-trap lines in *Arabidopsis* (26). The screening by Kan/NAM resulted in the isolation of 611 chromatin charting transposants (CCT lines) from seven CCP4 launchpad lines. Six CCP4 lines and 271 CCT lines have been mapped by tail PCR and further confirmed through locus specific PCR analyses. Sequence and functional characterizations of these lines can be obtained from the Cold Spring Harbor web site for our Chromatin Charting project (http://charting.cshl.org/).

Quantifying Luc activity in 277 CC lines dispersed throughout the five chromosomes of *Arabidopsis* generated the first transcription potential map and revealed insertion locations that may facilitate or repress transcription of an inserted gene. Focusing on a selected 100 kbp region on Chr. 2 that has a relatively high concentration of insertion lines represented in our CC collection, we characterized in depth the functional and physical properties of several insertion lines that showed interesting position effects. While DNA methylation is apparently required for maintaining the silencing observed in some of the insertion lines within this region, differential effects on nucleolar periphery association were observed for some of the CC lines examined. In general, however, increased locus mobility is well correlated with higher activity of the Luc reporter gene (24). The CC line collection, thus, provides a unique resource to begin a systematic comparison of the physical properties of the *Arabidopsis* genome at various tagged locations and to correlate the observed behavior with the associated gene expression potential at that position. In this chapter, we focus on describing the details for using these CC lines to tag and visually track the inserted element in live *Arabidopsis* plants, in addition to assaying Luc activity with live plants for correlation studies between structural parameters and transcription potential under similar conditions.

2. Materials

2.1. Seed Germination

1. 50% Bleach.
2. Solid Growth Medium: 0.5× Murashige and Skoog (MS) mineral salts, 1% sucrose, 0.8% agar, 50 μg/ml Kanamycin (for hemizygous CC lines).
3. 0.1% bacto-agar (Sigma Aldrich).
4. SterilGARD Laminar Flow Hood.

2.2. Luciferase Assays

1. D-Luciferin, potassium salt (Gold Biotechnology) (0.3 mg/mL D-Luciferin, 0.01% Triton X-100).
2. Biotek Synergy™ HT Multi-Detection Microplate Reader.
3. White opaque 96-well plate (Costar).
4. Lumazone FA bioluminescence/chemiluminescence/fluorescence macroscopic imaging system (MAG Biosystems) equipped with a Photometrics 1024B CCD camera.

2.3. In Vivo Visualization of Tagged Loci and Microscopy

1. Dexamethasone stock solution (3 mM dexamethasone in ethanol; Gold Biotechnology).
2. Ethanol (>95%).
3. *Agrobacterium tumefaciens* GV3101/mp11 strain transformed with pJM71 or pEL700.
4. Applied Precision DeltaVision image restoration microscope system Version 3.5 equipped with Nikon TE200 microscope and CH350 high-speed camera (Photometrics).
5. softWoRx 3.6.1 Suite software package (default analytical software of DeltaVision system).

3. Methods

3.1. Germination of Seeds

3.1.1. Germination of Seeds on 96-Well Plates

1. Dispense 200 μL of 0.5× MS solid media (kept at 60°C) per well in 96-well plates (plates are allowed to solidify at 4°C for several hours).
2. For surface sterilization, seeds of CCT or CCP lines are vortexed for 5 min in eppendorf tubes with 50% bleach and then rinsed with sterile water five times in a laminar flow hood.

3. Resuspend seeds in 0.1% agar, and deposit five surface-sterilized seeds in each well of a 96-well plate.
4. Synchronize seed germination by placing the plates at 4°C in the dark for 72 h, before transferring plates to a growth room set at 22°C under continuous light.

3.1.2. Germination of Seeds on Petri Dishes

1. Seeds of CCT or CCP lines are sterilized by vortexing for 5 min with 50% bleach and then rinsed with sterile water five times in a laminar flow hood.
2. Spread seeds on 0.5× MS plates or selection plates with Kanamycin (for hemizygous lines). Synchronize seed germination by placing plates at 4°C in the dark for 72 h, before transferring plates to a growth room at 22°C (as above) and set vertically for 7 days.

3.2. In Vivo Luciferase Assay

To quantify relative Luc expression level in whole plants, Luc activity can be measured with a microtiter plate reader or a high-performance CCD camera system (Lumazone Fluorescence Automated Imaging System, MAG Biosystems). The measurement of Luc activity with the microtiter plate reader allows one to screen transgenic plants in a high-throughput format, using 96-well plates. On the other hand, with the CCD camera technology, it is possible to quantify Luc expression from different tissues and organs of a single live plant. With spatial resolution at the macroscopic scales, the CCD camera approach enables the comparison of Luc expression at the whole plant level in real time. These two in vivo methods of Luc expression assay are, thus, complementary in their applications.

3.2.1. Luciferase Assay with the Microtiter Plate Format

1. Spray the plants directly on the 96-well plate with Luciferin working solution.
2. Leave the plate on the bench for 5 min to allow uptake of d-Luciferin into the plant tissues and cells.
3. Set up the program for the plate reader and load the assay plate onto the plate reader.
4. Read for at least 30 min.

3.2.2. Luciferase Assay with the Lumazone FA Imaging System

1. Spray the petri dish with Luciferin working solution and leave the plate for 10 min on the bench at room temperature.
2. Put the plate into the ultradark chamber of the Lumazone FA system and take a bright-field image of the sample for recognition of the plant outlines.
3. Keep the plate for 5 min in the dark to eliminate interference from delayed fluorescence of chlorophyll in the green tissues.
4. Acquire bioluminescence for 7 min in the dark at room temperature through the chemiluminescence detection channel

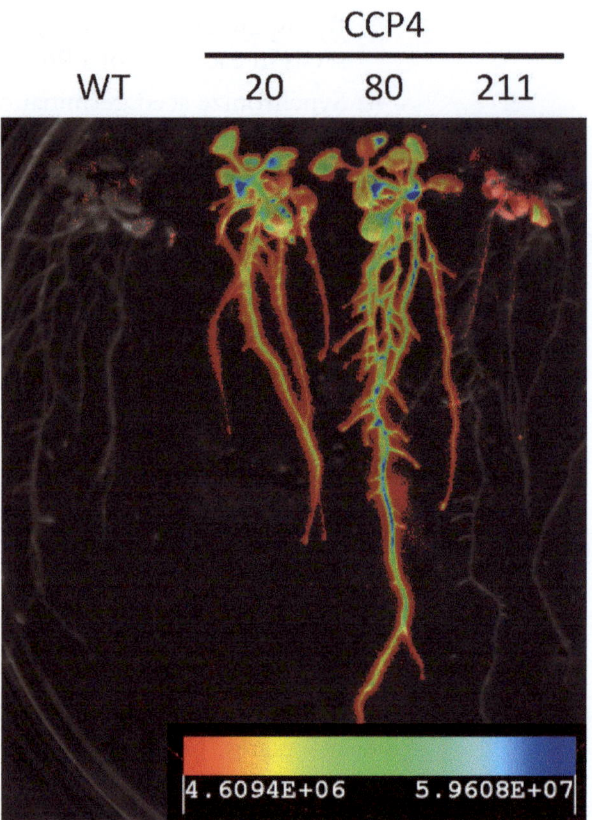

Fig. 2. Differential expression of luciferase activity in CC lines visualized by Lumazone FA. Bright-field and bioluminescence images were acquired separately and then merged. Acquisition was done with a MS agar plate containing wild-type (WT), CCP4.20, CCP4.80, and CCP4.211 plants. The color scale of luminescence shown represents the range of 4.8×10^6 photons s^{-1} (*red*) to 4.8×10^7 photons s^{-1} (*blue*).

of the Lumazone FA. The bioluminescence can then be superimposed onto the bright-field image to facilitate subsequent quantification (Fig. 2).

5. Quantify Luc activity from live plants. Luc activity of a single plant was defined as the total pixel intensity divided by the area on the plate that is covered by the plant. The count/size function of Lumazone Analyzer 2.0 was applied to the bright-field images to identity the outline for areas covered with plant tissues. The measurement of pixel intensities was performed with the Measurement Tool from the Lumazone FA software. Average pixel intensities of wild-type plants were determined as system background and were subtracted from each measurement.

3.3. Microscopy Assay

Since the construct in the CC lines does not contain any cassette for LacI–GFP fusion protein expression, visualization of the *LacO* array tag requires the supply *in trans* of this protein. This can be achieved through crossing of the CC lines with a CCV (Chromatin

Charting Visualization) line that expresses a GFP–LacI–NLS protein. Ideally, the expression level of the GFP–LacI–NLS protein should be controlled carefully in order to maximize the signal to background ratio (GFP–LacI–NLS bound to the *LacO* array vs. diffuse free GFP–LacI–NLS in the nucleoplasm) for detecting the tagged loci. Each 2-kb-long *LacO* repeat in the CC lines can bind no more than ~128 GFP-LacI-NLS molecules (assuming about 64 copies of the *LacO* sequence). Therefore, a background fluorescence significantly weaker than that emitted by 128 GFP molecules will be essential for the detection of the tagged locus. In our studies, chemical inducible expression systems were chosen to express the GFP–LacI–NLS (or other spectral variants of GFP) protein under tight transcriptional regulation. In our lab, we have used successfully two visualization constructs (named CCV constructs) in combination with the CC lines, through which the expression of GFP–LacI–NLS protein can be induced by either ethanol (pJM71) or Dexamethasone (pEL700) treatment of the F1 progeny (Fig. 3).

3.3.1. Introducing CCV Constructs into Chromatin Charting Lines

The CCV construct can be introduced into Chromatin Charting (CC) lines using three approaches: (1) Transform CCV construct into wild-type *Arabidopsis* and screen for stable transgenic CCV lines. CC lines of interests are then crossed with selected CCV lines. (2) CC lines can be directly transformed with the pEL700 vector. pJM71 cannot be used for this method as it has the same selection marker (Kanamycin resistance) as found in the CC lines. (3) Transient expression of GFP–LacI–NLS protein in leaves of CC lines through infiltration of *Agrobacterium* containing CCV vectors and subsequently activated by chemical induction. We recommend the first approach if consistent inducible expression of GFP–LacI–NLS in multiple cell types and if quantitative comparison between different CC lines are desired. The direct transformation of CC lines with a CCV construct will also allow visualization in various cell types. However, one should expect

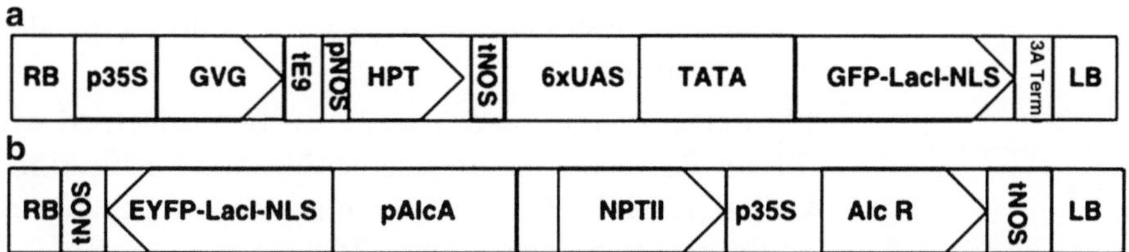

Fig. 3. Two vectors that can be used to generate CCV lines. (**a**) pEL700: *GVG* glucocorticoid binding domain/VP16 acidic activation domain/Gal4 DNA-binding domain fusion protein; *tE9* pea *rbcS-E9* terminator; *pNOS* promoter from the nopaline synthase (NOS) gene of *Agrobacterium*; *HPT* Hygromycin phosphotransferase; *tNOS* terminator from NOS; *6 X UAS* six tandem copies of the upstream activation sequence for yeast Gal4 binding; *TATA* TATA box; *GFP-LacI-NLS* green fluorescent protein/Lac repressor/nuclear localization signal fusion protein; *3A Term* pea *rbcS-3A* terminator; (**b**) pJM71: *EYFP–LacI–NLS* enhanced yellow fluorescent protein/Lac repressor/nuclear localization signal fusion protein; *pAlcA alcA* promoter from *Aspergillus*; *AlcR* alcohol-responsive *alcR* transcription factor from *Aspergillus*.

complexities such as transgene silencing of the CCV construct as frequently seen with *Agrobacterium* transformation. Thus, comparison between independent transformed lines with the CCV construct inserted at different locations may be problematic. The last method is suitable for working with a large number of CC lines. Chromatin dynamics can only be visualized conveniently in a few cell types (i.e., leaf epidermal cells and mesophyll cells) using this approach. We focus on describing methods relating to methods 1 and 2 in the subsequent sections, since method 3 has been described previously (27).

3.3.2. Visualization of the LacO Array in CC Lines with Stably Transformed CCV Lines

1. Cross selected CC lines with a stable CCV line that can express GFP–LacI–NLS under ethanol (JM71) or dexamethasone (EL700) induction respectively (24, 28).
2. The *LacO* array can then be visualized in either F1 or F2 plants generated from the cross. Quantifying chromatin dynamics will require at least two tagged alleles. Therefore, chromatin dynamics can only be measured in endoreduplicated cells of F1 plants, while diploid cells (e.g., guard cells) can be assayed with plants that are homozygous for the CC locus in the F2 generation. The coexistence of the CC construct and the EL700 construct can be selected by using plates containing both 50 μg/ml kanamycin and 10 μg/ml hygromycin. Since pJM71 as well as our Chromatin Charting construct confers kanamycin resistance, fluorescence microscopy after ethanol induction and PCR-assisted genotyping together will be required for characterizing the identity of individual F1 and F2 progeny.

3.3.3. Direct Transformation of Chromatin Charting Lines with pEL700

1. Transform selected Chromatin Charting lines with *Agrobacterium* containing pEL700 (28) by the standard floral dipping protocol (29).
2. Germinate T1 seeds on 0.5× MS solid media supplemented with 50 μg/ml kanamycin, 10 μg/ml hygromycin, and 200 μg/ml carbenicillin. Resistant seedlings should be transferred to a fresh plate with 0.5× MS solid media at ~10 days after germination.
3. Fluorescence microscopy can then be performed with either T1 or T2 seedlings after screening and verification of their GFP–LacI–NLS fusion protein's induction characteristics by epifluorescence microscopy (28).

3.3.4. Induction of GFP–LacI–NLS Expression by Ethanol or Dexamethasone

1. For inducing expression of GFP–LacI–NLS in EL700 transgenic plants, float 1- to 2-week-old *Arabidopsis* seedlings on 0.3 μM dexamethasone (stock solution diluted 10,000× with water) for 12–16 h at room temperature under identical illumination conditions as described above.

2. Expression of YFP–LacI–NLS fusion protein from JM71 transgenic plants can be induced by overnight treatment of seedlings with 1% ethanol vapor on 0.5× MS plates.

3.3.5. 3D-Restoration Microscopy of Chromatin Dynamics with LacO-Tagged Insertions in Live Plants

Three dynamic parameters of *LacO*-tagged insertion loci can be determined through in vivo 3D-restoration microscopy imaging, once an active CCV construct has been introduced into a CC line (24, 25): (1) the radial position of the tagged locus within the nucleus that is often represented as the distance of the GFP labeled "spot" to the nuclear or nucleolar periphery; (2) the confinement radius of the tagged locus, which may be correlated with high-order chromatin organization and transcription activities; (3) the diffusion coefficient of the tagged locus that correspond to the intranuclear mobility of chromatin at the location of the insert in the particular CC line.

1. Mount the induced seedling on a microscope slide with a drop of water and cover with a glass coverslip.
2. Locate the region of interest using lower magnification (40–200×). Mark the area with the "Mark Point" function of softWoRx software to allow the return to the same location at a later time.
3. Acquire fluorescence images of *Arabidopsis* nuclei using a 600–1,000× magnification. We generally acquire 40–60 Z-stacks with 0.2-μm Z-steps. Perform a trial Z-stack scan to ensure that the whole nucleus can be covered. The target nuclei can then be tracked for 10 min in order to extract dynamic information. The exposure time for each set of images needs to be empirically optimized to generate images with required quality while at the same time minimizing photobleaching of the sample.
4. The acquired images are then deconvolved by the deconvolution tool included in the softWoRx suite. An example from a 3D-restored image of an *Arabidopsis* nucleus rendered from the fluorescence microscopy is shown in Fig. 4 from two different angles.
5. In order to distinguish *LacO* array spots from random fluctuations in nuclear GFP background fluorescence, clusters of pixels with average intensity $o > m + 3.3S$ were determined as fluorescent spots corresponding to the inserted *LacO* array, where m is the average pixel intensity of the nucleus and S represents the standard deviation.
6. To define the radial position of the decorated *LacO* array spots within nuclei, the 3D coordinate of the spots is defined as the position of the brightest pixel. The distance between the center of the fluorescent spot and nuclear or nucleolar periphery is measured in several consecutive Z-stacks with the

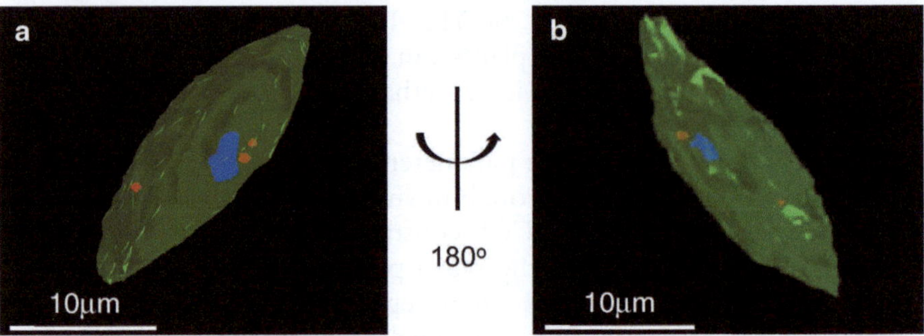

Fig. 4. 3D-model rendition of an endoreduplicated nucleus in CCP4.211 plants. *LacO* array spots were represented as *red* dots dispersed within the nucleoplasm (*green*). Nucleolus, an area that is underrepresented in free GFP fusion protein, is shown in *blue*. Two different tilted views for the same 3D-model reconstructed from the Z-stack of images collected are shown in adjacent panels. A size bar representing 10 μm is shown on each panel.

Distances tool of the softWoRx suite. The minimal values from these measurements were taken as the distance between the fluorescent spot and a particular periphery.

7. Derivation of both confinement radius and diffusion coefficient requires precise measurement of distance between two *LacO* array spots in the same nucleus as a function of time. We calculate distances between *LacO* array spots from their 3D coordinates using the Pythagorean theorem. The squared distance between two spots A and B, $d_{A,B}{}^2=(x_a-x_b)^2+(y_a-y_b)^2+((z_a-z_b)\times s)^2$, where (x_a, y_a, z_a) and (x_b, y_b, z_b) each refers to the 3D coordinate for spots A and B, respectively; s equals to the step distance between two consecutive Z-stacks.

8. Plot $\Delta d_t^{\,2}=(d_t-d_0)^2$ between two *LacO* array spots against the elapsed time t, where d_t and d_0 represent the distance between two spots at time points t and 0, respectively. The characteristic of confined diffusion is that $\Delta d_t^{\,2}$ will reach a plateau after a certain period of observation. In contrast, two freely diffusing particles will give rise to an infinitely increasing Δd^2. The confinement radius can be derived from the value of $\Delta d_t^{\,2}=(d_t-d_0)^2$ at the plateau. For each tagged insertion locus, an average confinement radius computed from multiple spot pairs can thus be used to describe the dynamic behavior.

References

1. Lanctôt, C., Cheutin, T., Cremer, M., Cavalli, G. and Cremer, T. (2007) Dynamic genome architecture in the nuclear space: regulation of gene expression in three dimensions. Nat Rev Genet. 8, 104–15.
2. Chen, H.K., Pai, C.Y., Huang, J.Y. and Yeh, N.H. (1999) Human Nopp140, which interacts with RNA polymerase I: implications for rRNA gene transcription and nucleolar structural organization. Mol Cell Biol. 19, 8536–46.
3. Rossi, F., Moschetti, R., Caizzi, R., Corradini, N. and Dimitri, P. (2007) Cytogenetic and molecular characterization of heterochromatin gene models in *Drosophila melanogaster*. Genetics. 175, 595–607.

4. Lam, E., Luo, C. and Watanabe, N. (2009) Charting functional and physical properties of chromatin in living cells. Curr Opin Genet Dev. **19**, 135–41.
5. Goetze, S., Mateos-Langerak, J., Gierman, H.J., de Leeuw, W., Giromus, O., Indemans, M.H., Koster, J., Ondrej, V., Versteeg, R. and van Driel, R. (2007) The three-dimensional structure of human interphase chromosomes is related to the transcriptome map. Mol Cell Biol. **27**, 4475–87.
6. Verschure, P.J., van Der Kraan, I., Manders, E.M. and van Driel, R. (1999) Spatial relationship between transcription sites and chromosome territories. J Cell Biol. **4**, 13–24.
7. Cremer, T., Cremer, M., Dietzel, S., Müller, S., Solovei, I. and Fakan, S. (2006) Chromosome territories-a functional nuclear landscape. Curr Opin Cell Biol. **18**, 307–16.
8. Fransz, P., De Jong, J.H., Lysak, M., Castiglione, M.R. and Schubert, I. (2002) Interphase chromosomes in Arabidopsis are organized as well defined chromocenters from which euchromatin loops emanate. Proc Natl Acad Sci USA. **99**, 14584–9.
9. Mersfelder, E.L. and Parthun, M.R. (2006) The tale beyond the tail: histone core domain modifications and the regulation of chromatin structure. Nucleic Acids Res. **34**, 2653–62.
10. Jackson, J.P., Johnson, L., Jasencakova, Z., Zhang, X., PerezBurgos, L., Singh, P.B., Cheng, X., Schubert, I., Jenuwein, T. and Jacobsen, S.E. (2004) Dimethylation of histone H3 lysine 9 is a critical mark for DNA methylation and gene silencing in *Arabidopsis thaliana*. Chromosoma. **112**, 308–15.
11. Fuchs, J., Demidov, D., Houben, A. and Schubert, I. (2006) Chromosomal histone modification patterns-from conservation to diversity. Trends Plant Sci. **11**, 199–208.
12. Xu, L. and Shen, WH. (2008) Polycomb silencing of KNOX genes confines shoot stem cell niches in Arabidopsis. Curr Biol. **18**, 1966–71.
13. Turck, F., Roudier, F., Farrona, S., Martin-Magniette, M.L., Guillaume, E., Buisine, N., Gagnot, S., Martienssen, R.A. Coupland, G. and Colot, V. (2007) Arabidopsis TFL2/LHP1 specifically associates with genes marked by trimethylation of histone H3 lysine 27. PLoS Genet. **3**, 86.
14. Towbin, B.D., Meister, P. and Gasser, S.M. (2009) The nuclear envelope – a scaffold for silencing? Curr Opin Genet Dev. **18**, 180–6.
15. Casolari, J.M., Brown, C.R., Komili, S., West, J., Hieronymus, H. and Silver, P.A. (2004) Genome-wide localization of the nuclear transport machinery couples transcriptional status and nuclear organization. Cell. **117**, 427–39.
16. Thompson, M., Haeusler, R.A., Good, P.D. and Engelke, D.R. (2003) Nucleolar clustering of dispersed tRNA genes. Science. **302**, 1399–401.
17. Heintzman, N.D., Hon, G.C., Hawkins, R.D., Kheradpour, P., Stark, A., Harp, L.F., Ye, Z., Lee, L.K., Stuart, R.K., Ching, C.W., Ching, K.A., Antosiewicz-Bourget, J.E., Liu, H., Zhang, X., Green, R.D., Lobanenkov, V.V., Stewart, R., Thomson, J.A., Crawford, G.E., Kellis, M. and Ren, B. (2009) Histone modifications at human enhancers reflect global cell-type-specific gene expression. Nature. **459**, 108–12.
18. Pickersgill, H., Kalverda, B., de Wit, E., Talhout, W., Fornerod, M. and van Steensel, B. (2006) Characterization of the *Drosophila melanogaster* genome at the nuclear lamina. Nat Genet. **38**, 1005–14.
19. Mateos-Langerak, J., Goetze, S., Leonhardt, H., Cremer, T., van Driel, R. and Lanctôt, C. (2007) Nuclear architecture: is it important for genome function and can we prove it? J Cell Biochem. **102**, 1067–75.
20. Dostie, J., Richmond, T.A., Arnaout, R.A., Selzer, R.R., Lee, W.L., Honan, T.A., Rubio, E.D., Krumm, A., Lamb, J., Nusbaum, C., Green, R.D. and Dekker J. (2006) Chromosome Conformation Capture Carbon Copy (5C): a massively parallel solution for mapping interactions between genomic elements. Genome Res. **16**, 1299–309.
21. Lieberman-Aiden, E., van Berkum, N.L., Williams, L., Imakaev, M., Ragoczy, T., Telling, A., Amit, I., Lajoie, B.R., Sabo, P.J., Dorschner, M.O., Sandstrom, R., Bernstein, B., Bender, M.A., Groudine, M., Gnirke, A., Stamatoyannopoulos, J., Mirny, L.A., Lander, E.S. and Dekker, J. (2009) Comprehensive mapping of long-range interactions reveals folding principles of the human genome. Science. **326**, 289–93.
22. Marshall, W.F., Straight, A., Marko, J.F., Swedlow, J., Dernburg, A., Belmont, A., Murray, A.W., Agard, D.A. and Sedat, J.W. (1997) Interphase chromosomes undergo constrained diffusional motion in living cells. Curr Biol. 7, 930–9.
23. Vazquez, J., Belmont, A.S. and Sedat, J.W. (2001) Multiple regimes of constrained chromosome motion are regulated in the interphase *Drosophila* nucleus. Curr Biol. 11, 1227–39.
24. Rosin, F.M., Watanabe, N., Cacas, J.L., Kato, N., Arroyo, J.M., Fang, Y., May, B.,

Vaughn, M., Simorowski, J., Ramu, U., McCombie, R.W., Spector, D.L., Martienssen, R.A. and Lam, E. (2008) Genome-wide transposon tagging reveals location-dependent effects on transcription and chromatin organization in Arabidopsis. Plant J. **55**, 514–25.

25. Kato, N. and Lam, E. (2003) Chromatin of endoreduplicated pavement cells has greater range of movement than that of diploid guard cells in *Arabidopsis thaliana*. J Cell Sci. **116**, 2195–201.
26. Sundaresan, V., Springer, P., Volpe, T., Haward, S., Jones, J.D., Dean, C., Ma, H., and Martienssen, R. (1995) Patterns of gene action in plant development revealed by enhancer trap and gene trap transposable elements. Genes Dev. **9**, 1797–810.
27. Luo, C. and Lam, E. (2009) Chromatin charting: global mapping of epigenetic effects. Methods Mol Biol. **553**, 127–39.
28. Kato, N. and Lam, E. (2001) Detection of chromosomes tagged with green fluorescent protein in live *Arabidopsis thaliana* plants. Genome Biol. **2,** 1–10.
29. Zhang, X., Henriques, R., Lin, S.S., Niu, Q.W. and Chua, N.H. (2006) Agrobacterium-mediated transformation of *Arabidopsis thaliana* using the floral dip method. Nat Protoc. **1**, 641–6.

Chapter 18

Transposable Elements as Catalysts for Chromosome Rearrangements

Jianbo Zhang, Chuanhe Yu, Lakshminarasimhan Krishnaswamy, and Thomas Peterson

Abstract

Barbara McClintock first showed that transposable elements in maize can induce major chromosomal rearrangements, including duplications, deletions, inversions, and translocations. More recently, researchers have made significant progress in elucidating the mechanisms by which transposons can induce genome rearrangements. For the *Ac/Ds* transposable element system, rearrangements are generated when the termini of different elements are used as substrates for transposition. The resulting alternative transposition reaction directly generates a variety of rearrangements. The size and type of rearrangements produced depend on the location and orientation of transposon insertion. A single locus containing a pair of alternative transposition-competent elements can produce a virtually unlimited number of genome rearrangements. With a basic understanding of the mechanisms involved, researchers are beginning to utilize both naturally occurring and in vitro-generated configurations of transposable elements in order to manipulate chromosome structure.

Key words: Transposable elements, Chromosome rearrangements, Deletion, Duplication, Inversion, *Ac/Ds*, Maize, Rice, Arabidopsis

1. Introduction

This chapter will focus on the maize transposable element system *Activator/Dissociation* (*Ac/Ds*) as a catalyst of chromosome rearrangements. The maize *Ac/Ds* elements comprise a classical two-component system of the *hAT* transposon family (1). The autonomous *Ac* element is 4,565 bp in length, whereas the non-autonomous *Ds* elements vary in length and internal sequence composition. *Ac* and *Ds* elements share 11 bp terminal inverted repeat sequences (TIRs), and they also contain multiple copies of subterminal hexamer motifs (AAACGG or similar) located within

James A. Birchler (ed.), *Plant Chromosome Engineering: Methods and Protocols*, Methods in Molecular Biology, vol. 701, DOI 10.1007/978-1-61737-957-4_18,

250 bp of the element termini (2–4). The *Ac* element produces a single predominant spliced mRNA that encodes a 102 kD transposase protein. The transposase binds to the AAACGG hexamer motifs in the *Ac/Ds* subterminal regions (5–7). Although the exact mechanism of *Ac/Ds* transposition is unclear, evidence to date indicates that *Ac/Ds* elements transpose by a cut-and-paste mechanism (8–10). *Ac/Ds* excision is presumed to generate a chromosomal double-strand break (DSB), which is then religated to generate an excision site (11, 12). The *Ac/Ds* system has been used extensively for gene tagging in maize (13, 14) and heterologous plants including *Arabidopsis* (15–19), petunia (20), tomato (21), and rice (22, 23). *Ac/Ds* transposition has also been demonstrated in yeast, zebrafish, and human cells (24, 25). These examples show that the basic biology of *Ac/Ds* transposition is shared with that of *hAT* transposons in a variety of organisms.

1.1. Sister Chromatid Transposition

McClintock reported that the *Ac/Ds* system could induce chromosomal rearrangements, such as deletions, duplications, inversions, and reciprocal translocations (26–32). Whereas, analyses of single *Ac* or *Ds* elements showed that standard *Ac/Ds* transposition changes only the position of the elements, but does not generate chromosomal rearrangements. By studying chromosome-breaking state I *Ds* elements, English et al. (33), Weil and Wessler (35), and Martinez-Ferez and Dooner (34) showed that a pair of *Ds* 5′ and 3′ termini in direct orientation can undergo transposition reactions leading to ligation of sister chromatids (33–35). The covalent joining of sister chromatids occurs when the 5′ and 3′ *Ac/Ds* termini excise from the flanking DNA; transposon excision is followed by ligation of the two flanking DNA segments (the donor sequences). When the transposon termini are located on sister chromatids, donor sequence ligation generates a chromatid bridge which will break in subsequent divisions. This basic mechanism has been confirmed by analysis of deletion/duplication alleles isolated from a maize ear twinned sector. Sequences of the rearrangement junctions proved that these alleles were the reciprocal products of a single transposition event involving *Ac* 5′and 3′ ends from different sister chromatids (36); this reaction was termed sister chromatid transposition (SCT, Fig. 1.)

In the SCT model, reinsertion of the excised transposon ends into different sites in the chromatid bridge is predicted to generate a series of nested segmental deletions and inverted duplications. This prediction has been confirmed by the identification of 35 interstitial deletions and 15 inverted duplications flanking the maize *p1* gene on chromosome 1S. The deletions extend from the *p1* gene to various proximal endpoints, and range up to >4.6 cM (3.6 Mb) (37). The inverted duplications also extend proximally, and range in size from ~70 kb to ~14.7 Mb (Zhang

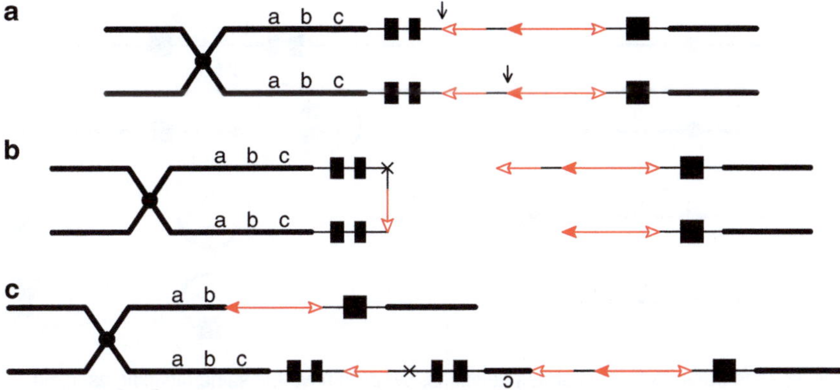

Fig. 1. Sister chromatid transposition generates reciprocal deletions and inverted duplications. The two lines indicate sister chromatids joined at the centromere (*black circle*). The boxes indicate exons 1, 2, and 3 (*left to right*) of the maize *p1* gene. *Red lines* with *arrow(s)* indicate *Ac* or *fAc* (terminally deleted 3′ fragment of *Ac*), and the *open* and *solid arrows* indicate the 3′ and 5′ ends, respectively, of *Ac*/*fAc*. The *small vertical arrows* indicate transposase cleavage sites. (**a**) *Ac* transposase cleaves at the 5′ terminus of *Ac* in one sister chromatid and the 3′ terminus of *fAc* in the other sister chromatid. (**b**) Cuts are made at the *Ac* and *fAc* termini. The two nontransposon ends join together to generate a chromatid bridge; minor sequence changes at the junction form a transposon footprint (×). (**c**) The excised transposon termini insert at a target site between *b* and *c*; the *fAc* 3′ end joins to the *c* side to generate one sister chromatid with an inverted duplication, while the *Ac* 5′ end joins to the *b* side to generate a second sister chromatid with a corresponding deficiency.

and Peterson, unpublished). Significantly, the rearrangement breakpoints are precisely bound by *Ac/Ds* termini, indicating that these are generated directly by transposon insertion (37). In summary, these results show that SCT can generate flanking deletions and inverted duplications of variable size. Reinsertion of the excised transposon termini into other chromosomes is also predicted to occur, but the products would be genetically unbalanced and not heritable from standard diploid parents.

1.2. Reversed-Ends Transposition

Alleles containing reversely oriented *Ac/Ds* 5′ and 3′ termini can undergo a second type of alternative transposition termed reversed-ends transposition (RET). In RET, transposition involves the apposed 3′ and 5′ termini of nearby *Ac/Ds* elements which are themselves in direct orientation. Excision of the transposon termini and ligation of the donor sequence is predicted to generate a covalently closed circular DNA molecule composed of the DNA between the two linked *Ac/Ds* elements (the intertransposon segment, or ITS). While direct evidence for the circularized ITS is lacking, its existence is inferred from rare cases in which the ITS becomes circularly permuted; these are proposed to result from insertion of the reversed-ends transposon into the circularized ITS (38). Otherwise, the ITS circle would be lost because it lacks a centromere. Insertion of the excised reversed-ends transposon

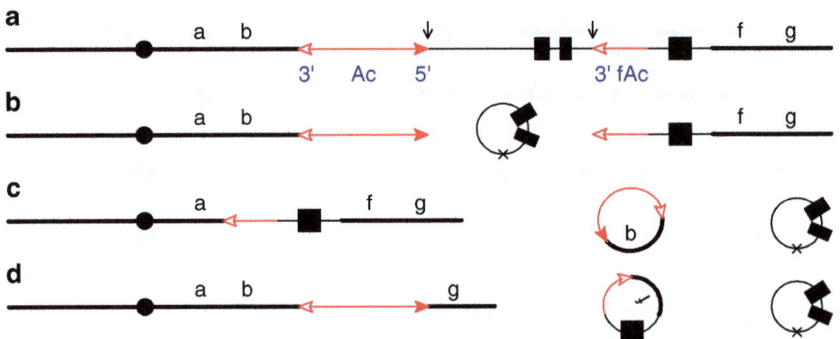

Fig. 2. Reversed *Ac* ends transposition generates deletions. Symbols have the same meaning as in Fig. 1. (**a**) *Ac* transposase cleaves at the 5′ end of *Ac* and the 3′ end of *fAc*. (**b**) Following transposase cleavage at the junctions of *Ac/p1* and *fAc/p1*, the internal *p1* genomic sequences are joined to form a circle. The "×" on the circle indicates a transposon footprint at the site of joining. (**c**) The excised transposon ends insert at a target site between *a* and *b*; the *fAc* 3′ end joins to the *a* side to generate a *p1* proximal deletion, while the *Ac* 5′ end joins to the *b* side to form a circle. (**d**) The excised transposon ends insert at a target site between *f* and *g*; the *Ac* 5′ end joins to the *g* side to generate a *p1* distal deletion, while the *fAc* 3′ end joins to the *f* side to form a circle (**c**, **d** are alternative outcomes).

into the flanking chromosomal DNA generates inversions or deletions, depending upon the insertion orientation (Fig. 2). Whereas, insertion into a different chromosome will produce (1) a pair of balanced reciprocal translocations, with the *Ac* termini precisely at the translocation junctions, or (2) a dicentric chromosome plus acentric fragment. Numerous cases of the former have been identified from alleles carrying *Ac* termini in reversed orientation on maize chromosome 1S (39). The latter class (comprising a dicentric plus an acentric fragment) would be expected to be unstable and nonheritable. The generation of large inversions and reciprocal translocations by RET has been confirmed by classical cytogenetic analysis and fluorescence in situ hybridization (FISH) (39). Moreover, the presence of excision footprints and 8 bp target site duplications at the inversion and translocation junctions confirm that these rearrangements are formed as direct products of *Ac/Ds* transposition (38–41). Similarly, deletions and inversions have also been reported as the result of RET at the maize *bz1* locus on chromosome 9S (42, 43). It is important to note that, unlike SCT that generates nested deletions on only one side of the transposon sequence, reversed *Ac* ends transposition can generate nested deletions on both sides of the transposon sequences.

1.3. Chromosome Rearrangements Induced by Engineered Ac/Ds Transposons in Maize, Rice, and Arabidopsis

To test whether alternative transposition-induced rearrangements could be reproduced in transgenic systems, we generated a series of transgene constructs containing *Ac/Ds* termini together with suitable reporter genes (Fig. 3). Constructs were transformed into Arabidopsis (44), rice, and maize (45) followed by selection for included marker genes. Plants were screened to identify those

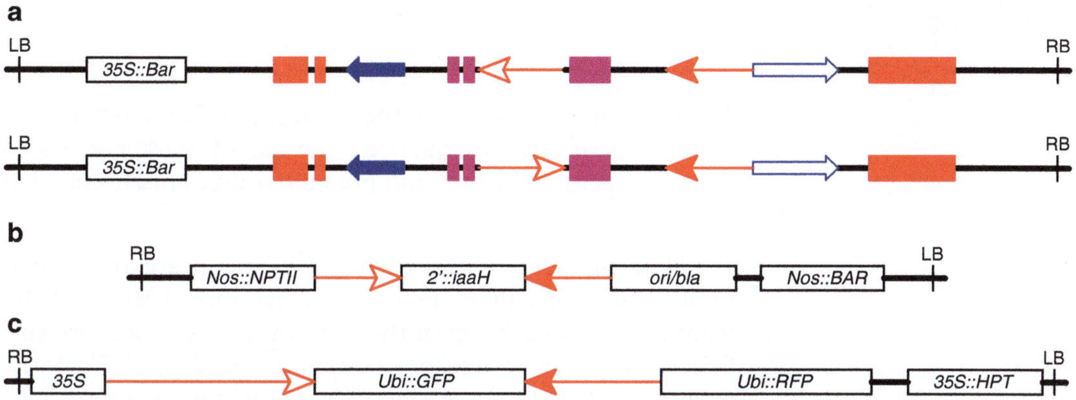

Fig. 3. Map of constructs used for maize, rice, and Arabidopsis transformation. In all constructs, LB and RB indicate the left- and right-border T-DNA sequence, respectively. The *open* and *solid red arrowheads* indicate the 3′ and 5′ ends, respectively, of *Ac*. The following lengths of *Ac* terminal sequences were used: For maize constructs, 307 and 239 bp from the *Ac* 5′ and 3′ termini, respectively; for Arabidopsis construct, 255 and 217 bp from the *Ac* 5′ and 3′ termini, respectively; for rice, 1,786 and 3,599 for the *Ac* 5′ and 3′ termini, respectively. (**a**) Maize constructs. The visual markers *p1* and *c1* are used to monitor transposition; *red boxes* indicate maize *p1* gene exons, and *purple boxes* indicate maize *c1* gene exons. The *open* and *solid blue arrowheads* are the 3′ and 5′ ends, respectively, of the *I/dSpm* element which is used to mobilize the *Ac* termini. Restoration of *p1* function indicates excision of *I/dSpm*, and loss of *c1* function indicates that alternative transposition events have occurred. (**b**) Arabidopsis construct. The *NPTII* gene confers resistance to kanamycin; *iaaH* gene is a negative selection marker that confers sensitivity to naphthalene acetamide; and the *BAR* gene confers resistance to bialaphos. The *ori/bla* segment (bacterial origin of replication and *beta-lactamase* gene conferring resistance to ampicillin) can be used for plasmid rescue of the T-DNA insertion locus. (**c**) Rice construct. *35S* indicates the CaMV 35S promoter driving the *Ac* transposase gene; *Ubi::GFP* and *Ubi::RFP* indicates the maize *ubiquitin1* promoter driving the GFP and RFP genes, respectively; *35S::HPT* indicates the CaMV 35S promoter driving the hygromycin phosphotransferase gene. The rice construct was modified from plasmid *pSQ5* (54).

containing intact single-copy insertions, and a source of *Ac* transposase was introduced by suitable crosses. Putative rearrangement events were identified by marker gene loss and confirmed by PCR analysis. Finally, rearrangement breakpoints were isolated by PCR methods, and sequenced. The results confirm that alternative transposition-induced rearrangements similar to those described above for natural maize alleles can be generated in transgenic maize, rice, and Arabidopsis (46).

1.4. Potential Uses of Transposon-Induced Chromosome Rearrangements

The chromosome rearrangements generated by transposable elements could have a variety of potential uses. For example, a set of local nested deletions can be used to functionally dissect a chromosome region or cluster of related genes (37). Duplications may be useful for increasing copy number of favorable genes whose expression is dose dependent. Inversions and translocations may provide favorable materials to study the position dependence of gene expression (position effect), and also as genetic tools for reproductive isolation. Finally, overlapping translocations and inversions can be used to generate segmental duplications with defined endpoints (47).

2. Materials

In this section, we describe the recommended lengths and configurations of *Ac/Ds* sequences; the use of marker genes to detect rearrangement events; and possible sources of *Ac*-encoded transposase.

1. *Ac/Ds* termini: Sequences of *Ac/Ds* elements are available in public databases. Previous research has shown that 238 bp from the 5′ and 209 bp of the 3′ terminal sequences are sufficient for element transposition (4). The 5′ and 3′ *Ac/Ds* termini can be cloned in the desired orientations using standard molecular techniques. For RET, the distance between the 5′ and 3′ termini (the ITS) appears to affect alternative transposition frequency. Highest transposition frequencies are observed for ITS lengths of ~0.4–4 kb; transposition frequency appears to drop dramatically for ITS lengths of <100 bp (Fig. 4). The maximum length of ITS which can still support RET is still unknown. Alleles with ITS of 13 kb can still undergo significant levels of RET, although at a lower

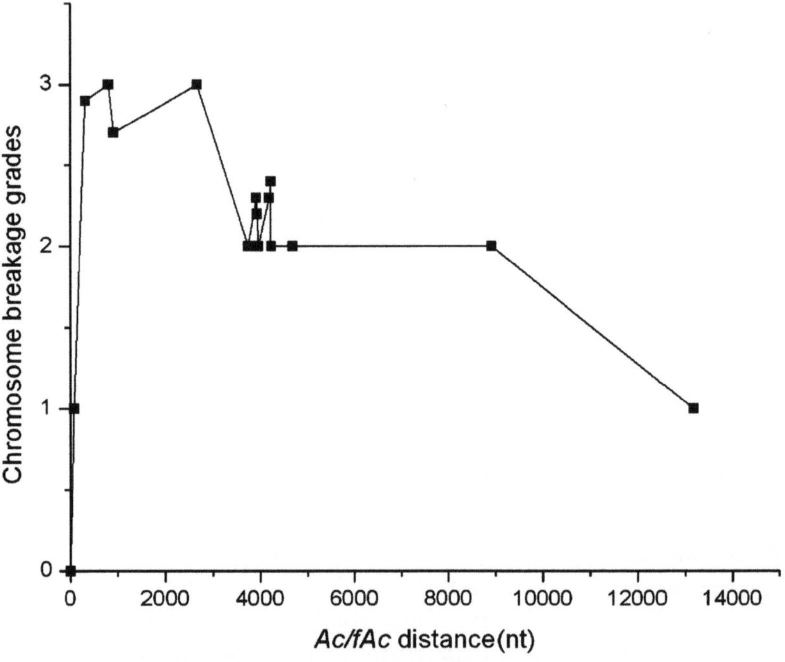

Fig. 4. The relationship between chromosome-breakage frequency and separation distance of reversed *Ac-fAc* termini at the maize *p1* locus. Chromosome-breakage frequency was measured by the loss of the *Dek1* marker distal to the *p1* gene. The frequency was evaluated by grades 0–3. Three indicates the greatest breakage frequency, and 0 shows no significant breakage frequency.

frequency than alleles with a smaller ITS. Interestingly, the Dooner laboratory has identified pairs of *Ac/Ds* elements separated by >100 kb distance which can still induce significant alternative transposition (43).

2. *Ac* transposase: A source of *Ac* transposase is essential for inducing rearrangements. Full-length *Ac* elements are not recommended due to their mobility. Several loci containing *Ac* elements with limited or zero mobility have been described in maize (48, 49). Stocks containing these alleles are available from the Maize Genetics Cooperation Stock Center, Urbana, Illinois. For experiments in other plants, e.g., Arabidopsis and rice, *Ac* transposase can be supplied from an *Ac* transgene containing an in vitro-generated deletion of sequences from one or both termini. The *Ac* transposase source can be incorporated into the rearrangement-inducing construct or can be integrated into a separate locus (46).
3. Selectable and scorable marker genes: A positive selectable marker gene (e.g., *bar, hpt, nptII*) is ordinarily required as part of the construct for selection of transformants (50). The incorporation of additional marker genes can facilitate the detection of rearrangement events. The most convenient markers are those that confer a visual pigmentation phenotype. Among the best known are the maize genes *bz1, c1, dek1,* and *p1*. Loss of function of any of these genes during maize plant development can result in a sector of nonpigmented or differentially pigmented cells (see Note 1). The size and number of sectors provides a useful indicator of the developmental timing and frequency of loss events. Loss events which are transmitted through the gametophytes may be recovered as mutant individuals. Thus, screening with visible markers enables convenient detection of chromosome rearrangements induced by alternative transposition events. Other useful visible markers include the fluorescent protein genes (e.g., *GFP, RFP*, and similar) which can function in many species. GFP appears to be more readily detected than RFP in plants, possibly because high levels of plant autofluorescence may obscure the RFP signal. Genes conferring herbicide resistance also provide good indicators of chromosome rearrangement events (e.g., loss of the ITS, or deletions to one side or other of the *Ac/Ds* termini; see Note 2). For Arabidopsis and rice, sensitive plants must be rescued from the herbicide-containing growth media. Maize plants can be screened for BAR sensitivity by leaf painting assays without killing sensitive plants. If alternative transposition frequency is sufficiently frequent (>1%), it may be possible to detect chromosome rearrangements directly via molecular screening (e.g., PCR, see below).

3. Methods

Configurations of transposable elements capable of undergoing alternative transposition reactions (SCT or RET) can be generated from naturally occurring (i.e., nontransgenic) genetic stocks containing active *Ac/Ds* transposable elements. The tendency of *Ac/Ds* elements to transpose during DNA replication and to insert into nearby sites can generate clusters of closely linked elements (51). If a pair of elements are in direct orientation, their apposed termini will be capable of undergoing RET (Fig. 2). Whereas, paired elements in inverted orientation will contain directly oriented 5′ and 3′ termini which are competent for SCT (Fig. 1). In either case, chromosome rearrangements can be detected via loss of expression of a nearby visible marker gene, e.g., *bz1* (43) or *p1* (38, 39). Putative rearrangements can be verified using diagnostic PCR as described below.

Where naturally occurring *Ac/Ds* pairs are not available, transgene constructs can be specifically designed to induce either SCT- or RET-induced chromosome rearrangements. Because production and analysis of transgenic plants requires considerable time and resources, great care should be taken to insure the integrity of the construct prior to plant transformation.

1. Transgene constructs containing *Ac/Ds* termini in either reversed or direct orientation can be prepared using standard molecular cloning techniques. The *Ac/Ds* termini sequences are inserted into a vector suitable for plant transformation. The frequency of alternative transposition reactions is affected by the physical distance separating the two termini. At the maize *p1* locus, the highest frequencies of chromosome breakage induced by RET occur when element termini are from ~300 bp to ~5 kb apart; with greater separation distances, chromosome-breakage frequency declines gradually (Fig. 4). More strikingly, separation distances of <100 bp show markedly reduced breakage frequency. Therefore, it is recommended that the *Ac/Ds* termini should be separated by at least 300 bp.
2. Marker genes are positioned with respect to the *Ac/Ds* sequences so as to provide an indication of the type of rearrangement induced. For example, an RET construct could contain three marker genes: one on each side of the *Ac/Ds* termini and one within the ITS. All RET events should show loss of the internal marker. Events that have lost in addition one or the other flanking markers indicate a deletion in that direction. Whereas, events that show loss of the ITS marker but retain both flanking markers are consistent with formation of inversions, translocations, or "fused-ends" (see Note 3).

3. The transgene construct is introduced into plant tissues and transformed plants are selected (44, 45). Transformed plants are screened for those carrying single-locus, single-copy insertion events, as these are generally required in order to detect rearrangements via marker gene loss (the presence of multiple transgene copies containing functional marker genes will mask transposition-induced loss of any single copy). In addition, multicopy transgene insertions containing multiple transposon termini could undergo a potentially large number of possible alternative transposition events, which would complicate subsequent detection and analysis.
4. An *Ac* transposase source is introduced by appropriate crosses or by subsequent transformation (see Note 4).
5. The progeny of plants containing the transgene construct together with an *Ac* transposase source are screened for marker gene loss (putative rearrangement events). Candidate rearrangement-containing plants are then analyzed by PCR. The choice of PCR primers will vary depending on the type of construct (whether directly oriented *Ac* termini for SCT, or reverse-oriented *Ac* termini for RET). As an example, the following strategy has been useful for screening natural alleles and transgene constructs containing reverse-orientated *Ac* termini for RET-induced rearrangements: Screening involves four sets of PCR primer pairs covering the following sequence junctions: (1) *Ac* 5′ sequence and external flanking sequence, (2) *Ac* 5′ end and ITS, (3) *Ac* 3′ end and ITS, (4) *Ac* 3′ sequence and external flanking. The pattern of PCR results produced by each allele gives an indication of the type of rearrangement. For example: (++++), no rearrangement; (++−+ or +−++), internal ITS deletion; (+−−+), inversion, translocation or fused-end; (+−−−), right-side deletion; (−−−+), left-side deletion; (−−−−), absence of construct, possibly due to segregation. Structures inferred by PCR patterns are compared with marker gene expression patterns to assess the validity of the rearrangement type.
6. The sequences at the chromosome rearrangement breakpoints can be cloned by several means, including I-PCR (53) or *Ac*-casting (52). Comparison with the genome sequence databases can precisely position the rearrangement.

4. Notes

1. In some cases, loss of marker gene expression may result not from chromosome rearrangements but from spontaneous

silencing, which can occur frequently in transgenic plants. Different marker genes appear to exhibit differing susceptibility to silencing. In transgenic maize, the *p1* gene undergoes frequent spontaneous silencing, whereas the BAR and *c1* genes exhibit little or no silencing. In any case, silencing can be distinguished from deletion by a PCR assay for gene sequences.

2. Attempts have been made to employ the *iaaH* gene (50), as a negative selectable marker for RET in Arabidopsis. Plants carrying *iaaH* are reported to be sensitive to the herbicide NAM, whereas plants lacking *iaaH* are NAM-resistant (18). Insertion of the *iaaH* gene within the ITS should enable selection for plants in which RET has occurred, resulting in deletion of the ITS including the *iaaH* gene. In practice, however, the identification of NAM sensitive and resistant plants has proved unreliable (46).
3. A frequent outcome of RET in Arabidopsis is the generation of "fused-ends"; i.e. deletion of the ITS and ligation of the 5′ and 3′ *Ac/Ds* termini. Fused-ends appear to occur much less frequently in rice, and very rarely in maize. It is not known whether the variation in occurrence of fused-ends is due to differences in the transgene constructs used, or species-specific differences in the transposition mechanism. Nevertheless, the generation of fused-ends products can result in a significant background of noninformative events. Where suspected, fused-ends can be detected by a simple PCR screen using oligonucleotide primers complementary to the 5′ and 3′ *Ac/Ds* termini. Once formed, fused-ends appear to be refractory to further transposition events in tobacco (9).
4. *Ac* dosage is known to affect the frequency of conventional *Ac/Ds* transposition. In maize, higher levels of *Ac* expression result in delayed timing and reduced transposition frequency. This is known as the *Ac* negative dosage effect (1). If needed, the dosage of *Ac* can be manipulated by using *Ac* transposase source in either homozygous or heterozygous condition.

Acknowledgments

We thank Kan Wang and Bronwyn Frame of the Iowa State University Plant Transformation Facility for maize and rice transformations. This research was supported by NSF Awards 0450243 and 0923826 to TP and JZ.

References

1. Kunze R, Weil CF. (2002) The hAT and CACTA superfamilies of plant transposons. In: Craig NL, Craige R, Gellert M, Lambowitz AM (eds), Mobile DNA II. ASM Press, Washington, DC, pp. 565–610.
2. Bravo-Angel AM, Becker HA, Kunze R, Hohn B, Shen WH. (1995) The binding motifs for Ac transposase are absolutely required for excision of Dsl in maize. *Mol Gen Genet* **248**, 527–34.
3. Coupland G, Baker B, Schell J, Starlinger P. (1988) Characterization of the maize transposable element Ac by internal deletions. *EMBO J* 7, 3653–9.
4. Coupland G, Plum C, Chatterjee S, Post A, Starlinger P. (1989) Sequences near the termini are required for transposition of the maize transposon Ac in transgenic tobacco plants. *Proc Natl Acad Sci USA* **86**, 9385–8.
5. Becker HA, Kunze R. (1997) Maize Activator transposase has a bipartite DNA binding domain that recognizes subterminal sequences and the terminal inverted repeats. *Mol Gen Genet* **254**, 219–30.
6. Becker HA, Kunze R. (1996) Binding sites for maize nuclear proteins in the subterminal regions of the transposable element Activator. *Mol Gen Genet* **251**, 428–35.
7. Kunze R, Starlinger P. (1989) The putative transposase of transposable element Ac from *Zea mays* L. interacts with subterminal sequences of Ac. *EMBO J* **8**, 3177–85.
8. Gorbunova V, Levy AA. (2000) Analysis of extrachromosomal Ac/Ds transposable elements. *Genetics* **155**, 349–59.
9. Gorbunova V, Levy AA. (1997) Circularized Ac/Ds transposons: formation, structure and fate. *Genetics* **145**, 1161–9.
10. Kunze R. (1996) The maize transposable element Activator (Ac). *Curr Top Microbiol Immunol* **204**, 161–94.
11. Rinehart TA, Dean C, Weil CF. (1997) Comparative analysis of non-random DNA repair following Ac transposon excision in maize and Arabidopsis. *Plant J* **12**, 1419–27.
12. Rommens CM, van Haaren MJ, Buchel AS, et al. (1992) Transactivation of Ds by Ac-transposase gene fusions in tobacco. *Mol Gen Genet* **231**, 433–41.
13. Levy AA, Walbot V. (1990) Regulation of the timing of transposable element excision during maize development. *Science* **248**, 1534–7.
14. Kolkman JM, Conrad LJ, Farmer PR, et al. (2005) Distribution of Activator (Ac) throughout the maize genome for use in regional mutagenesis. *Genetics* **169**, 981–95.
15. Bancroft I, Bhatt AM, Sjodin C, Scofield S, Jones JD, Dean C. (1992) Development of an efficient two-element transposon tagging system in *Arabidopsis thaliana. Mol Gen Genet* **233**, 449–61.
16. Dean C, Sjodin C, Bancroft I, et al. (1991) Development of an efficient transposon tagging system in *Arabidopsis thaliana. Symp Soc Exp Biol* **45**, 63–75.
17. Fedoroff NV, Smith DL. (1993) A versatile system for detecting transposition in Arabidopsis. *Plant J* **3**, 273–89.
18. Sundaresan V, Springer P, Volpe T, et al. (1995) Patterns of gene action in plant development revealed by enhancer trap and gene trap transposable elements. *Genes Dev* **9**, 1797–810.
19. Grevelding C, Becker D, Kunze R, et al. (1992) High rates of Ac/Ds germinal transposition in Arabidopsis suitable for gene isolation by insertional mutagenesis. *Proc Natl Acad Sci USA* **89**, 6085–9.
20. Chuck G, Robbins T, Nijjar C, Ralston E, Courtney-Gutterson N, Dooner HK. (1993) Tagging and cloning of a Petunia flower color gene with the maize transposable element Activator. *Plant Cell* **5**, 371–8.
21. Meissner R, Chague V, Zhu Q, Emmanuel E, Elkind Y, Levy AA. (2000) Technical advance: a high throughput system for transposon tagging and promoter trapping in tomato. *Plant J* **22**, 265–74.
22. Chin HG, Choe MS, Lee SH, et al. (1999) Molecular analysis of rice plants harboring an Ac/Ds transposable element-mediated gene trapping system. *Plant J* **19**, 615–23.
23. Izawa T, Ohnishi T, Nakano T, et al. (1997) Transposon tagging in rice. *Plant Mol Biol* **35**, 219–29.
24. Emelyanov A, Gao Y, Naqvi NI, Parinov S. (2006) Trans-kingdom transposition of the maize Dissociation element. *Genetics* **174**, 1095–104.
25. Weil CF, Kunze R. (2000) Transposition of maize Ac/Ds transposable elements in the yeast *Saccharomyces cerevisiae. Nat Genet* **26**, 187–90.
26. McClintock B. (1953) Induction of instability at selected loci in maize. *Genetics* **38**, 579–99.
27. McClintock B. (1950) The origin and behavior of mutable loci in maize. *Proc Natl Acad Sci USA* **36**, 344–55.

28. McClintock B. (1949) Mutable loci in maize. *Year B Carnegie Inst Wash* **48**, 142–54.
29. McClintock B. (1953) Mutation in maize. *Year B Carnegie Inst Wash* **52**, 227–37.
30. McClintock B. (1948) Mutable loci in maize. *Year B Carnegie Inst Wash* **47**, 155–69.
31. McClintock B. (1951) Chromosome organization and genic expression. *Cold Spring Harb Symp Quant Biol* **16**, 13–47.
32. McClintock B. (1952) Mutable loci in maize. *Year B Carnegie Inst Wash* **51**, 212–9.
33. English J, Harrison K, Jones JD. (1993) A genetic analysis of DNA sequence requirements for Dissociation state I activity in tobacco. *Plant Cell* **5**, 501–14.
34. Martinez-Ferez IM, Dooner HK. (1997) Sesqui-Ds, the chromosome-breaking insertion at bz-m1, links double Ds to the original Ds element. *Mol Gen Genet* **255**, 580–6.
35. Weil CF, Wessler SR. (1993) Molecular evidence that chromosome breakage by Ds elements is caused by aberrant transposition. *Plant Cell* **5**, 515–22.
36. Zhang J, Peterson T. (1999) Genome rearrangements by nonlinear transposons in maize. *Genetics* **153**, 1403–10.
37. Zhang J, Peterson T. (2005) A segmental deletion series generated by sister-chromatid transposition of Ac transposable elements in maize. *Genetics* **171**, 333–44.
38. Zhang J, Peterson T. (2004) Transposition of reversed Ac element ends generates chromosome rearrangements in maize. *Genetics* **167**, 1929–37.
39. Zhang J, Yu C, Pulletikurti V, et al. (2009) Alternative Ac/Ds transposition induces major chromosomal rearrangements in maize. *Genes Dev* **23**, 755–65.
40. Chopra S, Brendel V, Zhang J, Axtell JD, Peterson T. (1999) Molecular characterization of a mutable pigmentation phenotype and isolation of the first active transposable element from *Sorghum bicolor. Proc Natl Acad Sci USA* **96**, 15330–5.
41. Zhang J, Zhang F, Peterson T. (2006) Transposition of reversed Ac element ends generates novel chimeric genes in maize. *PLoS Genet* **2**, e164.
42. Dooner HK, Weil CF. (2007) Give-and-take: interactions between DNA transposons and their host plant genomes. *Curr Opin Genet Dev* **17**, 486–92.
43. Huang JT, Dooner HK. (2008) Macrotransposition and other complex chromosomal restructuring in maize by closely linked transposons in direct orientation. *Plant Cell* **20**, 2019–32.
44. Clough SJ, Bent AF. (1998) Floral dip: a simplified method for Agrobacterium-mediated transformation of *Arabidopsis thaliana. Plant J* **16**, 735–43.
45. Ishida Y, Saito H, Ohta S, Hiei Y, Komari T, Kumashiro T. (1996) High efficiency transformation of maize (*Zea mays* L.) mediated by *Agrobacterium tumefaciens. Nat Biotechnol* **14**, 745–50.
46. Krishnaswamy L, Zhang J, Peterson T. (2008) Reversed end Ds element: a novel tool for chromosome engineering in Arabidopsis. *Plant Mol Biol* **68**, 399–411.
47. Gopinath DM, Burnham CR. (1956) A cytogenetic study in maize of deficiency-duplication produced by crossing interchanges involving the same chromosomes. *Genetics* **141**, 382–95.
48. Xiao YL, Peterson T. (2002) Ac transposition is impaired by a small terminal deletion. *Mol Genet Genomics* **266**, 720–31.
49. Conrad LJ, Brutnell TP. (2005) Ac-immobilized, a stable source of Activator transposase that mediates sporophytic and gametophytic excision of Dissociation elements in maize. *Genetics* **171**, 1999–2012.
50. Sundar IK, Sakthivel N. (2008) Advances in selectable marker genes for plant transformation. *J Plant Physiol* **165**, 1698–716.
51. Chen J, Greenblatt IM, Dellaporta SL. (1992) Molecular analysis of Ac transposition and DNA replication. *Genetics* **130**, 665–76.
52. Ochman H, Gerber AS, Hartl DL. (1988) Genetic applications of an inverse polymerase chain reaction. *Genetics* **120**, 621–3.
53. Singh M, Lewis PE, Hardeman K, et al. (2003) Activator mutagenesis of the *pink scutellum1/viviparous7* locus of maize. *Plant Cell* **15**, 874–84.
54. Qu S, Desai A, Wing R, Sundaresan V. (2008) A versatile transposon-based activation tag vector system for functional genomics in cereals and other monocot plants. *Plant Physiol* **146**, 189–99.

Index

James A. Birchler (ed.), *Plant Chromosome Engineering: Methods and Protocols*, Methods in Molecular Biology, vol. 701,
DOI 10.1007/978-1-61737-957-4, © Springer Science+Business Media, LLC 2011

V

Y

Z

MIX
Papier aus verantwortungsvollen Quellen
Paper from responsible sources
FSC® C105338

If you have any concerns about our products,
you can contact us on
ProductSafety@springernature.com

In case Publisher is established outside the EU,
the EU authorized representative is:
Springer Nature Customer Service Center GmbH
Europaplatz 3, 69115 Heidelberg, Germany

Printed by Libri Plureos GmbH
in Hamburg, Germany